The World Energy Book

An A-Z, Atlas and Statistical Source Book

The World Energy Book

An A-Z, Atlas and Statistical Source Book

Consultant Editors and Principal Contributors
David Crabbe and Richard McBride
Open University Energy Research Group

The MIT Press
Cambridge, Massachusetts, and London, England.

Contributors:
Jennifer Clark, Michael Gabb, David Hemming, Norman Hill, Ray Oddy, Cyril Parsons

Editorial and Design:
Pauline Wingate, David Young, Merilyn David, Natasha Djuk, John Doyle, Anne Heat, Sandra Hosking, Rita Jacobs, David Kerby, Anita Kogan, Susan McCabe, Pamela Poulter, Tony Selina, Pat Thomas, Michael Tout

First MIT Press paperback edition, 1979

First published in Great Britain in 1978 by Kogan Page Limited; first published in the USA in 1978 by Nichols Publishing Company.

Printed in Great Britain

Library of Congress catalog card number: 78-71558
ISBN 0-262-53036-8

Contents

Statistical Appendices

Acknowledgements

The publisher would like to acknowledge assistance, direct and indirect, in the compilation of this book provided by:

British Gas Corporation
British Petroleum
Central Electricity Generating Board
Confederation of British Industry
Electricity Council
Friends of the Earth
Hydraulics Research Station
National Coal Board
United Kingdom Atomic Energy Authority
United Kingdom Timber Trades Federation

The Energy Technology Review Series, published by Noyes Data Corporation of Park Ridge, New Jersey, provided much source material which is also gratefully acknowledged.

Considerable efforts have been made to ensure the accuracy of the information contained in this book and every attempt has been made to trace and accredit copyright material. Source acknowledgements are to be found in the book where relevant.

Introduction

The World Energy Book is intended as a comprehensive reference guide to energy sources, energy related terminology, economics and all factors related to the search for, extraction of, production and utilization of the major and alternative sources of energy.

There are three main sections of the book, an *A-Z* containing some 1000 terms and definitions, the *Energy Atlas* and the *Statistical Data Section*.

A-Z

The text in this section is presented in the form of alphabetically ordered entries (under bold entry headings) to enable immediate and easy reference to be made. Cross-referencing (italics) is used throughout to direct the reader to related subjects if he wishes to delve further. Many other cross-references will be isolated under other related entries so that the inquisitive reader will discover a network of facts stemming from his original inquiry. Additionally, further data can often be pin-pointed in the atlas and statistical sections.

Although primary chemical, primary physical, secondary chemical, secondary physical and nuclear sources of energy are not headings under which energy sources are listed in the A-Z, some idea of the extent of the terminology included can be gained from the subjects listed below under these headings.

Primary chemical sources of energy

Solid fuels (coals, peat), petroleum/crude oil, oil shale, bituminous sands/tar sands, natural gas and photosynthates (cellulose and wood). Also included (where relevant): origins and nature, the state of relevant technology, methods of discovery, extraction (or growth) and production, research and development, geology, preparation, refining and extraction, transportation or transmission, conversion to heat, storage, energy economics and analysis, uses, the importance of each source to the overall energy picture, environmental and health factors, consumption, production and reserves, energy equivalents and conversion factors. (These last statistical areas are briefly covered in the A-Z then reference is made to the extensive tables in the statistical section.)

Primary physical sources of energy

Solar, hydroelectric, geothermal, wind, tidal, waves, and ocean thermal gradients. Also includes: nature and origins, climatic and geological conditions, harnessing technology and developments, uses, advantages and disadvantages, potential uses, economic and environmental considerations, legal factors. Production, consumption, energy equivalents and conversion factors and the proportion of the overall energy supply are dealt with by reference to the statistical section.

Nuclear sources of energy

Nuclear energy introduction, nature, nuclei and the release of energy. *Fission energy:* conditions of controlled reaction, heat transfer and transmission, research and development, energy equivalents and conversion factors, environmental and economic factors, radiation hazards, world demand, production and distribution, the proportion of total energy supply. *Nuclear reactors:* burner reactors, breeder reactors, present state of technology, future prospects and risks, nuclear materials. *Fission fuels:* uranium, thorium, plutonium, ancillary nuclear materials (including deuterium, zirconium), location/extraction (if applicable) and isolation, reprocessing, forms in use, future importance of each fuel. *Fusion* (thermonuclear energy): nature, importance, economic and environmental factors, thermonuclear reactors, laser reactors, safety and development. *Fusion fuels:* tritium, lithium, deuterium, physical properties, reserves, extraction and isolation, potential uses.

Secondary chemical sources of energy

Alcohols, manufactured gases, synthetic petroleum, waste materials, sewage, heat of fermentation, acetylene, hydrogen and hydrogen economy. Also included: nature and origins, present use and potential value, conversion factors, the state of technology, economic and environmental considerations, production and consumption.

Secondary physical sources of energy

Fuel cells, photoelectric energy, high temperature boilers, pumped storage, electrolysis, magnetohydrodynamics; technical, economic and environmental factors, conversion factors, energy value and specific advantages and disadvantages.

The Energy Resources Atlas

The atlas section is cross-referred to from the A-Z. It covers certain information which, as in the statistical section, is beyond the range of the A-Z. Maps provide detailed information on, for example, the location of oil and gas fields offshore and onshore in North America, South America, the Middle East and the North Sea; on coal, tar sands, oil shales, uranium and other fuel deposits; on possible tidal energy, sea wave energy and ocean thermal gradient sites; on geothermal areas, solar energy levels, hydroelectric sites; and on data on energy production, consumption and trade. The atlas section is also able to illustrate geographical distribution and movement of fuel sources.

Statistical Appendices

This section comprises tables, classification scales and conversion charts which provide statistical background information relating to all the present-day energy sources and some of the alternative sources. This section is cross-referred to and from the A-Z and atlas sections.

A

absolute scale of temperature

A scale of temperature based on the concept of an absolute zero of temperature, ie a point at which atoms and molecules possess no kinetic energy and at which an ideal gas would exert zero pressure. In practice, temperatures very close to absolute zero have been attained, and at these very low temperatures interesting physical phenomena such as *superconductivity* have been observed.

On the *centigrade scale* absolute zero is calculated to be -273.15° (approx) and on the absolute scale of temperature this point corresponds to zero degrees of temperature. Degrees absolute are commonly refered to as degrees Kelvin ($^\circ$K) and the general relationship between degrees Kelvin and degrees centigrade is:

$$^\circ K = {}^\circ C + 273.15.$$

absorption

In chemistry, the process in which one substance is taken up by another or penetrates it. Also the giving of energy, eg heat energy, to an insulating material by electromagnetic waves.

Absorption has many applications in industry. For example, by applying an absorption process to *wet gas* the higher *hydrocarbons* can be separated to provide a *dry gas* suitable for use as a fuel or *chemical feedstock*. The wet gas is passed through an absorption tower under slight pressure and the higher hydrocarbons are absorbed by a light *gas oil* flowing against it. The solvent gas oil is then heated and the dissolved products removed by steam.

See also *adsorption*.

absorption cycle heat pump

A particular type of *heat pump* which exploits the latent heat of vaporization of a working fluid to transport heat from a low temperature source to a high temperature sink. Compression of the working fluid is achieved by absorption in a liquid or solid. This type of heat pump is used for gas-powered refrigerators and for *solar cooling*.

See also *refrigeration*.

absorption gasoline

Natural *gasoline* obtained by *absorption* from *wet gas*. The gasoline is extracted from the *absorption oil* (used to dry the gas) by *distillation*.

absorption oil

A petroleum or coal tar oil that is used to absorb heavy components from a gas stream, as in the recovery of natural *gasoline* from *wet gas*.

abundance ratio

In a mixture of *isotopes*, the ratio of the number of *atoms* of one isotope to the total number of atoms present.

accelerator

A machine used to accelerate charged nuclear particles to very high energies. During the early 1930s accelerators using *protons* and deuterons were first employed to investigate nuclear reactions and this subsequently led to the discovery of *neutrons* and nuclear *fission*.

accumulator

Also known as a storage battery. An *electric cell*, or an assembly of such cells, that is capable of storage. An accumulator can store energy on charging and release energy on discharging.

acetylene

The usual term for the simplest member of the *alkyne* group of *hydrocarbons*, more correctly termed ethyne, and having the molecular formula C_2H_2($HC{\equiv}CH$). It is a gas at room temperature and atmospheric pressure. Acetylene may be manufactured from natural gas or petroleum feedstocks by high temperature cracking or partial oxidation and from coal via carbide production. Acetylene is used as a *chemical feedstock* and has a widespread, if small scale, use as a fuel employed in gas welding, the gas having a high calorific value of 1455 Btu/ft^3 (54.2MJ/m^3).

Acetylene has a high *energy requirement* for manufacture. This is largely because of the energy required to form the triple carbon-to-carbon bond of the molecule. Also acetylene prepared from carbide tends to contain impurities and therefore has to be cleaned before further conversion.

acid gas

Natural gas containing *hydrogen sulphide* is usually referred to as acid gas or sour gas. The term is also applied to gas containing a large proportion of carbon dioxide. Gas from the Lacq field in France contains 15 per cent hydrogen sulphide by volume, while gas from Retiden in Germany contains 18 per cent carbon dioxide by volume.

Hydrogen sulphide present in small amounts can be removed by washing with a strong alkali (eg 10 per cent aqueous solution of NaOH). If the hydrogen sulphide is present in large amounts, it is more economical to use a regenerative process in which the hydrogen sulphide is absorbed in an *amine*, such as mono- or di-ethanolamine. Hydrogen sulphide is an undesirable impurity in any fuel or feedstock since it causes corrosion of plant and pollution when finally burnt. Carbon dioxide may be removed by *absorption* in an aqueous alkali.

Acid gases can be an important source of *sulphur*. Sulphur can be generated from hydrogen sulphide by partial *oxidation* with air, using a bauxite *catalyst*.

acidizing

Method of stimulation of *petroleum* wells by injecting hydrochloric acid into the oil-bearing formation. The acid enlarges the porous passages to allow the oil to flow through.

actinide series

The group of elements with *atomic numbers* between 89 and 103. They are named after the first member of the series actinium.

The first four members of the series (actinium, *thorium*, protactinium and *uranium* and possibly also *plutonium*) occur naturally on earth: the remainder have been produced by nuclear reactions, either in nuclear reactors or in particle accelerators.

All the actinides are radioactive, and none of them has any stable isotopes.

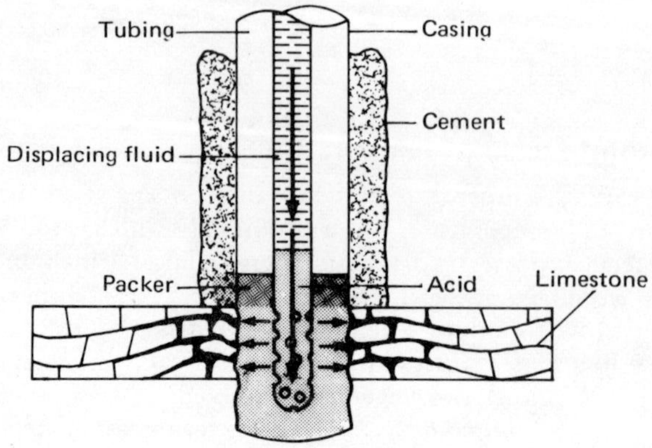

Outline of the acidizing process

activated carbon

A form of amorphous *carbon* characterized by a very large surface area per unit volume, approximately 1,000,000 m^2/kg, due to its porous structure, and capable of adsorbing considerable quantities of matter. Also known as activated

Element	Symbol	Atomic number	Atomic weight*	Earthly abundance (ppm)	Present and potential uses:
Actinium	Ac	89	(227)	3×10^{-10}	Neutron source
Thorium	Th	90	(232)	11.5	Gas mantles: potentially important fission power source.
Protactinium	Pa	91	(231)	8×10^{-7}	Important only as research material
Uranium	U	92	(238)	4	Important fission power source; also used in chemical analysis and photography to small extent
Neptunium	Np	93	(237)	–	Component in neutron detection instruments
Plutonium	Pu	94	(244)	–	Important fission power source (both in reactors and as a portable power source, eg heart pacemakers)
Americium	Am	95	(243)	–	Used in thickness gauges in float glass manufacture
Curium	Cm	96	(247)	–	Potentially useful portable power source
Berkelium	Bk	97	(247)	–	No commercial or technical use at present
Californium	Cf	98	(251)	–	Important neutron source in moisture gauges; also used for well logging and neutron activation analysis
Einsteinium	Es	99	(254)	–	Presently available only in minute amounts
Fermium	Fm	100	(257)	–	Presently available only in minute amounts
Mendelevium	Md	101	(258)	–	Presently available only in minute amounts
Nobelium	No	102	(255)	–	Presently available only in minute amounts
Lawrencium	Lr	103	(256)	–	Presently available only in minute amounts

* ( ) *Mass No of longest lived isotope*

Actinide series

charcoal, it is obtained from a variety of carbonaceous materials, among them *wood, petroleum coke, peat, coal* and bones. Activated carbons for decolorizing are made from sawdust or *lignite*, while those used for *adsorption* of vapours are best made from briquetted coal or *charcoal.*

The most widely used method of activation is treatment with air, steam or carbon dioxide at a high temperature (approximately 800°-900°C). Activated carbon is used to purify gases (eg the removal of heavy components from *wet gas*), and liquids, (eg the purification of water supplies and pharmaceutical preparations).

activation

1. The process by which certain molecules in a body acquire an amount of energy, called the *activation energy*, above the average energy of that body of molecules. The process of activation is a necessary initial step in reactions between molecules, and the magnitude of the activation energy for a reaction is an important determinant of the rate of the reaction.
2. Rendering substances (ie their nuclei) radioactive by neutron bombardment. The activation process can be used as a sensitive method of chemical analysis for the detection of trace quantities of elements.

activation energy

In chemistry, the energy that must be given to the reactants to enable a reaction to take place. The concept of an activation energy, or 'energy barrier', is applicable to many other physical processes. The rate of a process is related to the magnitude of the energy barrier, and to the temperature.

activity

A numerical parameter characteristic of the intensity of *radiation* emitted by a *radioactive isotope*.

The rate of breakdown of such an isotope is proportional to the amount present in a sample, ie:

$$\frac{dn}{dt} = -kn$$

(where n is the number of nuclei of the isotope present).

The constant of proportionality, k, is known as the activity of the isotope, and is related to the *half-life*, $t_{1/2}$, by the formula:

$$k = \frac{\ln 2}{t_{1/2}}$$

additives

In *petroleum* technology substances added to fuels to improve their combustion characteristics or other properties. The term is also used to describe any substance added to a product with the purpose of enhancing its usefulness.

Some representative additives to fuels are detailed below. Concentrations of additives in fuels range from one or two to several hundred parts per million. Many are now essential ingredients of *gasolines* and other motor fuels.

Additive	Function
Anti-knock compounds	Reduce susceptibility to detonation
Scavengers	Remove combustion products of anti-knock compounds
Anti-icing agents	Reduce icing in carburettor
Antioxidants	Enhance storability of fuel
Upper cylinder lubricants	Lubricate upper cylinders and pistons, and control inlet system deposits
Dyes	Indicate visually type and grade of gasoline

Additives to automotive gasoline

Additive	Function
Anti-icing agents	Prevent icing at altitude and suppress growth of bacteria and fungi in fuel
Anti-static additives	Decrease resistivity of fuel and so reduce hazard of sparking under high flow rate conditions
Antioxidants	Prevent formation of gums and coke in fuel lines and nozzles
Corrosion inhibitors	Reduce rust formation in fuel. Copper chelating agents are also used to suppress corrosion by this metal

Additives to aviation jet fuel

Additive	Function
Cetane improvers	Increase Cetane number
Detergents	Maintain fuel injection nozzle cleanliness
Rust inhibitors	Suppress fouling of injection system by rust
Antioxidants	Enhance storability of fuel
Pour-point depressants	Lower temperature at which a fuel will flow readily through pipe

Additives to diesel fuel

adsorption

The adhesion of the molecules of one substance, in the gaseous or liquid phase, to another substance in the solid phase, resulting in a relatively high concentration of the adsorbed component at the surface.

The concentrating effect of adsorption is thought to be an important factor in the heterogeneous catalysis of

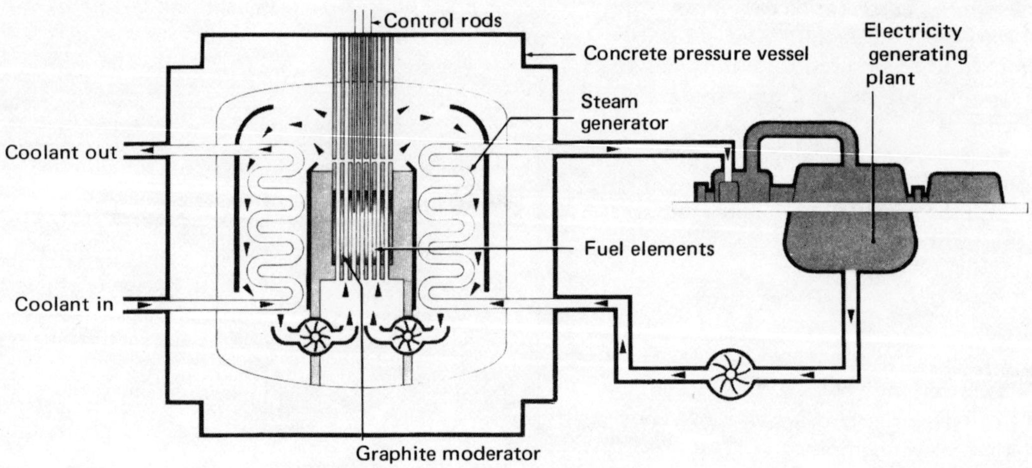

Advanced gas-cooled reactor (AGR)

Reproduced by kind permission of the UKAEA

chemical reactions and, as such, adsorption plays an important part in the energy industries. The phenomenon is also exploited in the removal of high boiling point *hydrocarbons* from *wet gas* by passing the gas over particles of *activated carbon.* Adsorption is a surface phenomenon and the very high surface area of activated carbon enables significant quantities of the hydrocarbons to be retained. The adsorbed hydrocarbons can be subsequently recovered by passing steam over the saturated adsorbent, so regenerating the carbon for further use.

advanced gas-cooled reactor (AGR)

A type of nuclear reactor in which heat extraction is by carbon dioxide gas under pressure (40 atmospheres at 634°C). The heat from the gas is transferred to water in a heat exchanger and converted into steam for a *turbo-generator.* The design specifies a heat output of 1494MWt giving about 621MWe with an efficiency of 41.6 per cent. The fuel is *uranium dioxide* pellets, 2 per cent enriched, clad in thin stainless steel. The moderator is *graphite.*

The AGR was chosen for the second generation of gas-cooled reactors in the UK, but problems have been encountered in their construction. Higher gas temperatures leading to more efficient production of electricity are achieved by using uranium dioxide, which melts at 2800°C, as fuel rather than the *uranium metal* of the *Magnox reactor.* Also the *burn-up* of uranium dioxide is greater, 18,000 megwatt-days per tonne.

advance mining

A method of underground mining in which extraction proceeds from the base of the shaft outwards towards the boundary of the deposit.

Advanced gas-cooled reactor data*

Peak power density	4.5 kW/l
Proposed heat output	1494MWt
Proposed electrical output	621MWe
Proposed efficiency	41.6%
Fuel	Uranium oxide pellets, 2% enriched, clad in stainless steel
Weight of fuel	113.7 tonnes
Fuel burn-up	18,000MWD/te
Moderator	Graphite
Coolant	Carbon dioxide gas
Coolant pressure	40 atmospheres
Coolant outlet temperature	634°C
Refuelling	On load

**Data based on Hinkley Point B (UK)*

aerodynamic uranium enrichment

The enrichment of uranium fuels by the aerodynamic separation and concentration of the required *isotopes.*

See also *uranium enrichment.*

aerogenerators

General name given to devices which extract useful energy from wind.

See also *wind energy.*

afterdamp

The major gaseous product formed by the combustion of *coal gas.* It is nonflammable, occurs in coal mines after an explosion and comprises mainly *carbon dioxide.*

See also *black damp; firedamp; white damp.*

afterheat

The heat from fission products remaining in a nuclear reactor after shut down.

AGR

Initials for *advanced gas-cooled reactor*.

Airco-Hoover process

A method of 'sweetening' gasoline by means of alkali and water washes, followed by passage through a diatomaceous clay/copper chloride catalyst. The process converts corrosive and troublesome mercaptans to the less harmful disulphides.

air cycle heat pump

A particular type of *heat pump* which uses air as the working fluid.

See also *refrigeration*.

air gas

Alternative name for *producer gas*.

air pollution

The presence in the atmosphere of contaminants such as dust, fumes, gas or smoke in such quantities and of such a nature as to be injurious to plant, animal and human life or an unreasonable interference with people's welfare and property.

Air pollution arises from many energy producing and processing technologies, from the burning of wood to the operation of nuclear power stations.

See the table *Sources of Pollution and Means of Control* (pages 16-17) which briefly summarizes the major air pollution sources associated with energy production and consumption.

See also *environmental effects; sulphur emissions*.

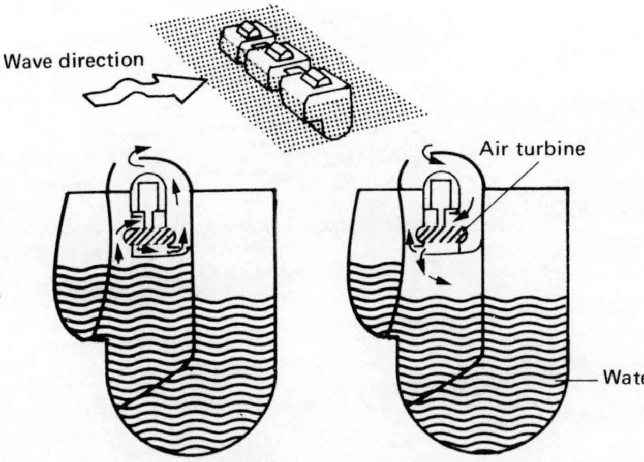

Sectional views of two air pressure ring units together with part of a row of units

air pressure ring buoy

A *wave energy* device which tilts backwards and forwards with the movement of waves thus distributing water inside from compartment to compartment and back again. Air inside is made to flow by the changing distribution of water and this air movement operates an air turbine within the buoy.

The air pressure ring buoy was initially developed from Japanese studies on floating breakwaters.

alcohol

The common name for *ethanol*, C_2H_5OH. In chemistry the term is applied to a large number of organic compounds containing the hydroxyl $-OH$ group: the simplest alcohol is *methanol* CH_3OH.

Both methanol and ethanol are manufactured from *hydrocarbon* sources: methanol from *natural gas* and ethanol from *ethylene* obtained from *petroleum*. Ethanol is additionally obtained by *anaerobic fermentation* of natural products, usually *carbohydrates*. Although such production is largely used for human consumption as beverages, it has been suggested that fermentation alcohols (both ethanol and methanol) could be important sources of fuel and *chemical feedstock* when oil becomes scarce.

alcoholic fermentation

Also known as *anaerobic fermentation*, the process by which certain yeasts decompose sugars in the absence of oxygen to form *alcohols* and *carbon dioxide*, eg the production of *ethanol*, wine and beer from *carbohydrate* sources.

algae

An important division of non-flowering plants (thallophytes) having a considerable size range, including *seaweed* and unicellular plants. The majority of the *photosynthesis* in the sea is due to algae which play an important part therefore in the reoxygenation of the atmosphere and the overall energy cycle. Algae provide a possible alternative energy source as a *cellulose* material.

alicyclic hydrocarbons

Hydrocarbons in which the carbon skeleton is arranged in

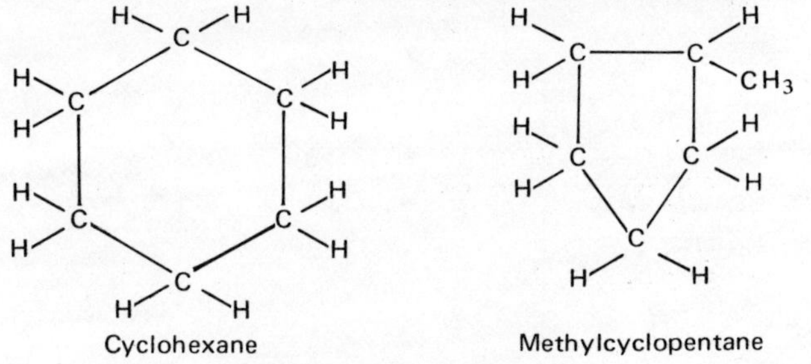

Typical alicyclic hydrocarbons
(These are both *cycloalkanes*.)

Air Pollution: Sources of Pollution and Means of Control

Pollutant	Produced by:	Important Sources	Control Methods
SMOKE	Combustion of any fuel with high-ash content	Coal and oil-fired power plant using high-ash fuel	Use of low-ash fuel Electrostatic precipitators on stacks
		Open fires burning coal	Use of smokeless solid fuels (manufactured)
	Incomplete combustion of hydrocarbon fuel	Diesel engines	Careful maintenance and adjustment
		Refuse incineration	After burners to complete combustion Filters and precipitators on stacks
DUST	Handling of solid fuel	Coal mining	Water spraying
		Coal processing	Precipitation and/or filtration
HYDROGEN SULPHIDE AND OXIDES OF SULPHUR SO_x	Burning of any fuel containing sulphur SO_x	Coal and oil-fired power plant using sulphur-containing fuel	Use of low-sulphur fuel Desulphurizing fuel before use Dispersion by use of high stacks Catalytic oxidation or reduction Absorption in dolomite or limestone Wet scrubbing of stack gases
		Domestic and industrial heaters and boilers	Use of low-sulphur or desulphurized fuel
		Petroleum refineries	Desulphurization prior to combustion or cracking Chemical absorption Catalytic oxidation or reduction
		Refuse incineration	Chemical absorption or wet scrubbing of flue gases
	Release of natural sources H_2S and SO_x	Geothermal energy plants	Control of and absorption of sulphur oxides from exit gases
HYDROGEN SULPHIDE AND MERCAPTANS	Processing of sulphur-containing hydrocarbons	Petroleum processing	Wet scrubbers Catalytic oxidation or reduction
OXIDES OF NITROGEN NO_x	Burning of any fuel in air at high temperature	Internal combustion engines	Lowering of flame temperatures Restricting of oxygen supply by enriching fuel/air mixture After burning of exhaust gases in combustion chamber Catalytic re-conversion to N_2 and O_2
		Fossil fuel power plants	Use of two-stage combustion Exhaust gas recycling
		Petroleum processing	Lowering of flame temperatures in crackers
CARBON MONOXIDE CO	Burning of carbonaceous fuel with restricted oxygen supply	Internal combustion engines	Maintenance of correct air/fuel ratio Design to ensure complete combustion Catalytic oxidation of exhaust gases
		Fossil fuel burners	Design to ensure complete combustion Operate at correct air/fuel ratio
ORGANIC COMPOUNDS **HYDRO-CARBONS**	Evaporation/losses of hydrocarbon fuels	Evaporative losses from fuel tanks	Use of unvented tanks with activated carbon vapour adsorption
		Crankcase 'blowby' in piston engines	Vent crankcase to inlet manifold
		Petroleum processing	Careful maintenance of seals and valves Absorption/adsorption of off gases Cryogenic removal from exit streams

Pollutant	Produced by:	Important Sources	Control Methods
ORGANIC COMPOUNDS HYDRO-CARBONS	Incomplete combustion of hydrocarbon fuels	Internal combustion engines	Combustion chamber design to ensure complete combustion Use of weak fuel/air mixture Catalytic converters to oxidize exhaust gases
ORGANIC COMPOUNDS ALDEHYDES		Internal combustion engines	Maintenance of correct air/fuel ratio Catalytic oxidation of exhaust gases
BENZO PYRENE		Internal combustion engines	Use of gasolines low in aromatic content
LEAD (in particulate form)	Burning of fuels containing lead-based anti-knock additives	Internal combustion (spark ignition) engines	Filtration of exhaust gases Use of low lead fuel (with high aromatic content) Use of lower compression ratios
		Refuse incineration	Efficient precipitation of fly/ash
HYDROGEN CHLORIDE	Burning of hydrocarbon fuel containing chlorine	Refuse incineration (source of chlorine mainly PVC)	Wet scrubbing of exit gases
RADIOACTIVE NUCLIDES: **Kr-85** **Kr-87** **Kr-88** **Xe-133** **Xe-135** **Xe-138** **H-3**	Fission products of radioactive decay	Leaking fuel rods in nuclear reactors	Rapid detection and removal of leaking fuel rods (gas cooled reactors) Adsorption of emissions on activated carbon to delay release beyond several half-lives of decay products Detention of gaseous emission products in gas storage (pressurized water reactors)
Kr-85 **H-3**		Fuel reprocessing	Cryogenic removal Adsorption in activated carbon
Ar-41	Product of neutron activation of argon in air	Nuclear reactors using air cooling in high-neutron areas	Cryogenic removal Adsorption in activated carbon
Ra-226 **Ra-228** **U-238** **Th-230** **Th-232** **K-40**	Burning of fossil fuels	Fly-ash from fossil-fuelled power plants	Efficient precipitation of fly/ash Dilution of effluent by use of tall stacks
HEAT	Use of any non-ambient energy source	Fossil-fuelled and nuclear power stations Geothermal power plants	Dispersion of power plant reduces local effects of thermal pollution Reduction in waste heat output in energy conversions Effective mixing of waste heat streams reduces ecological effects at a local level

one or more closed rings, but without the characteristic structure of alternating bonds that leads to aromatic properties. (The word is an abbreviation of 'aliphatic cyclic' and the compounds are so called because of their chemical similarities with *aliphatic hydrocarbons*.)

Alicyclic hydrocarbons occur naturally in crude oils as a mixture of many different molecular species, generally known as *naphthenes*. Examples include cyclohexane and methylcyclopentane.

See also *aromatic hydrocarbons*.

aliphatic hydrocarbons

Hydrocarbons in which the carbon skeleton is in the form of an open chain, as opposed to the *alicyclic hydrocarbons* and the *aromatic hydrocarbons* in which the carbon atoms form closed rings. They are the major constituents of petroleum.

alkaline accumulator

Also known as alkaline storage cell or battery. A secondary *electric cell* in which the electrolyte consists of an aqueous solution of an alkali, usually potassium hydroxide.

See also *nickel-cadmium accumulator; nickel-iron accumulator*.

alkaline primary cell

An *electric cell* in which a negative *electrode* of zinc is separated from the manganese dioxide positive electrode by a potassium hydroxide *electrolyte.* The energy-producing reaction is irreversible so that the cell cannot be recharged. This cell offers high discharge rates and although more costly to manufacture than the similar *Leclanché cell* is nevertheless an important source of small-scale, portable electricity.

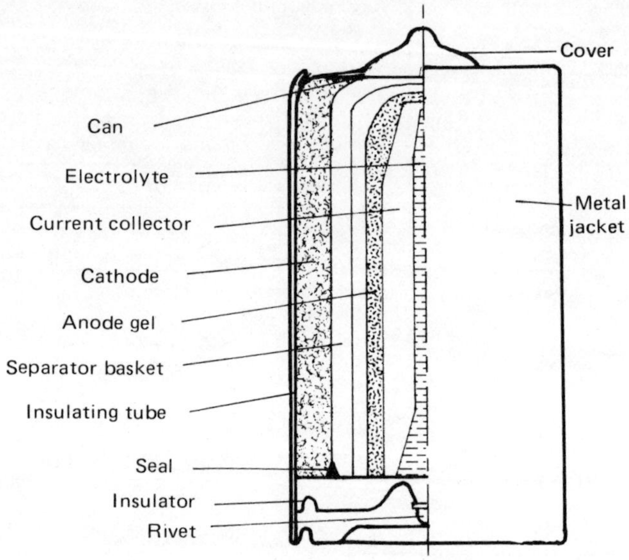

Diagram of a primary alkaline cell

alkaline storage cell or battery

An alternative name for *alkaline accumulator.*

alkanes

Members of the series of saturated *aliphatic hydrocarbons* with the general formula C_nH_{2n+2}. Also known as paraffins, alkanes occur naturally in *petroleum* and *natural gas*, and include *methane, ethane, propane, butane* (all shown below) and *pentane.* Butane and higher alkanes (ie those with a larger number of carbon atoms) can occur in different molecular forms, depending on whether the carbon skeleton is straight or branched. The form which has the straight carbon chain is known as the normal alkane.

See also *branched chain hydrocarbons.*

CH_3	methane	
CH_3CH_3	ethane	CH_3
$CH_3CH_2CH_3$	propane	$CH_3-CH-CH_3$ isobutane
$CH_3CH_2CH_2CH_3$	normal-butane	

alkenes

Members of the group of unsaturated *aliphatic hydrocarbons* containing one double bond between carbon atoms (generally written as C=C in chemical formulae). They have the general formula C_nH_{2n}, and the simplest member is ethene, C_2H_4, which is generally known as *ethylene.* The names of the alkenes can be derived from those of the corresponding *alkanes* by replacing the -ane suffix by -ene (eg pentane, pentene).

The three lowest alkenes, *ethylene, propene* and *butene* are all important *chemical feedstocks* or intermediaries.

alkylation

In petroleum processing, a process by which a *branched chain hydrocarbon*, usually isobutane, is reacted with an *alkene* (ethylene, propene, butene, pentene are all used) to form compounds for blending to give *gasoline* with a high *octane number.*

Also the process by which *benzene* and other aromatics are converted to aromatic blending agents by the addition of an *alkene* to the benzene ring.

Alkylation may be carried out by thermal means but the use of a *catalyst* results in less severe conditions and hence cheaper process plant. Sulphuric acid or hydrofluoric acid are used as catalysts.

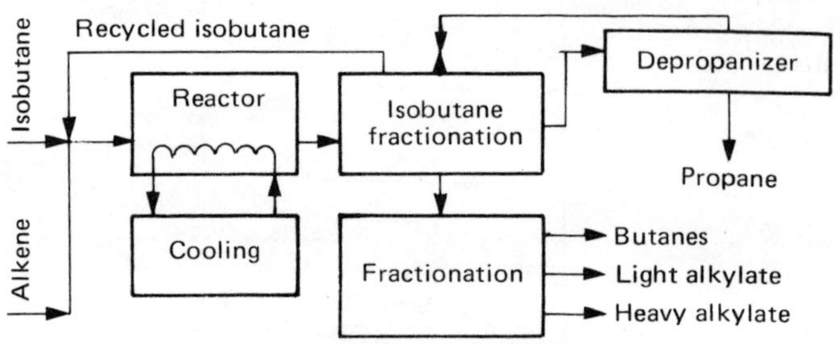

Alkylation — flow diagram

alkynes

Members of a group of unsaturated *aliphatic hydrocarbons* with one triple carbon-carbon bond (usually written C≡C in chemical formulae). They have the general formula: C_nH_{2n-2}. The most important member commercially is the simplest member of the series, acetylene, C_2H_2 (or HC≡CH). The name of any alkyne can be derived from the name of the corresponding alkane by replacing the -ane suffix by -yne, so that acetylene is more correctly called ethyne. (Alkynes are also known as acetylenes.)

Alnico

American trade name for an alloy of *aluminium*, nickel, *cobalt*, iron and *copper* forming a high-energy permanent magnet material.

alpha particle (α particle)

A helium *nucleus* (2 *protons* and 2 *neutrons*) emitted at very high speed by a radioactive substance. This happens because the heavy nuclei of radioactive substances are unstable with an excess of protons and will eventually emit alpha particles and thus reduce the number of protons. As a result the *atomic mass* of the nucleus and its *atomic number* (Z) will change.

alpha radiation

Strongly ionizing *radiation* consisting of a stream of *alpha particles* emitted at high velocity by radioactive substances. They can penetrate a few centimetres of air but most are stopped within the ordinary thickness of paper or a very thin layer of aluminium foil.

The emission of an alpha particle from a *nucleus* may be followed by a burst of electromagnetic radiation, a *gamma ray*.

alternating current (AC or ac)

An electric current that reverses direction periodically, usually many times per second. Many *electricity transmission* networks use alternating current because of the relative ease with which voltage may be controlled.

alternative energy sources

Sources of heat and power that do not rely on *fossil fuels* or nuclear *fission.* These include *solar energy, hydroelectric power, wind energy, wave energy, tidal energy, biological energy sources*, nuclear *fusion, geothermal energy* and *ocean thermal energy conversion.* The term may also be understood to include nuclear fission and such fossil fuels as are presently little used, such as *oil shales* and *tar sands.*

aluminium (aluminum)

A chemical element, aluminium is a light, ductile, grey metal. It is produced by the electrolysis of a molten mixture of bauxite and cryolite, a process which consumes large amounts of electric power. For this reason aluminium smelters are usually sited near cheap sources of power, such as hydroelectric schemes.

symbol	Al
atomic number	13
atomic weight	26.98
melting point	660.37°C
boiling point	2350°C
density	2,698 kg m^{-3} (2.70 g/cm^3)

The metal is a good conductor of heat and electricity and is resistant to atmospheric corrosion. It is used for high voltage *electricity transmission* lines where, for equal conductance, the weight of aluminium required is half that of copper, and the greater surface area of aluminium reduces corona discharges. Aluminium and its alloys are extensively used for heat-transfer equipment, such as *heat exchangers.* As a reflective foil the metal is used for the thermal insulation of *cryogenics* equipment, furnaces and other high temperature plant and in buildings. Deposited as a reflecting layer on glass the metal is used in some types of focusing solar collectors. The powdered metal finds application as a rocket propellant.

aluminum

American spelling of *aluminium.*

ambient energy sources

A term often used to describe sources of power that arise from the harnessing of natural forces and, in particular, those whose utilization leads to no net input of heat into the earth. For example, the exploitation of *hydroelectric power, wind energy* or *wave energy* entails the diversion of a source of mechanical power that would otherwise be degraded to heat at the temperature of the surroundings. In contrast the use of *fossil fuels* may lead to *thermal pollution.* Certain other energy sources, including *solar energy* and *tidal energy*, are often included under the banner of ambient sources.

Interest in these renewable resources is on the increase at present with the growing realization that the reserves of fossil fuels are finite and fears that large scale exploitation of *nuclear energy* may carry unacceptable risks.

The term 'alternative energy sources' is commonly used to describe the whole spectrum of energy sources that do not rely on fossil fuels or nuclear *fission.*

See also *biological energy sources; environmental effects.*

amines

Organic derivatives of *ammonia* obtained by replacing one or more hydrogen atoms of the organic radical with an amino group. An important use is in the treatment of *acid gas.* Monoethylamine in solution is used to remove corrosive substances such as *hydrogen sulphide* and *carbon dioxide* from natural gas.

ammonia

A pungent colourless gas (formula NH_3) produced through the combination of *hydrogen* with atmospheric *nitrogen.* Although nitrogen is relatively simple to obtain, hydrogen must be obtained either from hydrocarbon sources or by the chemical dissociation of water. In the past *coke* was the main source of hydrogen, via the *water gas* reaction, but now hydrogen is made by the *steam reforming* of hydrocarbons, largely from *petroleum fractions* and *natural gas.* Ammonia is also a by-product of coal *carbonization.* (Outside the USSR, Eastern Europe and China, some 40 million tonnes a year is currently produced mainly for the manufacture of fertilizers.)

Ammonia has a *calorific value* of approximately 400 Btu/ft^3 (14.9 MJ/m^3). Its melting point is −78°C and its boiling point is $-33.3°C$. It can be liquified easily and is a possible secondary fuel. *Ammonium nitrate* (mixed with an oxidizer) is used as a rocket propellant. Anhydrous ammonia is used in *ammonia-air fuel cells.* Ammonia is an efficient working fluid for heat exchange apparatus.

See also *cryogenics; heat pump; refrigeration.*

ammonia-air fuel cell

A *fuel cell* using compressed liquid *ammonia* as fuel which has an efficiency of 30 per cent at an output of approximately 300 watts. (Larger plants are expected to have higher efficiencies.)

ammonium nitrate

Low cost, medium performance, smokeless fuel (formula NH_4NO_3) used in gas generators for turbine pumps. It is also used as a slow burning propellant for rockets and missiles, commonly combined with liquid oxygen as an oxidizer.

anaerobic fermentation

The breakdown of sugars, proteins (when it is also known as putrefaction) and other organic substances by micro-organisms in the absence of oxygen to yield *alcohol* and *carbon dioxide* or other simple molecules. For example, the fermentation of glucose:

$$C_6H_{12}O_6 \rightarrow 2C_2H_5OH + 2CO_2$$

Other substances including fatty acids, glycerol and aldehydes remain in small quantities. The chemical reactions are facilitated through the catalytic intervention of enzymes in the living organisms.

The process has applications in *sewage* plants where *methane* released during the fermentation of organic substances in sludge is often used to provide power for running the sewage plants. This process is known as 'digestion'. During World War II *carbohydrates* and most root crops were fermented to produce *ethanol* and *methanol* which were constituents of *aviation spirit.*

Serious studies are currently in hand in the USA, Australia and various EEC countries to investigate the feasibility of energy farming by which, for example, crops are grown for their total energy content for food and fuel, or for fuel alone. For example, *sugar cane* and *cassava* are carbohydrate-rich plants which yield alcohol by fermentation at a price competitive with other methods of producing industrial alcohol.

animal wastes

Biological waste materials of animal origin, such as *dung,* which are *alternative energy sources.* Digestion of animal wastes anaerobically yields *methane.*

See also *anaerobic fermentation; biological energy sources; methanol; waste materials energy.*

anode

The positive *electrode* in an *electric cell.*

anthracene oil

A crude *fraction* obtained by the *distillation* of *coal tar* between 300°-400°C and containing 5 to 10 per cent anthracene and other hydrocarbon oils, eg phenanthrene, chrysene and carbazole. Relatively pure anthracene is obtained by fractional crystallization and is an important precursor in the dye industry.

anthracite

A high-grade *coal* having a lustrous appearance, a high carbon, low ash content, high density and burning slowly with a blue flame, producing little smoke or odour.

It has a *calorific value* of 320 therms/tonne (33.8 GJ/tonne) and is most generally employed as a domestic heating fuel.

anticline

An arch-like folded structure in which rock strata dip down outwards on each side away from the axis of the fold. The diagram under the entry for *caprock* shows the typical features of an anticline. In *sedimentary rocks* anticlines are common oil traps and as such are geological features investigated geophysically. The top and sides of the arch form impermeable barriers to trap and hold *hydrocarbon* materials in the underlying permeable strata. A large proportion of the oil discovered to date occurs in anticlines.

anti-knock component

A substance such as tetraethyl lead added to *gasoline* to reduce the fuel's tendency to detonate or 'knock' on ignition in a petrol engine. Knocking or pinking can damage an engine. The addition of an anti-knock component to a fuel raises its *octane number.*

API

Initial letters for American Petroleum Institute.

apiezon oils

The residue left by the vacuum distillation of *petroleum* products which has an almost zero vapour pressure.

API gravity

An arbitrary scale expressing the specific gravity or density of liquid petroleum products, and calibrated in degrees API:

$$\text{Deg API} = \frac{141.5}{\text{specific gravity } 60°\text{F}/60°\text{F}} - 131.5$$

appraisal well

Also known as outstep well or step out well. A well drilled close to another (already proven to have oil or gas producing potential) to gain information about a reservoir's characteristics and help assess its commercial potential by removing and analyzing cores and undertaking production tests. A *delineation well* is a special kind of appraisal well drilled to determine the boundaries of the reservoir.

See also *petroleum.*

arch dam

A type of dam whose arch design enables the overall cross-section to be reduced as water pressure on the face of the dam is transmitted to the abutments. It is particularly well suited to narrow gorges. Examples of arch dams are the Boulder Dam, Colorado, and in the UK the Galloway hydroelectric scheme.

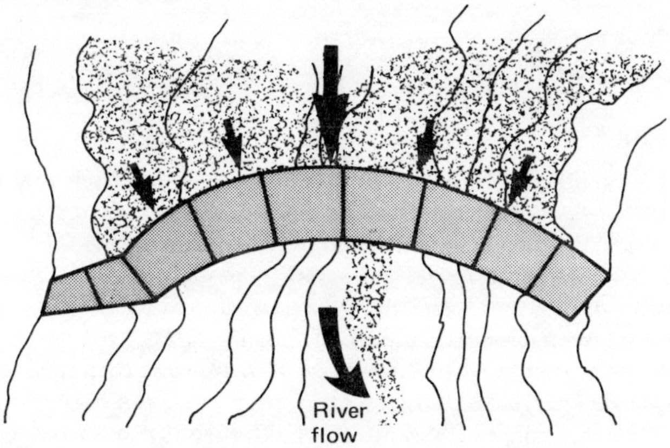

Arch dam

argon

An inert gaseous element, traces of argon are found in sea water, geothermal steam and natural gas and it constitutes about 1 per cent by volume of the atmosphere. It is isolated by the *fractionation* of liquid air.

Argon is used for filling incandescent electric lamps and fluorescent tubes; and in radiation counters. It may also be used in the experimental development of rocket fuels as a frozen solid in which free hydrogen atoms could be held.

symbol	Ar
atomic number	18
atomic weight	39.948
density	1656 kg m^{-3} at 233°C
melting point	−189.2°C
boiling point	−185.7°C

In liquid-metal-cooled *nuclear reactors*, the inert argon is used as a 'blanket' on open surfaces of the *coolant* which is strongly reactive sodium or sodium and potassium alloy. In carbon dioxide or air-cooled nuclear reactors, any argon present in the coolant forms radioactive isotopes on the absorption of neutrons and thus a radioactive gas which may pollute the atmosphere in the neighbourhood of the reactor.

aromatic hydrocarbons

Compounds, or mixtures of compounds, containing *benzene rings* in their molecules. The simplest aromatic is *benzene,* C_6H_6 ; more complex aromatics consist of combinations of this basic molecular block with others, such as alkyl groups.

Aromatics are found in considerable quantity in the low boiling distillates of *crude oils*, and can also be manufactured by *catalytic reforming* of *naphthenes* and certain *alkanes.*

Aromatics are useful anti-knock additives for *gasolines,* but adversely affect the burning properties of *kerosines* and *gas oils.* Certain aromatics, notably benzene, *toluene* and xylene are prepared in large quantities for use as solvents and chemical intermediates.

artificial oil

Oil that is manufactured from a carbon-containing energy source other than petroleum, such as coal.

See also *coal liquifaction.*

ash

The non-volatile inorganic residue resulting from the burning of an organic compound or mixture of compounds. For example, when *coking coal* is burnt for the production of *coke* large quantities of ash are formed.

asphalt

The term given in the USA to the solid or semi-solid residues from the distillation of certain *crude oils.* In the UK this is known as *bitumen* and asphalt is used to describe natural or artificial mixtures of bitumen with mineral matter.

asphaltenes

Complicated *hydrocarbons* with molecular skeletons of the graphite type linked together by long *carbon* chains. Asphaltenes or 'asphaltic matter' are present in almost all *crude oils*, but being effectively non-volatile they appear only in the heavier fractions, heavy oils and *bitumen.* Asphaltenes are generally removed from lubricating oils during refining.

asphaltic

A petroleum base classification indicating oil at the heavy end of the range (15° API and lower). Asphaltic petroleums are rich in *hydrocarbons* of high boiling point.

See also *API gravity.*

associated gas

Natural gas occurring with *petroleum* deposits which is either overlying the oil-bearing strata or dissolved in the oil. Also known as gas cap gas and solution gas.

The presence of associated gas means that the oil is under pressure. This aids extraction and in many areas the gas is reinjected for the same purpose. Until recently associated gas was regarded as a hindrance and vast quantities were wasted through flaring off at the well head. The value of such natural gas as a premium fuel and *chemical feedstock* is now fully appreciated and wastage has been dramatically reduced. In offshore situations especially, activities on the production platforms are often fuelled by associated gas.

ASTM

Initial letters for American Society for Testing and Materials.

Atgas process

An American *coal gasification* process in the development stage which produces a *pipeline quality gas* with a calorific value of up to 900 Btu/ft^3 (33.5 MJ/m^3). Crushed, dried *coal* is injected into a molten iron bath along with steam, oxygen and limestone at a temperature of approximately 2,600°F (1,430°C).

The coal dissolves and its volatiles are cracked into *carbon monoxide, methane* and *hydrogen.* Further carbon monoxide and hydrogen are produced as the *carbon* reacts with the steam and oxygen. (The ash and slag are continuously withdrawn; the slag is desulphurized and a portion is recycled into the gasifier.)

The raw gas from the furnace is cleaned and upgraded to pipeline quality gas by the *shift reaction* and *methanation* processes. By using air instead of oxygen a *low-Btu gas* of about 145 Btu/ft^3 (5.4 MJ/m^3) is produced. The process is still undergoing tests to establish its feasibility as a large scale conversion method.

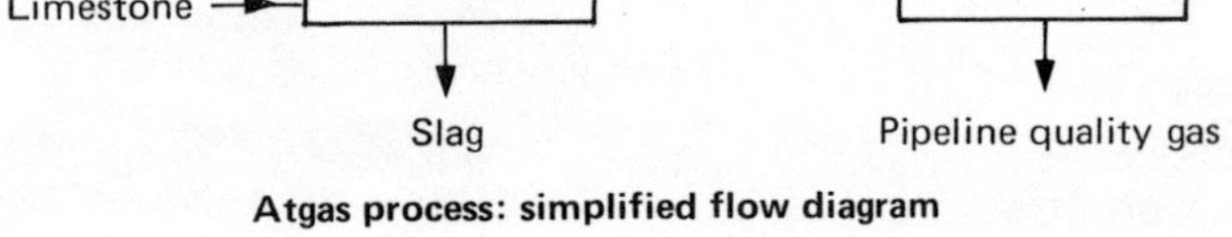

Atgas process: simplified flow diagram

Athabasca tar sands

The world's largest known *tar sands* deposits cover an area of 34,000 km^2 in the province of Alberta, Canada. Estimates of reserves suggest an upper limit of approximately 300 thousand million barrels of artificial oil. (This roughly compares to 5/6 of the total known Middle East conventional oil deposits.)

The quantity of *oil in place* is certainly gigantic, but until the technology of *in-situ mining* is substantially improved, a major portion of these reserves remains irrecoverable. Prevailing environmental conditions are extremely hostile, and winter temperatures can reduce metal machinery to an unworkably brittle state.

The Great Canadian Oil Sands enterprise has been operating there since 1967. After initial difficulties the company produces in the region of 45,000 barrels of oil per day. Other projects include the *Syncrude* venture, established in 1973, which has a projected target of 125,000 barrels per day by the 1990s.

See also *tar sands.*

ATK

UK abbreviation for aviation turbine kerosine.

See also *aviation turbine fuel.*

atom

The smallest amount of a chemical element which retains its chemical identity. The smallest amount of matter which cannot be sub-divided by purely chemical means. Neutral atoms consist of *nuclei* about 10^{-12}cm diameter, of density about 10^{24}g cm^{-3} and charge Z surrounded by Z *electrons,* the whole having a diameter of approximately 10^{-8}cm. The charge Z is due to the presence of Z *protons* each having a positive charge of 1.6021 x 10^{-19}coulombs, and each of the Z electrons carries the same charge though it is negative. The number of electrons and their arrangement determine the chemical properties of an atom and hence of a chemical element. Nearly all (99.9 per cent) of the mass of an atom is concentrated at the nucleus. Z is known as the *atomic number.*

atom bomb

A bomb, whose power is provided by nuclear fissile material, eg *plutonium-239* or *uranium-235* (it is also known as A bomb and fission bomb).

See also *fission energy; hydrogen bomb.*

atomic energy

An alternative name for *nuclear energy,* this term was initially widely used to mean the energy obtained from *fission* as in the *atom bomb* and *atomic reactors*: hence the United Kingdom Atomic Energy Authority, the United States Atomic Energy Commission, or Atomic Energy of Canada Ltd.

atomic mass

The mass of an atom of an *isotope* relative to the mass of

an atom of carbon-12, the most abundant isotope of carbon. See also *atomic weight.*

atomic number

Symbol Z, the atomic number is the number of *protons* in the *nucleus* of an *atom*, equal to the number of *electrons* around the nucleus of the atom in a neutral state. Elements are chemically classified by their atomic numbers.

atomic pile

Outdated general name for a *nuclear reactor.* It arose from the first reactor – CP-1 (Chicago Pile No 1) – set up by Enrico Fermi which consisted of a stack or pile of graphite bricks in which were embedded balls of uranium metal and uranium powder. The whole roughly spherical structure was contained in a wooden frame. Strips of neutron-absorbing *cadmium* were placed in the pile to control the possibility of a premature *fission chain reaction.* The pile reached a power level of 200 watts but was shut down for safety reasons. It was rebuilt as Chicago Pile No 2 in 1943. Other types of nuclear reactor have been derived from these primitive experimental beginnings and for some years nuclear reactors were called atomic piles.

atomic reactor

Popular name for a *nuclear reactor.*

atomic weight

The atomic weight of an element, which may in nature exist as a mixture of isotopes of different *atomic mass*, is the ratio of the statistical average weight of the atoms of the element (ie allowing for the differing natural abundances of the isotopes), to one twelfth of the weight of one atom of carbon-12.

Originally atomic weights were measured relative to the weight of one atom of hydrogen-1: the carbon-12 scale (which leads to very similar numbers for atomic weights), was chosen for ease of experimental determination.

auger mining

A method of extracting *coal* by the drilling of large-diameter holes into the seam, using an auger. It is used for seams of moderate depth when the overburden is too deep to be removed economically.

autofining

A preliminary process in *petroleum refining* to remove the *sulphur* from light *petroleum* products. Partial *cracking* produces *hydrogen* which catalytically removes sulphur from the remaining stream.

availability factor

1. A measure of the maximum theoretical efficiency with which heat can be converted to work. It is a function of the temperature at which the heat can be supplied, T, and the temperature of the heat sink to which heat must be rejected, T_0, and takes the form:

$$\text{availability factor} = \frac{T - T_0}{T}$$

(T_0 will generally be ambient temperature.)

Also known as Carnot fraction, the availablity factor can be taken as a measure of the quality, or grade, of the heat; a high availability factor indicates high quality heat.

2. The term is also used as a measure of the reliability of an uncertain source of power, such as an aerogenerator or wave energy device. It is calculated as the ratio of the total energy produced in a given period to that which would have been produced if the maximum rated power output had been maintained throughout that period.

For example, an annual availability factor would be defined, as a percentage, as:

$$\text{annual availability factor} = \frac{100 \times \text{total energy output (MWh)}}{\text{(number of hours in year)} \times \text{rated power output (MW)}}$$

In this sense, the term is analogous to the *load factor* of a power station. See also *energy.*

available energy

Energy which is, in principle, capable of being transformed completely into mechanical work.

AVGAS

Abbreviation for aviation gasoline (*aviation spirit*).

aviation spirit

Also known as aviation gasoline. A *petroleum fraction* in the *gasoline* range used as fuel in piston-engined aircraft. There are five grades of aviation spirit in the UK whose octane ratings fall in the range 73-115/145; in the United States a similar range is covered by four grades.

aviation turbine fuel

A straight-run *kerosine* used as a fuel in jet-engined aircraft. Two fractions may be used, aviation kerosine with a boiling range 150°-250°C and the so-called 'wide-cut gasoline', also known as aviation turbine gasoline, with a broader boiling range (30°-260°C).

AVTAG

UK abbreviation for aviation turbine gasoline.

See also *aviation turbine fuel.*

AVTUR

UK abbreviation for aviation turbine kerosine.

See also *aviation turbine fuel.*

B

background radiation

Radiation from naturally occurring, rather than man-made, sources to which the population as a whole is exposed. It is usually less than 0.1 *rem* per year but in Aberdeen, Scotland, a city built of uranium-containing granite, this background level is as high as 0.15 rem per year, 30 per cent of the ICRP *dose limit*.

See also *dose level*.

back-pressure steam turbine

A type of *steam turbine* in which all the exhaust steam is used for heating purposes. Steam at low pressure for heating purposes or process work must be free from contaminants. Turbine exhaust finds use in industrial areas such as textiles, laundries, sugar refining and paper manufacture.

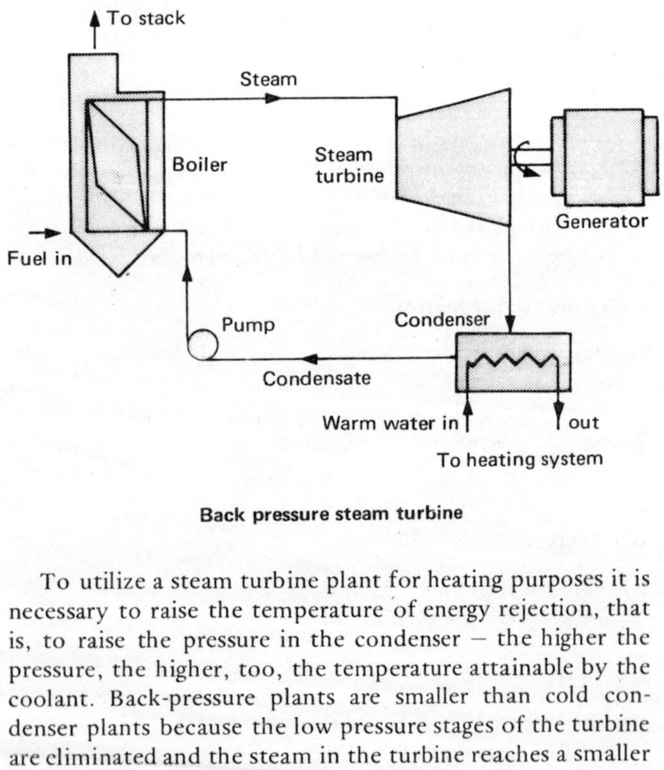

Back pressure steam turbine

To utilize a steam turbine plant for heating purposes it is necessary to raise the temperature of energy rejection, that is, to raise the pressure in the condenser – the higher the pressure, the higher, too, the temperature attainable by the coolant. Back-pressure plants are smaller than cold condenser plants because the low pressure stages of the turbine are eliminated and the steam in the turbine reaches a smaller volume. An additional advantage is the elimination of the cooling towers of conventional power stations. On the other hand, the power output is reduced. For modest back pressures this amounts to some 1 per cent per 3°C increase in condensing temperature.

See also *district heating; geothermal energy; total energy systems.*

bagasse

The dried fibrous refuse of *sugar cane*. Bagasse, a *cellulose* material, has a possible use as a fuel. It is currently burnt as fuel on a small scale in some sugar processing factories.

barrel (bbl)

The standard unit of volume for petroleum. One barrel is equivalent to 42 US gallons, 34.97 Imperial gallons, 35 Canadian gallons or 159 litres.

barrels per day (b/d)

A measure of the average daily petroleum output from an oil well or oil field. It is also used as a measure of the rate of consumption of crude petroleum, or petroleum products, on a national or regional basis.

barrels per stream day (BPSD)

The rate per day in barrels of *chemical feedstock* passed through a continuously operating petrochemical process.

barrels per tonne

The number of barrels of oil (a barrel = 42 US gallons) yielded by a tonne of crude. Depending on the specific gravity of the crude this might vary from 5.8 to 9.1 barrels per tonne. An average yield of 7.4 barrels per tonne is used to convert tonnes to barrels and vice versa.

base load

The minimum load of an electrical supply system over a given time period, usually expressed in watts or multiples thereof, such as MW, GW.

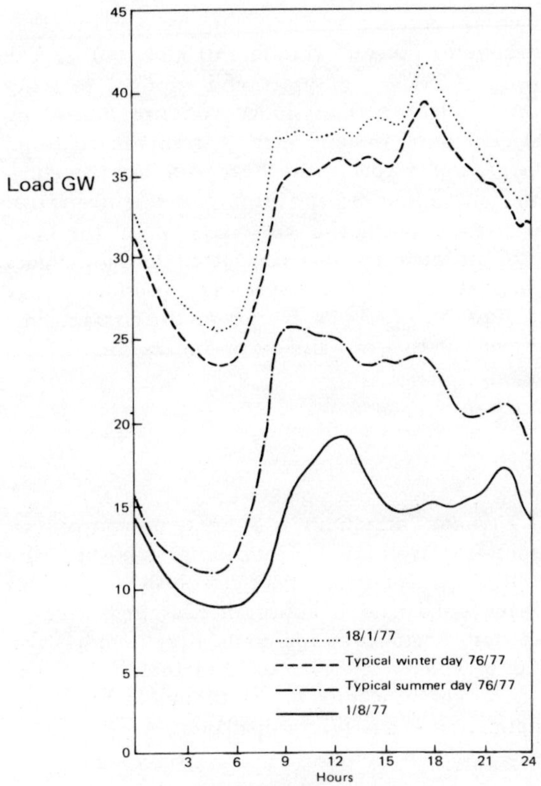

Base load

Load curve for the Central Electricity Generating Board (England and Wales) for two days of maximum and minimum demand and two typical winter and summer days.

CEGB UK 1966/1977 Annual Report

The diagram shows the load curve for the system of the Central Electricity Generating Board of England and Wales for the two days of maximum and minimum demand and two days of typical summer and winter demand in the (financial) year 1976/77. The base load was approximately 9000MW in that year.

base load plant

Electrical generating plant which is kept running to meet the *base load* demand of an electrical supply system. As demand rises above the base load more generating capacity must be brought into use. The order in which this is done is carefully controlled for the system by a hierarchy of generating costs (called a *merit order* in the UK). The stations having the lowest marginal generating costs are highest in the merit order and supply base load. Conversely, those stations having highest marginal generating costs supply *peak load*.

For systems having nuclear as well as *fossil fuel* stations, the nuclear stations are higher in the merit order since these have low fuel costs and very small marginal running costs. The larger and more efficient fossil fuel stations will be next in the merit order and will also contribute to meeting the base load.

base oil

A refined or untreated oil which is used with other oils and additives to produce lubricants.

battery

1. A term often used to describe an *electric cell* or an assembly of such cells.
2. A series of distillation columns used for *petroleum refining* and operated as a single unit.

In the energy industries the term is sometimes used to describe a *direct-current* voltage source made up of several units that convert thermal, chemical or *solar energy* into *electrical energy*.

bbl

Abbreviation for barrel.

bd, b/d

Abbreviation for barrels per day.

Becker nozzle process

A method of separation of isotopes of *uranium.* Jets of *uranium hexafluoride* vapour are mixed with helium and subjected to pseudo-gravitational fields. Isotopes of different mass are separated.

Becquerel rays

Penetrating *alpha, beta* and *gamma radiations* from *uranium*, named after the discoverer.

benign energy sources

Another term for *ambient energy sources.*

benzene

The simplest and first known *aromatic hydrocarbon*, chemical formula C_6H_6, obtained as a by-product both in refining petroleum and in producing *coke* from *coal.* Discovered by Faraday in 1825, benzene is a colourless, highly flammable liquid, insoluble in water and soluble in organic solvents. It is itself a valuable solvent but has carcinogenic properties and the maximum permissible concentration is less than 10 ppm. It has a specific gravity of between 0.71 to 0.79, boiling point 80.1°C, and melting point 5.53°C.

The benzene molecule has a ring structure known as the *benzene ring.* This basic structure characterizes the entire aromatic class of organic compounds. The major uses for benzene lie in producing high grade motor fuels — it increases the octane rating — and as an intermediary in numerous chemical processes.

benzene ring

Name given to the basic characteristic atomic grouping in aromatic hydrocarbons. The ring comprises six carbon atoms joined to one another, with a single hydrogen atom linked to each carbon atom; the illustration shows two ways of depicting the benzene ring. A wide range of chemicals can be produced by substituting one or more groups for the hydrogen atoms around the ring. For example, *toluene* $C_6H_5CH_3$, *styrene* $C_6H_5CHCH_2$, and the *xylenes* $C_6H_4(CH_3)_2$.

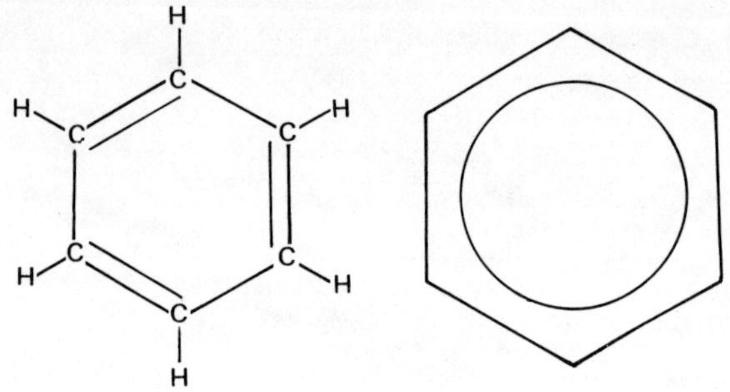

Two ways of representing the benzene ring symbol

benzine

A mixture of *aliphatic hydrocarbons* in the *paraffin* series with boiling points between 80°-130°C. Formerly the term itself was used synonymously with *gasoline* and petrol. Benzine is used in producing motor spirits and as a solvent for cleaning and in resin manufacture.

benzol

Alternative name benzole. A mixture of crude *benzene* and other *aromatic hydrocarbons* produced from *coal tar*. Benzol is used as motor spirit, in nitration and for general industrial purposes.

See also *benzol oils*.

benzol oils

Alternative name for benzol. Such oils are rich in benzol and derived by petroleum *cracking* or *coal tar* distillation. Light oil distillate from coal tar processing is fractionated to yield a mixture of benzols, toluols and xylols. *Benzene*, *toluene* and *xylene* (so-called btx chemicals) are derived from these alcohols.

Bergius process

A process for hydrogenating carbonaceous matter, primarily *coal*, but also *cellulose* materials, to produce a liquid fuel similar to petroleum.

The process comprises four stages: 1. Powdered coal together with a *catalyst* (usually iron oxide) is made into a paste with heavy oil or tar. 2. The preheated paste together with hydrogen gas enters a series of converters maintained at 250 atmospheres pressure (minimum) and 450°C. The reaction products consist of gases and vapours together with a heavy sludge. 3. The sludge contains heavy oil, unconverted coal dust, mineral matter from the coal and catalyst. *Distillation* recovers the heavy oil for recycling. 4. Passing the gas/vapour stream through a *heat exchanger* preheats the paste entering the converters and at the same time cools the product stream. The latter then undergoes fractional distillation under pressure. The products range from volatile liquids to fuel oils. Increasing the contact time reduces the yield of fuel oils and so optimizes the yield of liquid fuels.

See also *coal liquifaction*.

beryllium

A chemical element, beryllium is a light, non-corroding, silver-grey metal. It is obtained from the mineral beryl by chemical extraction followed by electrolysis. Alloys of the metal have considerable importance as high-strength corrosion-resistant engineering materials. Beryllium has also been used as a *moderator*, *reflector* and fuel *cladding* in nuclear reactors. The powdered metal, though highly toxic, may find application as a rocket propellant.

symbol	Be
atomic number	4
atomic weight	9.01
melting point	1285°C
boiling point	2470°C
density	1,846 kg m^{-3} (1.85 g/cm^3)

beta-alumina solid electrolyte battery

Alternative name for *sodium-sulphur battery*.

beta-particle (β-particle)

An electron emitted at very high speed by the nucleus of a radioactive substance. The effect is to increase the *atomic number* of the nucleus by one.

See also *beta-radiation*.

beta-radiation

Ionizing *radiation* consisting of a stream of *beta particles* emitted by radioactive substances at high velocity. They can penetrate 700–800 cm of air or several millimetres thickness of aluminium.

The emission of a beta particle from a *nucleus* may be accompanied by a flash of electromagnetic radiation, a *gamma ray*.

bi-gas process

A process for converting *coal* into a gaseous fuel. Pulverized coal and steam are introduced (under pressures of between 70 and 100 atmospheres) directly into a column of hydrogen-rich gas (*synthesis gas*) which is produced in the bottom of the gasifier by the slagging gasification of *char* with oxygen and steam at temperatures of 2500°F (1370°C) and above. The coal is converted partly into *methane* and further synthesis gas and this raw gas is drawn off for purification (primarily the removal of *hydrogen sulphide* and *carbon dioxide*). Further *methanation* yields a *high-btu gas*.

The principal feature of the bi-gas process is the direct production of methane from coal in larger quantities than can be achieved in conventional steam-oxygen gasifiers. No tar or oils are formed in the process and residual char is driven off through the top of the gasifier to be recycled as solid char for further slagging. The two stage gasifier is relatively simple in design and large scale operations are believed feasible.

See also *coal gasification*.

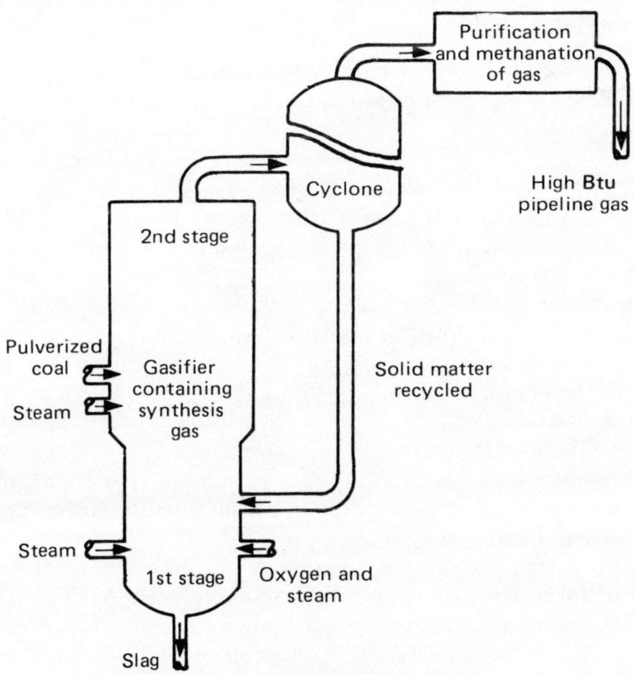

Bigas process: simplified flow diagram for a two-stage gasifier

biogas

A term generally used to describe any gas, usually combustible, produced by the enzymatic breakdown of organic materials. An example is *methane* produced by the digestion of sewage sludge, which represents a source of power for sewage disposal plants.

See also *biological energy sources; waste materials energy*.

biological energy sources

The major suppliers of fuel energy represented principally by *petroleum* and *coal* derived many millions of years ago from plant and animal remains. These are finite resources and attempts are currently being made to find alternative sources of energy including the application of natural processes such as *photosynthesis*. Apart from manpower and the use of animals to work machinery in primitive communities and as beasts of burden, little use is made of animals as suppliers of energy. In recent years attempts have been made to use *animal wastes* as energy sources. Various devices (*digesters*) are available in which the breakdown of farm wastes – eg chicken, pig and cattle manure – can be achieved with the production of *methane*. It is possible to fit converters to any type of internal combustion engine so that motor vehicles can operate using methane gas rather than *gasoline*.

See also *alcohol; algae; anaerobic fermentation; biogas; biomass; carbon cycle; cellulose; energy crops; fossil fuels; methanol; wood*.

biological shielding

The protective wall of material (concrete, lead, water) surrounding a source of radiation to reduce its intensity and especially to prevent the escape of gamma radiation.

biomass

The total weight of plants and aminals living in a particular area. Also a general term for living matter.

Biomass can provide food, fuel or fibre. In practical energy terms, plant production is of most interest, as plants have the ability to utilize the energy of sunlight to produce *carbohydrates* from carbon dioxide and water by *photosynthesis*. Further organic substances can be formed from the carbohydrate intermediates. For fuel requirements the organic products of photosynthesis can be processed by *anaerobic fermentation* to yield *alcohols*. It is also suggested that *hydrogen* could be produced using *chloroplast* membranes.

It is estimated that the amount of *carbon* fixed every year into stored energy through photosynthesis is ten times the world's total energy use. Only half a per cent is used by the world's population, and various ideas are proposed for utilizing biological energy sources as alternatives to conventional fuels.

See also *biogas; cellulose; energy crops; methane; methanol; sewage; waste materials energy; wood*.

bitumen

(Known as *asphalt* in the USA.) In the UK, the term applied to the solid or semi-solid residues from the distillation of certain *crude oils*. Bitumen also occurs naturally, alone or mixed with mineral matter. Substantially pure bitumen deposits are found in Utah and Colorado, USA. In the so-called 'pitch lake' in Trinidad, the deposit contains about 39 per cent bitumen, 32 per cent minerals and 29 per cent water and gas by weight. A tonne of bitumen yields from 5.8 to 6.4 barrels of oil.

In the UK natural or artificial mixtures of bitumen with mineral matter are known as asphalt. Common uses for this material are for road surfaces, roofing and waterproofing. A similar range of substances, usually called *pitch*, is obtained by the distillation of *coal tar*, or by the distillation of kraft pulping process residues from paper manufacture.

bituminous coal

A *coal* of medium to high rank, normally compact, dark and banded. It burns with a long, smoky flame because it contains a high proportion of volatile matter (*hydrogen, carbon monoxide, methane, ethanol, tar,* and *ammonia* liquor).

Bituminous coals are advanced in the stage of coalification which gives them a high carbon and low oxygen content, 5 to 15 per cent oxygen, 78 to 86 per cent carbon. Calorific value is 275 therms/tonne (29 GJ/tonne). This is the most abundant rank of coal and usually carboniferous in origin. Bituminous coals are widely used as *coking coals* because on heating they swell and give off water vapour, gas and tar, leaving behind *coke*. Classification of the major coal types is slightly varied between countries. For example, the ASTM standard includes five groups within the general class of bituminous coal.

bituminous shales

An alternative name for *oil shales*.

black damp

A term used in mining, which refers to air in which the oxygen has been replaced by *carbon dioxide*. This may occur following an explosion or *combustion*.

black oil

Also known as dirty oil. Dark-coloured heavy petroleum products such as heavy diesel oil and fuel oils. It is used when referring to storage and transportation of these fuels, for tanks which have contained black oil need to be cleaned out before they are used to store lighter petroleum fractions.

black shale

An alternative name for *oil shale*.

blanket

The layer of *fertile material*, usually *uranium-238* or *thorium-232*, placed around or within the core of a *breeder reactor* to be converted into *fissile material* by *neutron* capture. See also *fast breeder reactor*.

blast furnace gas

A low energy-content gas produced in the smelting of iron by *coke* in a blast furnace and containing about 30 per cent *carbon monoxide*. Calorific value is typically 90 Btu/ft^3 (3.4 MJ/m^3).

blending naphtha

A *petroleum* distillate used to thin or dilute heavy *petroleum fractions* (such as lubricating oils) to ease handling and processing.

blue water gas

A fuel consisting of approximately equal amounts of *carbon monoxide* and *hydrogen* produced by passing steam over heated *coke*. It has a calorific value of 350 Btu/ft^3 (13.0 MJ/m^3). When blue water gas is passed through a carburettor into which gas oil is sprayed *carburreted water gas* is formed.

See also *water gas*.

boghead coal

Also known as parrot coal. A non-banded, translucent *coal* yielding a high proportion of tar from the resins, waxes and spores it contains.

boilers

Closed vessels in which pressurized water is converted to steam by the action of heat. The term is also used more generally for equipment where boiling occurs and to describe water heaters connected to distribution systems. Boilers form an important element in many energy conversion technologies, including the generation of electricity by use of *steam turbines*, and the distribution of process heat in an industrial plant.

Boiler design aims to produce maximum transfer of the energy contained in the heating medium to the steam. The heating medium may be hot gases from fuel combustion or *coolant* from a *nuclear reactor core*. Boilers may be classified into two main types: 1. fire-tube boilers, in which the hot fluids pass through pipes immersed in the water and 2. water-tube boilers, in which the hot fluids circulate around pipes containing water. Water-tube boilers are inherently safer and, though more costly to make, are preferred for the majority of applications.

To ensure maximum use of the heat contained in the flue gases, incoming feedwater may be passed through a *heat exchanger* placed in the exit gas stream. This 'economizer' preheats the water and reduces the energy required for heating to boiling point. The air for combustion may similarly be treated in an 'air heater'.

If dry steam is required, as in the majority of applications, the output steam may be passed through a 'separator' to remove water. This may be of cyclone form,

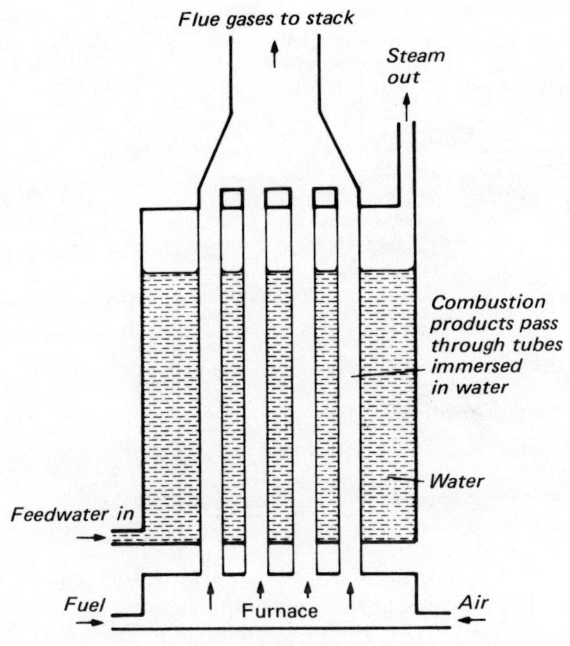

Fire-tube boiler

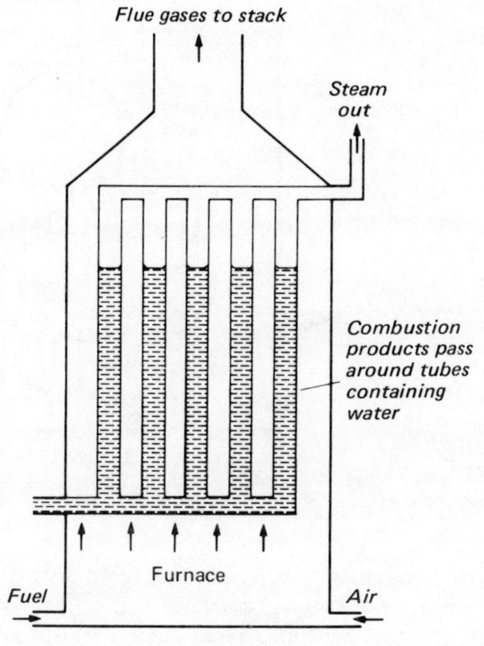

Water-tube boiler

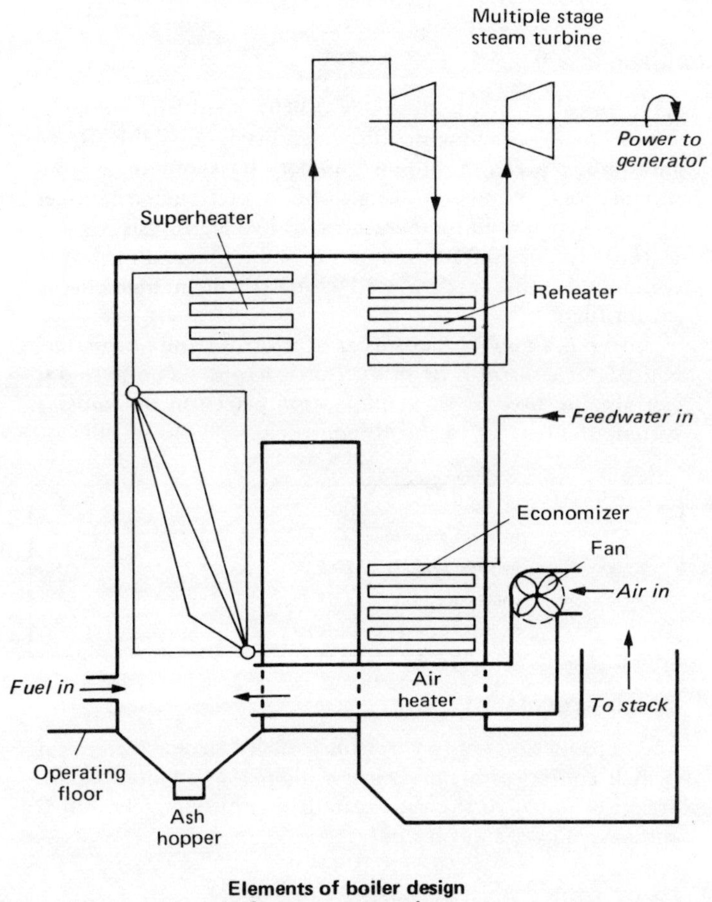

Elements of boiler design for power generation

in which the steam is made to move in a circular path, spilling water to the outside. More usually the steam is 'superheated' in a separate part of the boiler.

Turbines achieve highest efficiencies when expansion is closest to isothermal. For this reason multiple-stage turbines are used, with steam reheating cycles between stages. The steam is tapped from each stage, passed through a 'reheater' and passed on to the next.

boiling water reactor

A type of light water *nuclear reactor* in which steam is generated directly in the reactor core under pressure (typically around 68 atmospheres) and at a temperature of less than 300°C. Water circulated through the hot reactor core produces saturated steam. This is separated from the recirculated water, dried and piped to the turbo-generator unit. Such a direct-cycle steam generation system is similar to that used in fossil-fuelled thermal power plants. Numerous plants of the BWR type are in use. It is the simplest type of reactor. The steam generated is radioactive, but fortunately most of the *radioactivity* produced is due to N-16, an isotope of nitrogen with a half-life of only seven seconds. *Control rods* are inserted and withdrawn at the bottom of the reactor vessel where *reactivity* and *moderator* density are highest. Fuel is *uranium dioxide* pellets (2.2 per cent enriched is a typical figure) and clad in zirconium alloy (*zircaloy*).

A BWR operates at approximately half the pressure of a *pressurized water reactor (PWR)* but produces steam at a similar temperature. The reactor vessel is made of low alloy steel at the base and stainless steel elsewhere. Special jet

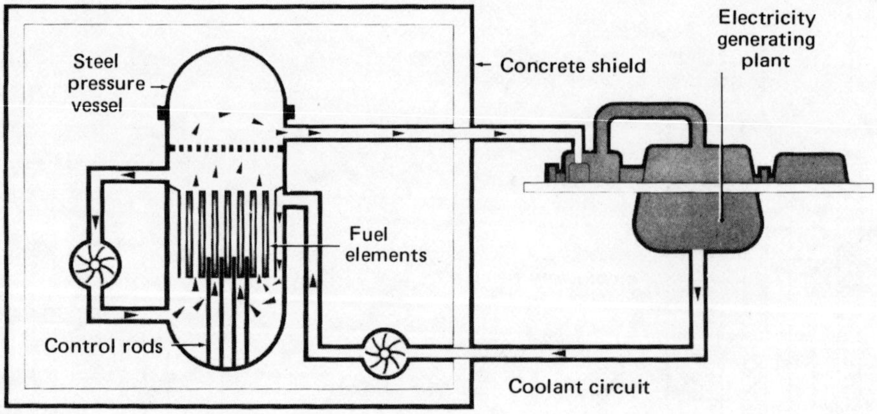

Reproduced by kind permission of the UKAEA

Boiling water reactor

pumps recirculate water in the primary system. This separate recirculation system will still operate should there be an external *coolant* system failure. It also permits rapid start-up. Control rods are boron carbide contained in stainless steel tubes: working life is expected to be fifteen years of full power production. A containment and clean-up pool is provided to absorb heat from *beta* and *gamma radiation* emitted from fission products remaining in the spent fuel in addition to heat from the dry well surrounding the reactor vessel.

Boiling water reactor data*	
Peak power density	49 kW/l
Heat output	3293 MWt
Electrical output	1065 MWe
Efficiency	32.3%
Fuel	Uranium dioxide, 2.2% enriched, clad in zircaloy
Weight of fuel	169 tonnes
Fuel burn-up	19,000 MWD/te
Moderator	Light water
Coolant	Light water
Coolant pressure	68 atmospheres
Coolant outlet temperature	285°C
Refuelling	Off load

* *Browns Ferry 1 — USA*

bord and pillar

An underground coal mining technique for deep mining operations. This method involves 'blocking out' areas of *coal* by a roadway network of squares, rectangular blocks or pillars and extracting coal through conventional or continuous processes. The coal-pillars are efficient supports but significant proportions of coal are left behind unless these are replaced by steel or timber supports.

Bord and pillar is the major technique employed in the USA, where it is called room and pillar, although *longwall mining* (widely used in Europe) is being introduced to reach deeper coal reserves.

See also *hydraulic mining.*

boron

A chemical element, boron is a light, hard non-metal. It appears as a yellowish brown crystalline solid or an amorphous greenish brown powder. It is obtained from several ores by reduction using carbon metals, metal carbides or hydrogen at high temperatures; or by electrolysis.

It has found limited use in semiconductors and the boron compounds, boranes, are used as ingredients in high-energy rocket fuels.

Boron is a powerful absorber of *neutrons* and boron steel is used for *control rods* in *nuclear reactors.* As boric acid it can also be used as a *chemical shim* placed in the *coolant* system of a reactor to absorb neutrons.

symbol	B
atomic number	5
atomic weight	10.811
melting point	2300°C
boiling point	2550°C

Bosch process

The production of *hydrogen*, usually from *water gas* (which comprises mainly *carbon monoxide* and hydrogen). Steam is added to the gas, heated to approximately 500°C and passed over an iron *catalyst.*

$$(CO + H_2) + H_2O \xrightarrow{Fe} CO_2 + 2H_2$$

Carbon dioxide removal is effected by absorption in aqueous solution.

bottled gas

The popular name for *liquified petroleum gas* stored under pressure in liquid form in portable steel 'bottles'. It consists mainly of *butane* and *propane.* The former is burned to provide heat for cooking, domestic heating and low temperature metalwork. Bottled propane has higher temperature applications such as for cutting metal.

BPSD

Abbreviation for *barrels per stream day.*

bradenhead gas

An alternative name for *casing head gas.*

branched chain hydrocarbon

A *hydrocarbon* in which the carbon skeleton is in the form of a chain with one or more branches, rather than a continuous line, as in a straight chain hydrocarbon. Many hydrocarbons can occur in branched or straight chain forms, eg *butane.*

The proportions of straight chain and branched chain hydrocarbons in an engine fuel are of great importance. For the compression ignition, or diesel engine, a fuel with a high proportion of straight chain hydrocarbons is required. On the other hand, for the spark ignition engine the fuel must have resistance to detonation, and the proportion of branched chain hydrocarbons should be high.

See also *anti-knock component; cetane number; octane number.*

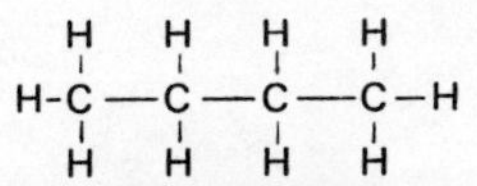

Normal butane
(n-butane)

A straight chain hydrocarbon

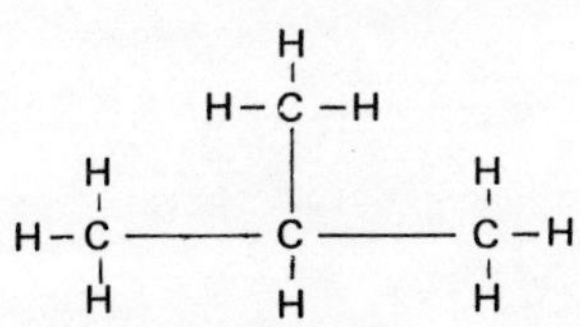

iso-butane

A branched chain hydrocarbon

breeder reactor

A *nuclear reactor* whose *breeding ratio* is greater than 1, that is, a reactor that produces by the *neutron* irradiation of *fertile material*, more *fissile fuel* than it consumes.

See also *fast breeder reactor.*

breeding ratio

Also known as the breeding factor, conversion ratio or conversion factor. The ratio between the *fertile fuel* converted into fissile by neutron irradiation in a *nuclear reactor* and the *fissile fuel* consumed. If, for every 10 fissile nuclei which undergo *fission* in the *fission chain reaction*, 10 fertile nuclei are converted into fissile nuclei, the breeding ratio is said to be 1; if only 5 are converted for 10 consumed, the breeding ratio would be 0.5.

bright coal

The general description given to *coal* which includes shiny bands of clean, bright vitrain or clarain.

briquettes

A combustible fuel, consisting mainly of low-grade powdered *coal* or *coke* compacted in pressure moulds into convenient shapes for transport and use. Commonly the moulds are prismal, ovoid or cylindrical in shape.

British thermal unit

Abbreviation Btu. The heat required to raise the temperature of 1 lb of water through 1°F at 60°F. Equivalent to 252 calories, 1055.79 joules.

brown coal

Another name for *lignite*, though some forms of lignite are black. Brown coals are useful sources of synthetic fuels and oils which are obtained by *hydrogenation.* Its calorific value is between 120 and 140 therms/tonne (12.7-14.7 GJ/tonne).

Btu

Abbreviation for *British thermal unit.*

bunker fuels

A general term covering heavy diesel oil and fuel oil stored in ship's bunkers.

burner reactor

A *nuclear reactor* that consumes more *fissile fuel* than it produces. Such a reactor has a *breeding ratio* of less than 1. All commercial power reactors are of this type, though various types of *breeder reactor*, which could produce more fissile fuel than they consume, are under development.

See also *thermal reactor.*

burning oil

A term used to describe a *petroleum fraction* at the heavy end of the *light distillate* range used in room heaters and oil-fired domestic heating boilers. More familiarly known as *paraffin* (kerosine in the US). In blocks of flats and similar premises equipped with larger oil-fired boilers a slightly heavier *petroleum fraction, gas oil*, is usually employed as boiler fuel.

burn-up

A measure of the degree to which fuel is consumed in a *nuclear reactor* calculated either as the percentage of *fissile nuclei* that have undergone *fission* or as the amount of heat energy produced per unit weight of fuel. It is usually measured in megawatt-days per tonne of nuclear fuel.

butane

An *alkane* containing four *carbon* atoms. Butane may exist in two forms – n-butane and iso-butane – and the commercial grade is usually a mixture of the two. A gas at room temperature and pressure, butane can be conveniently stored under pressure as a liquid and is so distributed as *bottled gas* for cooking and domestic heating. When blended into *gasoline*, butane improves *volatility* and *octane number*. Butane is obtained as a gaseous product from the distillation of crude *petroleum*, and as a by-product from *reforming* and *cracking* processes.

Butane also occurs in *natural gas* from some areas to a small extent, and in *associated gas* from oil fields.

See also *branched chain hydrocarbon; liquified petroleum gas.*

BWR

Initial letters for *boiling water reactor.*

C

C_4 fraction

Also known as C_4 and lower fraction. The fraction of *hydrocarbons* (mainly gases) occurring in *petroleum* deposits that is composed of C_4 compounds (those with four or less carbon atoms). Examples are *butane* (four), *propane* (three), *ethane* (two) and *methane* (one).

cadmium

A chemical element, cadmium is a rare, silver-white, soft metal always occurring in nature with zinc which it resembles. It is produced as a by-product of zinc smelting. It is used in electroplating steel and also for alloying with other metals to improve their properties.

symbol	Cd
atomic number	48
atomic weight	112.40
melting point	321°C
boiling point	770°C
density	8647 kg m^{-3} (8.65 g/cm^3)

Cadmium is a powerful absorber of *neutrons* and is used in steel for *control rods* in *nuclear reactors.* It is also used in *nickel-cadmium accumulators.*

The nickel-cadmium cell is an excellent energy source but costly so it is generally confined to small units. The *cadmium sulphide ceramic cell* is a photovoltaic energy converter still under development.

cadmium sulphide ceramic solar cell

A continuous low power *solar cell* primarily used in space craft as a small scale energy converter. Although expensive at present, the ceramic CdS cell shows promise as a terrestrial solar energy converter which would be considerably cheaper than the conventional single crystal *silicon cell* (ie mass production is more feasible for the CdS cell than for the silicon cell).

The required photovoltaic properties are derived from a barrier layer of copper doped cadmium sulphide deposited in a thin layer on the main body of treated CdS powder.

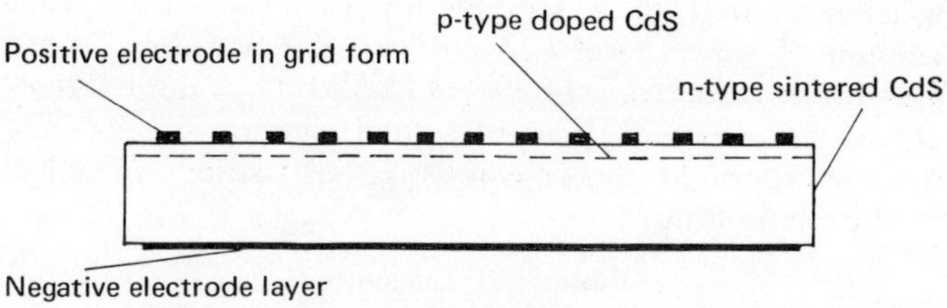

Structure of cadmium sulphide ceramic solar cell

The open-circuit voltage of the cell depends on the temperature and time of the activation heat treatment.

The ceramic CdS cell has a conversion efficiency from 5 to 8 per cent and under optimum conditions produces a current of 30 mA/cm^2.

caesium (cesium)

A chemical element, caesium is a soft, silver-white reactive metal. It is prepared by electrolysis of the fused cyanide.

It finds application in certain photoelectric devices and has been suggested for use in ion-propulsion low-thrust space craft.

symbol	Cs
atomic number	55
atomic weight	132.9
melting point	28.6°C
boiling point	686°C
density	1900 kg m^{-3} (1.90 g/cm^3)

Caesium-137, a radioactive *isotope* which decays by *beta particle* emission, is used as a source of *gamma radiation.* It has a half-life of 30 years. It is one of the *fission products* of uranium in nuclear reactors and is a potential source of radioactive pollution either by accidental release from an operational reactor or by routine releases during nuclear fuel reprocessing.

calandria

In nuclear engineering, a sealed vessel comprising a honeycomb of tubes used in the core of some kinds of *nuclear reactors*, eg *CANDU.* In *heavy water* moderated reactors, it contains the heavy water *coolant* that circulates through the reactor and which passes into large diameter 'header' pipes to the steam generators. In the *SGHWR* the calandria is filled with heavy water *moderator* and the light water coolant is contained in separate pressure tubes. In *PWR*s the calandria protects the control element fingers, and fits over the control element grid tubes aligning the fuel assemblies.

calorie

A unit of heat defined as the quantity of heat required to raise the temperature of one gramme of water by one degree Centigrade. A more rigorous definition requires the specification of the actual water temperature, and the 'calorie' usually referred to is strictly the '15°C calorie', ie the amount of water needed to raise the temperature of one gramme of water from 14.5°C to 15.5°C. The calorie is one of the cgs (centigrade gramme second) system of units and is related to the SI (Système International) system of units by the relationship:

$$1 \text{ calorie} = 4.185 \text{ joules}$$

where the *joule* is the SI unit of heat measurement.

The 'calorie' unit used when referring to the subject of human carbohydrate requirements is more strictly referred to as the kilocalorie or Calorie (with a capital C) and is equivalent to 1000 calories, as defined above.

calorie intake and requirements

The daily *Calorie* (energy) requirements of humans vary according to physical activity, age, sex, body size and structure and environmental temperature. The idea of 'reference' man and woman is a useful concept to illustrate average energy requirements. If reference man is 25 years old, weighs 150 lb, (70 kilograms), lives in a temperate environment with an average temperature of 68°F (20°C) and performs moderate physical activities, then the daily allowance should be in the region of 2,880 kcal/day.

For the reference woman, 25 years old, 128 lb (58 kilograms), of the same environment and similarly moderate activity, the allowance is around 2,100 kcal/day.

As the metabolic rate decreases with age (usually physical activity is also lessened) then calorie requirements are decreased. (It is suggested that the reduction is by 5 per cent per decade between the ages of 35 and 55, 8 per cent per decade between 55 and 75 and 10 per cent per decade following 75.)

calorific value

The amount of heat evolved when a given mass of fuel is completely burnt. It may be expressed in a variety of units, for example: calories per gramme, British Thermal Units (Btu) per pound, *therms* per tonne, and the SI unit *joules* per kilogram.

In this way, the energy content of different types of fuel may be directly compared and their relative usefulness and

Type of Activity	Time (hours)	Man Rate (kcal/min)	TOTAL	Woman Rate (kcal/min)	TOTAL
Sleeping and lying (essentially the basal metabolic rate)	8	1.1	540	1.0	480
Sitting (normal activities while sitting, such as clerical work, driving, etc.)	6	1.5	540	1.1	420
Standing (including unpurposeful walking)	6	2.5	900	1.5	540
Walking	2	3.0	360	2.5	300
Other (spasmodic sports, climbing stairs)	2	4.5	540	3.0	360
			2,880		2,100

Source: National Research Council (NRC publication No. 1146).

Energy expenditures by reference man and reference woman

convenience in particular applications evaluated.

See also *Appendix 2: Tables 2, 3, 4, 5.*

CANDU

Abbreviation for Canadian Deuterium Uranium reactor, a type of *nuclear reactor* in which pressurized *heavy water* (deuterium oxide) is used both as *moderator* and *coolant.* The fuel, natural *uranium dioxide* pellets, is contained in zirconium alloy (*zircaloy*) cans and these are enclosed as bundles in individual pressure tubes, also made of zircaloy. These tubes conduct the coolant and act as a pressure vessel rather than the whole reactor core being enclosed in a pressure vessel. A pressure of 85 atmospheres is used. The hot, heavy water emerges from the pressure tubes at the ends of the *calandria* and feeds into larger diameter header pipes which carry it to steam generators. Heat output is around 1750 MWt giving about 508 MWe with an efficiency

CANDU data*	
Peak power density	16.2 kW/l
Heat output	1744 MWt
Electrical output	508 MWe
Efficiency	29.4%
Fuel	Natural uranium oxide clad in zircaloy
Weight of fuel	92.6 tonnes
Fuel burn-up	9300 MWD/te
Refuelling	On load
Moderator	Heavy water
Coolant	Heavy water
Coolant pressure	85 atmosphere
Coolant outlet temperature	293°C

**Data based on Pickering 1 (Canada)*

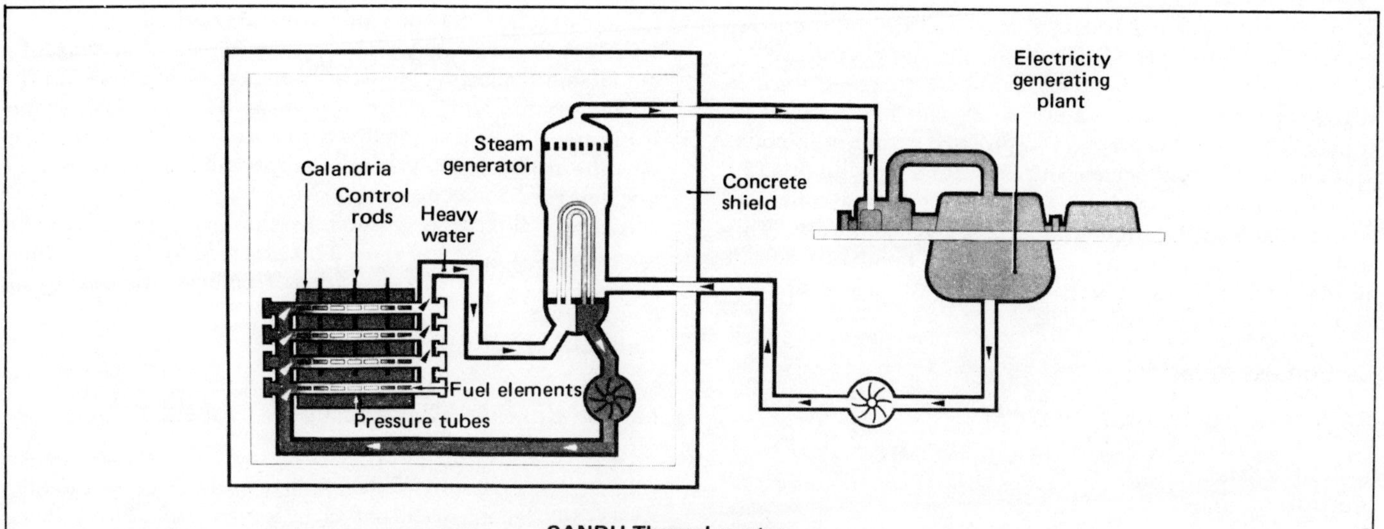

CANDU Thermal reactor
heavy water moderated (Canada)
Reproduced by kind permission of the UKAEA

of just under 30 per cent.

Using heavy water a higher natural uranium fuel *burn-up* is possible than with graphite-moderated reactors. A proportion of the energy output is produced by the *fission* of *plutonium* formed in the reactor.

caprock

An impermeable rock layer which stops the further upward migration of oil and gas so that it accumulates in a reservoir.

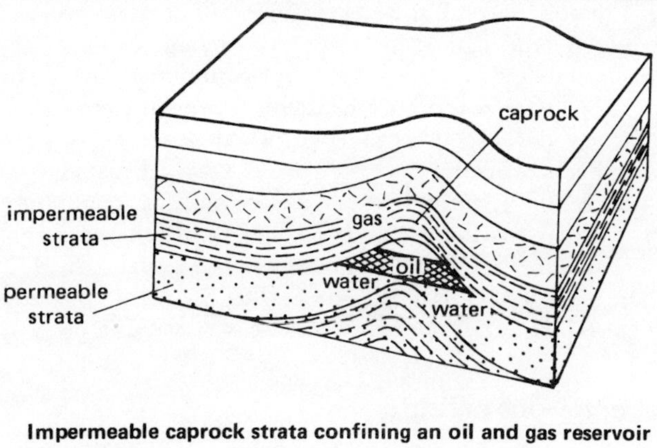

Impermeable caprock strata confining an oil and gas reservoir

carbohydrates

A large group of organic compounds composed of *carbon*, *hydrogen* and oxygen and widely found in plants and animals. Sugars, starches and *cellulose* are examples. They are important biochemically as energy stores and providers of energy. Carbohydrates are formed in nature through *photosynthesis* and are thus the precursors of the hydrocarbon energy stores in *fossil fuels*.

Anaerobic fermentation of carbohydrates produces *alcohols* which have been used as fuels and fuel additives, especially in World War II for *aviation spirit*. It has been suggested that the scientific farming both of agricultural and presently uncultivated plants could provide useful *alternative energy sources;* starch from potatoes, cellulose from marram grass, for example. Cellulose is a principal constituent of *wood* which is still an important fuel in primitive communities and some more advanced northern societies.

See also *energy crops*.

carbon

A chemical element, carbon is a non-metal that occurs naturally in three forms: diamond, a crystalline form with a density of 3.52 g/cm^3, the hardest of natural substances; graphite, a semi-crystalline, soft substance with a density of 2.25 g/cm^3; and the amorphous form with a density of 1.8-2.1 g/cm^3. All forms melt at approximately 3550°C.

Carbon is widely distributed in nature and is unique among chemical elements in forming compounds in which its atoms are linked together in long, stable chains. All life-forms contain carbon; the continuous circulation of carbon-containing materials in nature is termed the *carbon cycle*. Fossil fuels are the remains of ancient life-forms and owe much of their energy importance to their carbon content. On complete burning in air, carbon is converted to carbon dioxide with the release of energy:

$$C + O_2 \rightarrow CO_2 + 34\ MJ/kg \text{ of carbon.}$$

Coal consists largely of carbon. *Petroleum* and *natural gas* contain mainly *hydrocarbons*, in which carbon is chemically combined with *hydrogen*.

symbol	C
atomic number	6
atomic weight	12.01
graphite sublimes at 3700°C	
density graphite	2266 kg m^{-3} (2.27 g/cm^3)

Amorphous carbon, obtained from hydrocarbon sources, is used for the brushes of electric motors and *generators.*

Graphite is used as *moderator* in some designs of *nuclear reactor.* If *carbon dioxide* is used as *coolant* in such reactors it may react with the graphite to produce carbon monoxide.

Carbon-14, a *radioactive isotope* with a half-life of 5730 years, is produced by the irradiation of nitrogen in nuclear reactors and is a potential source of *air pollution.* This isotope is also produced in nature by bombardment of nitrogen nuclei in the upper atmosphere by cosmic rays.

carbon assimilation

An alternative term for *photosynthesis.*

carbon cycle

The circulation of carbon in nature. Carbon dioxide in the atmosphere is taken up by green plants and used in *photosynthesis* to form *carbohydrates.* Within the plant these are converted into other complex *hydrocarbons* – fats, proteins and their derivatives. Animals, unable to synthesize such compounds from the basic raw materials, are dependent on green plants for them. Plants and animals die and decay. The process of decomposition produces more carbon dioxide and, under anaerobic conditions, *methane* and other hydrocarbons. Marine planktonic organisms die and fall to the sea bed forming organic oozes. Ultimately they may become *petroleum* deposits following pressure and heat changes over a long period. In marshy areas trees and other plants die and decay. After millions of years such accumulated deposits may be changed into *coal. Peat* is an intermediary. Fossil fuels owe their existence to the photosynthetic activities of plants many millions of years ago.

Volcanic activity and the combustion of fossil fuels – either for space heating, heat production in thermal power stations or industrial processes – releases carbon dioxide into the atmosphere; the respiratory activities of animals and plants also liberate carbon dioxide. There is thus a continuous circulation of carbon-containing materials.

Some authorities have expressed concern at the growing output into the atmosphere of carbon dioxide from industrial activities. They feel that it is acting as a blanket in the upper atmosphere causing a rise in temperature and affecting the earth's heat balance.

See also *greenhouse effect; photosynthetic energy; thermal pollution.*

carbon dioxide

Colourless, odourless gas, chemical formula CO_2, formed by the complete combustion (in air or oxygen) of *carbon* and its compounds. It is one of the products when any fuel (other than *hydrogen*) or organic compound is burnt.

Certain *nuclear reactors* use carbon dioxide as a heat-transfer agent (*coolant*). Because the heat transfer is inferior to that of a liquid, higher operating pressures are necessary. The temperature at which it may be used in graphite-moderated reactors is limited by its chemical attack on the *moderator* to form *carbon monoxide.* The gas must be free from *argon* as this is readily turned into a radioisotope by the absorption of *neutrons.* Carbon dioxide absorbs very few neutrons and is not expensive.

Carbon dioxide is present in the air to the extent of 0.03-0.04 per cent and is used by plants in building up their structure. It is the main source of carbon essential to all living things.

See also *photosynthesis.*

carbon dioxide (CO_2) acceptor process

A coal gasification process that uses several pressurized *fluidized-bed reactors* to devolatilize coal or lignite and then to gasify the devolatilized char with steam. The mineral dolomite, a calcium-magnesium carbonate, has a unique role in the process. When calcined in one of the fluidized-bed reactors, dolomite releases CO_2 which is discharged with waste flue gas. The hot calcined dolomite, changed to the oxide form, serves as a heat carrier medium and the 'acceptor' material from which the process derives its name. It accepts or chemically combines with CO_2 and provides a way of removing CO_2 from the product gas stream. In reacting with CO_2, the exothermic heat released provides additional process heat.

Finely-divided, non-caking coal is fed into a pressurized fluidized-bed reactor for devolatilization. Hot calcined dolomite (the calcium-magnesium oxide 'acceptor') is fed into the top part of the devolatilizer, and, as it proceeds downward through the coal bed being devolatilized, degenerates and releases heat by combining chemically with CO_2 in the reactor atmosphere. Temperatures of about 800°C are developed in the bed, which is under 10 to 20 atmospheres pressure.

The spent acceptor, which has chemically recombined with CO_2, is re-calcined in a separate fluidized-bed reactor which operates at 1000°C and uses char as fuel.

The *synthesis gas* produced by this process is upgraded to *high-BTU gas* by *methanation.*

carboniferous period

The Upper Palaeozoic period (350-270 million years ago) when rocks of the carboniferous system were laid down, including the majority of the world's hard (*anthracite*) and *bituminous coals.* Much of the world's oil also come from carboniferous accumulations, particularly in the USSR, Argentina and eastcoast USA.

American geologists sub-divide the carboniferous system into the Mississipian, Pennsylvanian and Permian systems, but in Europe the Permian is regarded as a major system which followed the carboniferous.

See also *Appendix 3; Tables 1, 2.*

carbonization process

A general term covering the many processes in which *coal* is heated in the absence of air to form a variety of solid, liquid and gaseous products.

See also *coal gas; coke; low temperature carbonization.*

carbon monoxide

A colourless, odourless, and exceedingly toxic gas (chemical formula CO) formed when carbon and its compounds are incompletely burnt in air. It was a minor constituent of the old *town gas*, made from *coal*, and is present in the exhaust gases of petrol- and diesel-driven vehicles. Its characteristic absence of odour makes it highly dangerous, for it readily combines with the oxygen-carrying constituent in the bloodstream, haemoglobin, depriving vital organs of necessary oxygen, resulting in loss of consciousness and death. It burns with a blue flame to form the dioxide and has a calorific value of 150 Btu/ft^3 (5.6 MJ/m^3).

carbon-zinc cell

Another name for the *Leclanché cell.*

carburetted water gas

A *water gas* manufactured by passing steam over hot *coke* and enriched through passing *gas oil* over the coke as well. The resulting product, carburetted water gas, has a *calorific value* of approximately 500 Btu/ft^3 (18.7 MJ/m^3).

The gas is produced from a process originally developed in the USA in the last century, and was the first application of the use of a *petroleum* product in gas-making.

Carnot fraction

Another term for *availability factor.*

carnotite

Uranium potassium vanadate with small amounts of *radium*. A yellow metallic ore found in sandstones in the USA and West Africa. It is important as a source of *uranium*.

casing head gas

Also called Bradenhead gas. Low pressure gas produced at the well-head of a well that is primarily oil-producing, it helps to bring oil to the surface and is rich in heavy *hydrocarbons.*Casing head gas can be employed to power equipment on the rig; it can be compressed and reinjected to help oil flow, or it is sometimes flared.

casing head gasoline

Liquid *hydrocarbons* that separate naturally out of natural gas derived from oil wells as it rises to the surface and the pressure falls. It is a very volatile product, which is extracted at the well head.

cassava

The common name for two varieties of the genus *Manihot,* a member of the Euphorbiaceae. Cassava is a tropical shrub growing to a height of about two metres. Grown throughout the tropics, the roots of *M. ultisima* can be ground to yield cassava meal or specially prepared to make tapioca. Another variety *M. dulci aipi* has edible tuberous roots which are suitable for human and animal consumption. The plant is especially rich in *starch* – about 80 per cent in the tubers.

Cassava is a C_3 species (that is, it fixes solar energy into 3-carbon compounds such as triose phosphate). Experimentally it has been shown to use from 1.5 to 2 per cent of the total radiation (3 to 4 per cent of the incoming light energy). Fermenting of starch from the cassava tubers under experimental conditions to produce alcohol is competitive with the production of alcohol industrially. Cassava has been suggested (with several others) as an ideal plant for energy farming purposes.

See also *anaerobic fermentation; photosynthetic energy.*

catalyst

A substance that accelerates chemical change between other materials, without itself undergoing any apparent change. The catalyst may be in the same phase as the reactants (a homogeneous catalyst), or in another phase (a heterogeneous catalyst).

Heterogeneous catalysis is very important in the petrochemical industry, where catalysts are generally made to be porous, to increase the area of contact with the reactants. Silica, alumina and platinum are adsorptive substances used as catalysts in the petrochemical industry, eg in *catalytic cracking* of *hydrocarbon* fractions.

catalytic cracking

A *cracking* process in which a *catalyst* is used to facilitate the reaction. Use of the catalyst enables the reaction temperature to be reduced. It is used in the manufacture of *gasoline,* for example, and is much practised in the USA to satisfy the huge demand for gasoline.

catalytic dehydrogenation

The oxidation of a material by the removal of hydrogen using a *catalyst.* Catalytic dehydrogenation is usually carried out with less active catalysts than the corresponding hydrogenation reactions. Specially prepared copper, selenium and palladium are examples of such catalysts.

Catalytic dehydrogenation is widely used in hydrogenation processing. One example is the production of *synthesis gas* from naphtha or light distillate feedstock. A nickel-based catalyst is used.

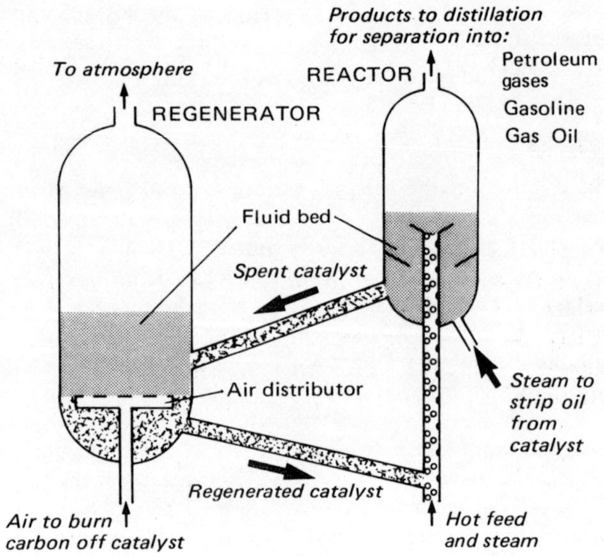

Catalytic cracking

catalytic reforming

An important process for the production of motor gasolines using a feedstock of material in the boiling range approximately 70°-190°C and raising the octane number from 40 to between 95 and 100. Catalysts used are platinum (0.3-0.75 per cent by weight) on highly purified alumina. Temperatures of 490°-540°C are used and pressures of 10-30 atmospheres. Four basic reactions are involved, *dehyrogenation, isomerization,* dehydrocyclization, (in which hydrogen is produced) and *hydrocracking.* Hydrogen is fed in to reduce the deactivation of the catalysts that occurs.

The hydrogen produced in the catalytic reforming is used in *desulphurization.* Catalytic reforming is also used to produce btx aromatics for use as *chemical feedstocks.*

catalytic rich gas process

An *oil gasification* process in which naphtha undergoes *steam reforming* over a catalyst at 500°C to give a gas rich in hydrocarbons and containing small amounts of hydrogen. A further reforming stage yields a *medium-Btu gas* which for some years was distributed as *town gas* in preference to using coal as a starting material in its manufacture.

cathode

The negative *electrode* in an electric cell.

cellulose

A complex polysaccharide *carbohydrate* consisting of numerous glucose units linked together in long chains. It is the principal constituent of plant cell walls and the basic structural material of plants. Its fibrous nature makes cellulose the basis of many textile products: cotton (cotton fibres are almost pure cellulose), jute, sisal, rayon (as cellulose acetate) and many more. When pure, it is highly insoluble, resistant to hydrolysis and chemical change generally. Cellulose has the chemical formula $(C_6H_{10}O_5)_n$.

When bound with lignin $(C_{12}H_{18}O_8)_n$ very tough tissue is formed. 25-30 per cent of the *wood* of trees consists of lignin. The proportions of carbon, hydrogen and oxygen in lignin, cellulose and wood are shown in the table.

	C%	H%	O%
Lignin	49.66	5.74	44.6
Cellulose	44.4	6.2	49.4
Wood	48.5	6.0	43.5

Like all organic substances in plants, cellulose is derived from the initial chemical products of *photosynthesis.* Wood is the principal tonnage source of cellulose, but other materials, so-called *energy crops,* are being investigated. Current energy farming research is aimed at the total utilization of photosynthetic materials as food, fuel and fibre. Cellulose-retaining materials can provide heat through direct combustion or they may first be converted into other substances, *alcohols,* by *anaerobic fermentation* for example, and these are suitable for fuels.

See also *algae; bagasse; biomass; cassava; ethanol; methane; methanol.*

Celsius scale

See *centigrade scale.*

centigrade scale

Also known as the Celsius scale. The temperature scale most widely used, in which each degree equals 1/100th the temperature interval between the melting point of ice and the boiling point of water. On the centigrade scale the melting point of ice is at 0°C, the boiling point of water is 100°C.

To convert to the Fahrenheit scale multiply by 1.8 and add 32. Degrees Kelvin, or absolute temperature, is related to the centigrade scale by simply adding 273.15.

centrifuge process

One of two principal methods available for enriching *uranium* for use in *nuclear reactors,* the other being the *diffusion* process. The centrifuge process takes advantage of the fact that *uranium-235* is slightly lighter than *uranium-238* so that when 'hex' gas (*uranium hexafluoride*) is passed into a spinning centrifuge, some separation of the isotopes takes place. The lighter molecules tend to stick closer to the axis of the centrifuge and this slightly richer uranium hexafluoride is drawn off and centrifuged in a similar manner, repeatedly, until a significant level of uranium-235 concentration is achieved. This method of enrichment promises great savings in electricity requirement as compared with the gaseous diffusion process and a plant is at present being constructed at Capenhurst in north-west England.

ceramic fuels

Nuclear fuels in which ceramic materials are mixed with the nuclear materials (eg *uranium oxide*) so that the fuel elements can withstand high temperatures. They are used in ceramic reactors.

See also *cermets*.

ceramic reactor

A *nuclear reactor* that has no metals in the reactor core, but uses refractory ceramic materials able to withstand high temperatures, several thousand degrees centrigrade. All ceramic reactors are versions of *high temperature gas-cooled reactors (HTGRs)*. Fuel and moderator are not kept separate as in other thermal reactors but are intimately associated. The fuel, usually uranium oxide or carbide enriched with as much as 93 per cent U-235, is processed into tiny spheres which are coated with layers of refractory carbon and silicon carbide in graphite. Thorium may be used in place of uranium. Because of the higher operating temperatures ceramic reactors produce higher quality heat for steam generating purposes.

See also *pebble-bed reactor*.

Cerenkov radiation

Blue light produced as a result of the emission of β-particles by radioactive materials and transmitted through a transparent, liquid medium, the particles initially moving at a speed greater than that of light in the medium. Such a phenomenon is often a feature of 'pool type' reactors and *cooling ponds* containing irradiated fuel.

cermets

Mixtures of materials ('ceramic' and 'metal') for use at high temperatures in *nuclear reactors*, and in the manufacture of *ceramic fuels* (eg uranium – plutonium cermets).

Cermets are excellent high-temperature materials and have a high resistance to radiation damage.

cesium

American spelling of *caesium*.

cetane number

A measure of the ignition characteristics of diesel fuels. The cetane number is the percentage volume of cetane in a mixture with alpha-methylnaphthalene that has identical ignition characteristics to the diesel fuel under test. Comparisons are made by using a standard test engine.

Diesel engines require fuels that have low self-ignition temperatures and little resistance to detonation or 'knocking'. A good diesel fuel therefore contains a high proportion of paraffinic hydrocarbons (*alkanes*), and the cetane number can be raised by the removal of *aromatic hydrocarbons* or by the addition of 'cetane improvers' – unstable compounds that promote chemical chain reactions. (The ignition requirements for diesel fuels are thus exactly the opposite of those for gasolines.) A cetane number of less than 40 would indicate an unsatisfactory diesel fuel; a rating of 50-60, a good one.

An alternative measure of fuel characteristics is the *diesel index*, which is generally calculated from laboratory measurements.

CFE

Initials for *controlled flash evaporation process* which is an open cycle *ocean thermal energy conversion* unit.

chain reaction

Chemical or atomic process in which the products of the reaction assist in promoting the process itself, such as ordinary *combustion* or nuclear *fission*. Chain reactions are characterized by an 'induction period' of comparatively slow reaction rate followed by a vastly accelerated reaction rate.

See also *fission chain reaction*.

char

A carbon rich product formed with oils and gas during *coal liquifaction* and *coal gasification* processes. The amount of char produced varies considerably with the type of coal used and the liquifaction or gasification process. Gas is produced from char and heat released during its combustion is also used for regenerative purposes.

charcoal

Often referred to as 'amorphous' carbon. Charcoal is the carbonaceous residue produced by heating vegetable matter (eg *wood*, sugar) or animal matter under airless conditions. (Concentrated sulphuric acid added to such materials also produces charcoal. The process essentially is one of dehydration.)

Early smelters converted wood to charcoal to provide sufficient temperature to isolate iron from its ore. (Wood burns with a comparatively cool flame, insufficient for the reduction of iron ore.)

Charcoal is sometimes used, instead of *coke*, in the production of pig iron and wrought-iron. Furthermore, as charcoal is highly porous and has a large surface area in relation to its mass, it is widely used in some forms of gas production as an adsorber. Forms of charcoal which are particularly effective in this role are known as active charcoals.

See also *activated carbon; cellulose; photosynthesis*.

chemical energy

Energy locked in chemical compounds and which can be released as other forms of energy. For instance, the chemical energy stored in an *electric cell* can be changed into electrical energy, and on combustion chemical energy is released from fuels (eg *hydrocarbons*) as heat energy. Solar energy, used in photosynthesis to convert carbon dioxide and water into *carbohydrates*, is effectively stored as chemical energy.

See also *energy storage*.

chemical feedstock

In petrochemicals, a chemical or a product of petroleum refining which is used as the starting point for other chemical syntheses. *Natural gas, hydrogen, benzene* and *naphtha* are common feedstocks.

chemical shim

A neutron absorber, usually boric acid, placed in the coolant system of water-cooled *nuclear reactors*, which helps to counteract the consumption of *fuel rods* and other materials bombarded by neutrons from the fission reactions; the build up and decay of various neutron-absorbing fission products; and the depletion of the fuel. It can also compensate for temperature changes in the coolant.

chlorophyll

A complex organic substance that plays a vital part in *photosynthesis*, enabling plants to form *carbohydrates* from carbon dioxide and water using the energy of sunlight. It has a crucial role, therefore, in providing both food energy and all organic fuels – *wood, peat, coal, petroleum* and *natural gas*. Attempts to understand completely the way in which solar energy is fixed in plants through the 'catalytic' intervention of chlorophyll are of considerable significance in enabling us to harness the sun's energy more effectively and thus aid appreciably in solving our increasing energy problems.

Efforts to culture 'concentrated' sources of chlorophyll such as unicellular *algae* have not met with much commercial success.

chloroplasts

Small, often disk-shaped bodies containing *chlorophyll* and present particularly in the cells of the leaves and young stems of plants and in algae. Experiments are taking place to try to produce *hydrogen* gas by splitting water using stabilized chloroplast membranes and hydrogenous enzymes derived from bacteria. It may also be possible to generate electricity by photoelectrochemical means using chlorophyll in chloroplast membrane systems.

CHP

Abbreviation for *combined heat and power*.

Ci

Symbol for *Curie*.

cladding

Material (eg magnox, stainless steel) used to contain the nuclear fuel and the products formed during nuclear *fission*. It is crucial that the cladding does not develop leaks, for the reactor *coolant* is then liable to become contaminated with radioactive products.

clark cell

A standard *primary cell* employed in measurements of *emf*. The cell consists of a zinc *anode* and mercury *cathode* covered with mercury sulphate paste. The anode and cathode are semi-submerged in a saturated solution of zinc sulphate.

CMU system

A system for closed cycle *ocean thermal energy conversion*, that uses *ammonia* as its working fluid, proposed at the Carnegie-Mellon University.

coal

Coal is a combustible stratified rock formed from the remains of decaying vegetation. It is the only organic rock, usually black or dark brown in colour, laid down principally in the *carboniferous period* on all continents. The various coals have *carbon* contents of between 60-95 per cent depending on the stage reached in coalification.

The major producers are North America, USSR and the EEC (China has large reserves but accurate figures are unavailable). World coal reserves are substantially in excess of those of oil and natural gas. Estimates suggest that there are sufficient reserves to meet projected demands for two to three hundred years. (Though this picture could change dramatically if coal were converted to synthetic liquid fuels, on a large scale, as replacement for oil.)

Coal is the dirtiest and least convenient of the fossil fuels. Environmental considerations figure increasingly in its extraction and use. Extraction is primarily through networks of underground mines or opencast surface works.

Origins. In the coal-forming ages the climate was hot and humid and the carbon dioxide level in the atmosphere was probably higher than at present. These conditions encouraged rapid forest growth (particularly of coniferous trees, tree ferns etc) providing large amounts of deposit vegetation in the swamps. Normally dead vegetation is converted to carbon dioxide and water but this process is prevented in a water logged environment. The damp climate assists the formation of acid bogs which are bacteriostatic. That is, slow bacterial decay took place until the bacteria could no longer survive under the acid conditions of the accumulated products of decomposition.

Peat is formed by this process. (Existing peat bogs were formed in the last million years.) Mature coal gradually develops as the products of bacterial decomposition are subjected to high temperatures and pressures resulting from the continuous accumulation of silt, soil and rock and land movements over time.

The succession of changes from the original vegetable matter to peat, then soft *brown coal* and *lignite*, to increasingly hard *bituminous coals* and finally *anthracite* is known as coalification. (Some classifications include low volatility steam coals immediately after bituminous coals.)

Periodically climatic changes and geographical disturbances result in the stratification of coal deposits into various sized seams, divided by roughly repetitive layers of sedimentary rocks and other deposited material.

Nature. Coal is composed of *carbon, hydrogen* and *oxygen,* chemically combined as the basic products of *photosynthesis.* (Also present are elements in small concentrations, nitrogen, chlorine, sulphur and traces of several metals.) The proportions vary according to the stage reached in the coalification process during which water and carbon dioxide are eliminated as *methane* (CH_4) evolves. (Methane is the main constituent of firedamp which makes an explosive mixture with air. It must be removed to avoid serious problems within underground mines.)

There is a progressive loss of oxygen and an increase in the proportion of carbon while the hydrogen content remains fairly constant. (However, coal has a low hydrogen content compared with oil and gas.) The rank of the coal indicates its carbon content and coalification stage. Brown coal is of low rank, anthracite high. Low rank coals are low in calorific value and have high proportions of volatile matter. The volatile matter and impurities determine the end use for coal.

The characteristics of various coals depend on the nature of the original vegetable matter, the kinds of silts washed concurrently into the deposit areas, the degree to which decay had advanced prior to the organic deposits being sealed under further earth and rock, and the subsequent conditions of heat and pressure to which the deposits are subjected. Classification of coal has to account for its economic value which depends on its end use, which in turn is determined by its characteristics. As international trade is of a relatively small scale there has been little impetus to adopt an internationally recognized system of classification. The systems that do exist usually consider rank; volatile matter; calorific value; coking properties; impurities. (Examples of classification systems include: British National Coal Board System, Economic Commission for Europe's hard and soft coal systems, the American Society for Testing Materials' system.) Commercial classification systems usually describe coal on an 'as received' basis to indicate the inorganic content. Further observations are based on the 'pure coal' substance, or the 'dry mineral matter free' portion.

Accountable factors should therefore include:

CARBON. Carbon content increases with rank. A knowledge of the carbon content is necessary when determining combustion conditions and air requirements.

MOISTURE. Moisture lowers calorific value and causes handling problems. The three types of moisture include: surface moisture from storage or washing, inherent moisture absorbed in the capillaries, chemically combined moisture in the form of hydrates.

SULPHUR. Varies from less than 0.5 per cent to approximately 5 per cent. A low sulphur content is often necessary to meet anti-pollution requirements and similarly metallurgical coke should be of less than 1 per cent sulphur to decrease contamination. (Sulphur dioxide causes problems through corrosion.)

ASH. Incombustible material left over after burning. It results from silts present in and around the original vegetable matter or introduced during mining operations. Ash causes problems during combustion; it lowers the calorific value and if large enough quantities are involved its presence leads to congestion and removal problems.

VOLATILE MATTER (combustible matter driven off at above 925°C). It has three main components: combustible gases (*hydrogen, carbon monoxide, methane, ethane* and hydrogen sulphide), *tar* and *ammonia liquor.*

OXYGEN. Oxygen content decreases with rank. The greater the oxygen content the lower the coke grade.

COKING PROPERTIES. Some indication is required in any classification system of the suitability of a given coal for the manufacture of *coke.* Coke is made by the *carbonization* process. Other useful products of this process are *benzol oils*, tar and also coal gas. These products can be sold separately as fuels or chemical intermediates.

COMPOSITION. Coal is not of homogeneous uniform structure; it is banded (petrographic). There are four banded constituents: 1. *Vitrain.* Brilliantly glossy and vitreous, occurring in thin bands from 2-10mm thick. It is brittle and coals containing a high proportion of it break up on handling. 2. *Clarain.* It is clearly strained, bright and has a satin lustre. Mechanically stronger than vitrain, on carbonization it yields a more compact coke. 3. *Durain.* Dull and hard, it occurs in bands 13-450 mm in thickness and has a matt granular surface. Mechanically very strong, it tends to form flat plates on breaking. Normally durain does not form a coke on carbonization. It contains more carbon and less hydrogen than the hard coals and yields a smaller amount of volatile matter. 4. *Fusain.* Powdery and dull, it occurs in thin patches in the bedding plane. It is non-coking irrespective of rank and tends to reduce the caking properties of the coal. Hydrogen content is usually less than 2.5 per cent and carbon greater than 90 per cent. Vitrain and clarain form the bulk of normal coal seams, durain between 5-30 per cent and fusain less than 3 per cent.

Production figures. For a comparison of energy production figures, area by area, covering coal alongside oil, natural gas, hydroelectricity and nuclear electricity, see *Appendix 1; Tables 9 and 10.*

World coal reserves & production

million metric tonnes of coal equivalent

Country	Economically Recoverable Hard Coal	Measured Reserves All Ranks	1976 Hard Coal Production
European and Asian USSR	165,800	273,204	546.0
North and Central America	128,320	377,912	603.0
Asia (including China)	114,668	240,377	674.3
Europe	41,292	300,918	470.7
Australasia	13,805	74,694	79.1
Africa	12,335	26,607	79.1
South America	60	3,762	8.7
Total World	476,280	1,297,474	2460.9

Primary source: World Energy Conference Survey of Energy Resources, 1974
Secondary source: World Power Conference Survey of Energy Resources, 1968
Production data: International Coal Trade, US Dept of the Interior, Vol 45, No 4, April 1976

Consumption figures. Appendix 1, Table 2 illustrates the yearly total consumption and change from 1966 to 1976 of the solid fuels (coal, peat, lignite) in a country by country comparison. Appendix 1, Table 11 illustrates world energy consumption (1956, 1966 and 1976) in a country by country analysis of solid fuel consumption levels alongside figures for the other primary energy sources.

Uses. Coal, once the major energy supplier and now superseded by oil, is primarily used in the electricity supply industry and the steel-making and metallurgical industries (together accounting for some 80 per cent of total coal consumption). The remainder is employed in industrial heating and steam-raising, domestic heating, in gasworks to produce town gas and in some countries still for powering railway transport. (The overall proportion of coal utilized in town gas production is steadily decreasing as natural gas supplants town gas in many countries.)

For metallurgical uses, coal which produces a high grade of coke is necessary. The coal must be reactive, able to withstand the high pressures of blast furnaces, and have a low impurity content. Research into methods of steel-making requiring less restrictive types of coking coal is receiving a great deal of attention, especially in Japan.

The electricity supply industry is less selective in its coal requirements although consistency of ash and moisture content is required. Environmental restrictions on the amount of sulphur dioxide emitted from power stations has meant that a low-sulphur content coal is also preferred. (*Fluidized-bed combustion* could provide one means of improving the quality of various coals by removing sulphur and other impurities.) With the possible exception of steel and electricity supply industries, world energy requirements have steadily moved toward more convenient forms of energy, away from solid fuels.

Coal needs to show a significant price advantage over gas, oil and electricity to persuade consumers to accept the additional problems and costs associated with its use. However, the upheaval of the world energy scene in 1973 further encouraged governments to maintain their coal industries for two major reasons: first to reduce dependence on imported supplies of energy and, second, to ensure that supply of coal in the longer term as other fossil fuels are gradually exhausted (that is, at a faster rate than coal).

Reserves. Estimating world reserves is limited by the lack of universally accepted methods of estimation. Early estimates simply suggested reserves 'in situ', giving no indication of ease of access or costs of exploitation. Market, socio-economic and technical factors will affect the definition of what is economically recoverable and therefore regular surveys are needed to update reserves.

The World Energy Conference defined three broad categories of reserves (but often interpretation is wide-ranging).

1. MEASURED RESERVES. Seams greater than 30cm thick, less than 1,200m deep on which there is reliable data. (At present the maximum coal mining depth is 1,200m but there are huge resources at greater depths. Only much cheaper and better ventilation and support systems would make it possible to reach them.)

2. INDICATED AND INFERRED RESERVES. Seams within the above limits which can reasonably be assumed to exist – but for which only approximate estimates can be prepared.

3. ECONOMICALLY RECOVERABLE RESERVES. Reserves physically workable at current costs (while considering comparative costs of alternative fuels.)

Research and Development. The realization that the world's coal reserves should last significantly longer than oil and gas (plus the considerations already mentioned) has encouraged R & D within the coal industry.

The first important area of R & D is in the actual mining of coal. Traditional methods of shallow mining, *bord and pillar* (or room and pillar) and in particular deep mining methods, notably *longwall mining* have to be considerably improved to reach and extract the coal economically and safely. Mechanization and automation on a large scale is favoured. Similarly, on the surface, reclamation of strip-mined land has become an increasingly important factor. In the UK effective reclamation of open-cast or strip-mined land is a legal requirement imposed on the mining sector.

The majority of coal is consumed within the country of origin, Japan being the major exception. Therefore international transport is not a major problem. However, the successful exploitation of coal reserves depends on the existence of an efficient transport infrastructure. (The absence of such has delayed the exploitation of known coal reserves in areas such as South America and India.) Transportation to date is of a 'discontinuous' nature, ie road, rail, barge and sea-going vessels. This contrasts with the 'continuous' movement of oil and gas. Some coal, in slurry form, is transported by pipeline, but this involves dewatering at destination, is expensive and difficult and therefore represents a very small proportion of the total movement. One development enabling quicker and cheaper coal transportation is the use of permanently coupled trains known as 'merry-go-round'.

The remaining major areas of R & D are:
(a) Pollutant Control, and improved usage of wider ranging coals as fuel; (b) Coal conversion to gaseous and liquid products. Because of environmental pressure, air polluting smoke stacks are undesirable. Regulations determining the height of stacks (for improved dispersion) have been introduced into many countries. However, decreasing *sulphur* content, and other impurities, has become an important part of coal R & D. Methods for using poorer quality coals, eg solvent refined coal, are being rapidly developed, but as yet no large scale processes of this nature are in operation. Improvements in this direction have greatest significance within the electricity generating and metallurgical industries.

As other fossil fuels run out, substitute coal-fuels could become as important as solid coal. (Conversion of coal to gas and oil would, of course, deplete coal reserves faster than traditional methods of consumption.) Gas from coal, in general, is less expensive to make than oil or liquid fuel from coal. Until the 1960s Germany carried out the majority of oil conversion research. The USA has since taken the lead and has committed large blocks of coal reserves to *coal gasification* and *coal liquifaction* purposes. Britain's programme comes under the *Coalplex* plans. There is collaboration between the National Coal Board and America where extensive coal conversion projects are in operation. In South Africa at Sasolburg a large oil from coal plant has been in commercial operation for roughly 25 years.

In the 1973/74 financial year Britain spent £5.6 million on coal R & D and in 1974/75 the USA spent $180 million.

See also *Appendix 1; Tables 1, 2, 9, 10 and 11 and Appendix 3; Tables 1 and 2.*

coal gas

The name given to the mixture of gases produced by heating coal to high temperatures in the absence of air. This process was once the basis of the now defunct *town gas* industry, but with the advent of cheap petroleum-derived gas and natural gas, it is now used almost exclusively for *coke* production.

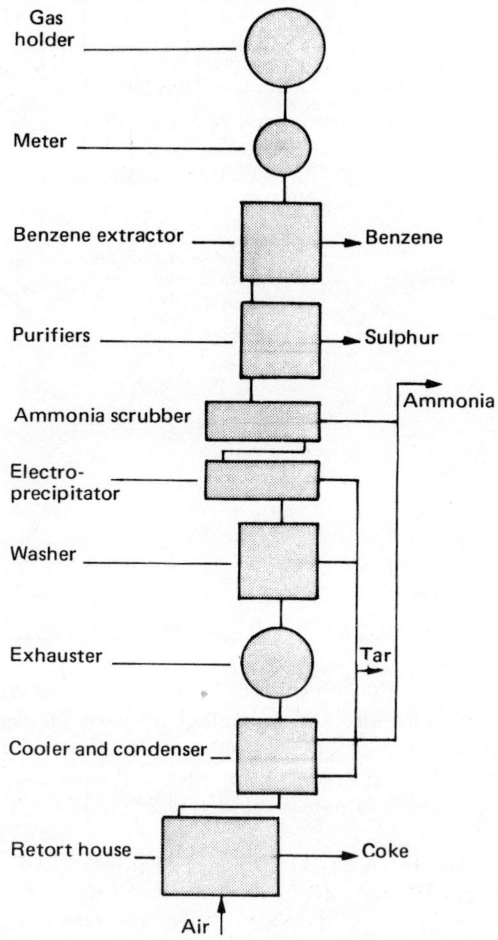

The manufacture of coal gas

Coal gas is a mixture of *hydrogen, carbon monoxide, methane, ethylene* and small quantities of other gases, obtained by removing the less volatile materials from the stream emerging from the coking retorts. The process was carried out at a 'gas works', the gas being stored in the familiar gasholders. Coal gas has a calorific value of approximately 500 Btu/ft^3 (18.7 MJ/m^3).

coal gasification

The conversion of coal into gaseous fuels. Different mixtures of gaseous products may be obtained from coal, depending on reaction conditions, but broadly these may be classified into three groups: low, medium and high-Btu gases. Their respective calorific value ranges are $<$350 Btu/ft^3 ($<$13.0 MJ/m^3), 350–700 Btu/ft^3 (13.0–26.0 MJ/m^3) and $>$700 Btu/ft^3 ($>$26.0 MJ/m^3). The old *town gas*, made from coal, was a medium-Btu gas consisting mainly of hydrogen and carbon monoxide. This has largely been superseded by natural gas, a high-Btu gas consisting principally of methane, and present research and development is aimed at producing a 'synthetic natural gas' with a high

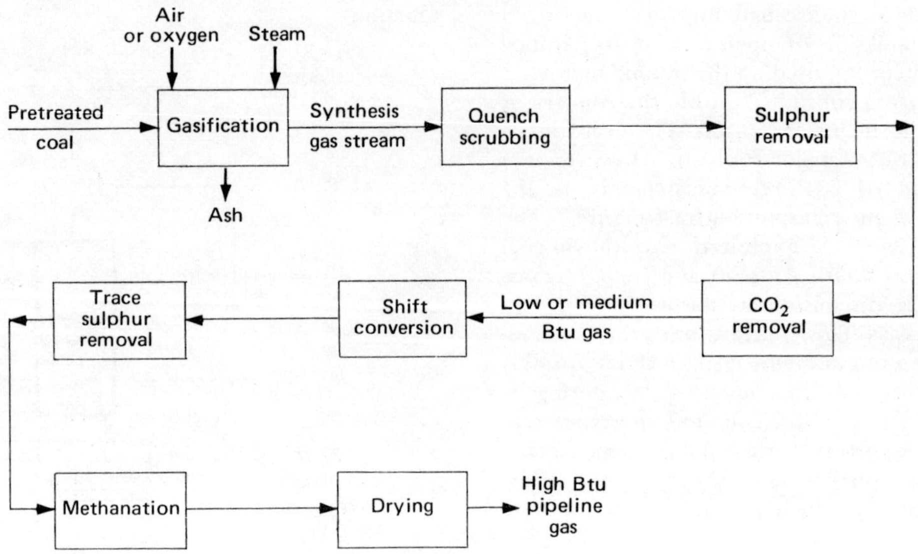

Schematic representation of the processing steps in coal gasification

calorific value. In the United States in particular, much work is being done, and a number of pilot plants are in operation. It is envisaged that coal gasification will play an important part in overcoming the problems caused by depleting reserves of natural gas in the United States.

See also *atgas process; bigas process; carbon dioxide acceptor process; COGAS process; hydrane process; hygas process; in-situ coal gasification; Kellogg process; Koppers-Totzek process; Lurgi process; methanation; shift reaction; synthane process.*

coal hydrogenation

A process for converting *coal* into gaseous and liquid products by treating it with *hydrogen.* Many coal hydrogenation processes have been developed, one of the best being the *Bergius process.*

See also *Coalplex; COED process; Consol process; Fischer-Tropsch process; H-coal process; Sasol; solvent-refined coal process; Synthoil process.*

coalification

The transformation of dead vegetable matter into *coal* by a succession of changes. Mature coal gradually develops as the products of bacterial decay are subjected to high temperatures and pressures resulting from the continuous accumulation of silt, soil and rock and land movements over time. The usual coalification sequence is from *peat* through soft *brown coal*, and *lignite* to increasingly hard *bituminous coals* and finally *anthracite.*

Coalite

Also known as *semicoke.* Trade name for a domestic fuel produced by partially *carbonizing* coal at a temperature of approximately 600°C. Heating the coal to this temperature drives off some, but not all, the volatile and tarry matter. Thus Coalite has a calorific value of 270 therms/tonne (28.5 GJ/tonne).

coal liquifaction

The conversion of coal into liquid hydrocarbon fuel. The various liquifaction processes fundamentally involve reducing the *carbon* to *hydrogen* weight ratio of the coal either by adding hydrogen or by removing part of the carbon as *coke* or *carbon dioxide.* The by-products of coal liquifaction range from waxes and heavy oils to light gasolines and gases, depending on the process and amount of hydrogen employed.

Although liquifaction technology has been understood for many years, the various processes are expensive and energy losses extremely high. Commercial experience in converting coal to liquid fuels has been limited. From the late 1920s and throughout World War II, Germany produced *gasoline* from coal through the *hydrogenation* of *coal tar* and by the *Fischer-Tropsch process.* The rise in world oil prices, the need for self sufficiency in oil and concern with pollution control have led to a renewed interest in coal liquifaction and *coal gasification* development, particularly in the USA. In South Africa, at the Sasol plant, gasoline and other liquid products are produced, through the Fischer-Tropsch process, from *synthesis gas* ($CO + H_2$) gen-

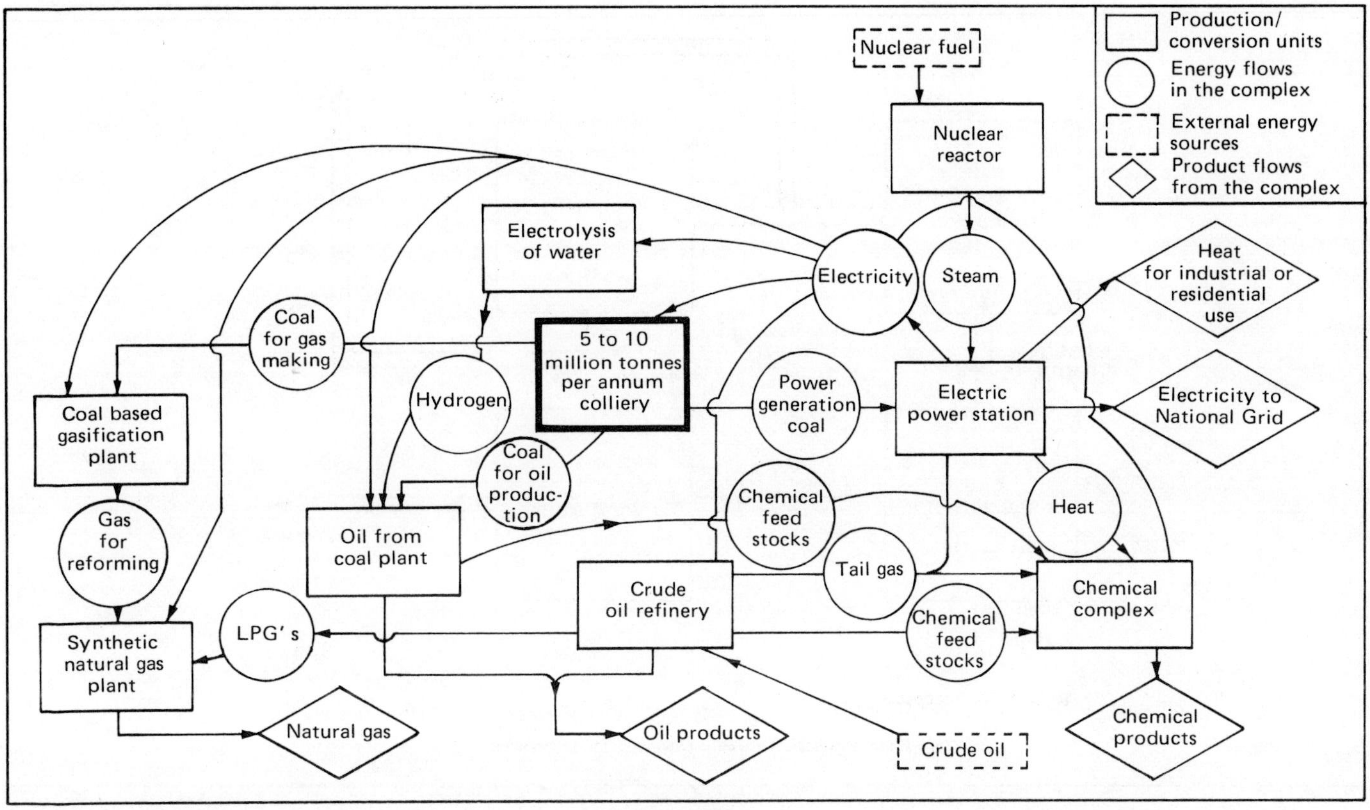

Coalplex

An energy complex operation in which a very large colliery could provide the basis for the integrated use of several types of energy input, and provide an output of usable energy in many forms. Chemical products are also obtained from the complex

erated in *Lurgi* coal gasifiers.

Present research efforts in the USA are directed towards improving the efficiency as well as the economic feasibility of coal liquifaction.

Four principal methods are available for coal liquifaction: 1. based on high pressure hydrogenation of coal; 2. *pyrolysis;* 3. dissolving the coal in a suitable oil solvent; and 4. the upgrading of synthesis gas as used in the Sasol plant.

See also *Bergius process; COED process; Consol process; H-coal process; solvent-refined coal process; Synthoil process.*

coal-oil gas (COG) refinery

US term for a *Coalplex* plant in which oil and gas is produced from coal in the same project.

Coalplex

The term given to conversion projects which combine oil from coal processes (*coal liquifaction*) and gas from coal processes (*coal gasification*). Employing both processes within one system reduces capital costs and increases the overall conversion efficiency. The American term for this complex is the Coal-Oil-Gas (COG) refinery.

coal tar

A by-product of the *carbonization* of coal, consisting of a variable mixture of hydrocarbon oils, phenols, cresols and nitrogen bases such as pyridine.

High-temperature carbonization is used for metallurgical *coke* production and produces a tar with a high content of *aromatic hydrocarbons* such as *benzene*, toluene, xylene and *anthracene oil.* Low-temperature carbonization, carried out at 500°-750°C, produces a volatile semi-coke that is used as a smokeless fuel in some countries, notably the UK. This process yields a tar containing a greater proportion of *aliphatic hydrocarbons.*

Improvements in blast furnace efficiency have reduced coke requirements in the steel industry so that, despite increasing world steel production, coal tar production has remained steady. As a source of organic chemicals coal tar is now secondary to petroleum.

COED process

The Char Oil Energy Development (COED) process is essentially a coal conversion process yielding an oil, a gas and a *char.* During conversion the coal feedstock is subjected to a four-stage *pyrolysis* process at temperatures increasing from 315°-815°C. Gaseous products are desulphurized and reformed to produce *hydrogen* which is

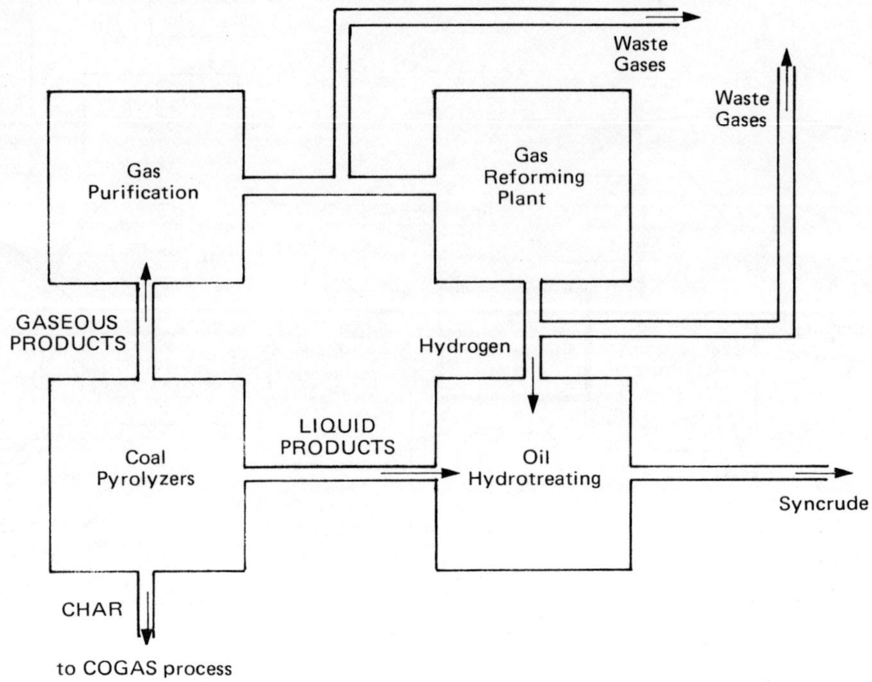

Diagrammatic represention of COED process

necessary to upgrade the liquid products of pyrolysis to a *synthetic crude oil. Syncrude* yields range from 0.7 – 1.3 barrels of oil for each ton of coal feedstock, depending on the type of coal used. The predominant product, however, is the char and this was originally seen as a possible feedstock for *electricity generation.*

Currently the *COGAS process* is being studied as a means of utilizing this char more effectively as a gasified product.

See also *coal gasification; coal liquifaction.*

coefficient of performance

For a *heat pump*, the ratio of the heat absorbed at the cold end to the work input to the heat pump.

COGAS process

A method of treating *char* (derived from the *COED process*) to produce gas. This is currently being developed with a view to making better use of the char which is at present the main product of the COED process (essentially aimed at manufacturing *Syncrude*). The char has high steam reactivity and because the raw oil and gas is produced in a low-pressure system it is economically comparable with other high-pressure gasification processes.

coke

Coke is the hard, solid residue left after *coal* has undergone *carbonization* and has its volatile matter removed. It has a calorific value of 270 therms/tonne (28.5 GJ/tonne). Removing the volatile matter both minimizes the serious pollution that coal burning can cause and makes available valuable raw materials for the chemical and other industries.

Coke is employed primarily within the metallurgical industries where it performs the dual function of providing heat and acting as a reducing agent. For this purpose coke must be reactive, contain little *ash* or *sulphur* and be capable of withstanding the high pressures inside blast-furnaces. (In a blast-furnace coke does not pack under pressure into an impervious mass, ie it allows air to pass through and supports the combustion and reduction of the ore. Although there is a great deal of pollution from the dust, electrostatic precipitation eliminates it.)

Industrial users also burn coke as one means of providing heat energy both for carrying out manufacturing processes and for *space heating.* Such coke need not meet such strict specifications as coke for the metallurgical industries. In particular it does not have to be so hard.

Although coke does not ignite as readily as coal it is finding an expanding domestic market as a result of the continuing introduction of smokeless zones.

World-wide coke production has remained fairly constant during recent years, while the proportion employed in the metallurgical industries has decreased. Although iron and steel production has expanded, other clean and convenient fuels, natural gas and petroleum, are successfully competing with coke as regards the heat requirement.

Furthermore, recent technology advances in steel-making processes have reduced the amount of coke required per tonne of steel produced.

Water gas and *producer gas* are both manufactured from coke.

See also *coking coal.*

coke breeze

A solid fuel with a calorific value of approximately 220 therms per tonne (23.0 GJ/tonne) separated from *metallurgical coke* by screening. It is mainly used to provide heat at coke ovens or at the sinter plants of the iron and steel industry.

coke oven gas

Gas produced at coke ovens as a result of the removal of volatile material from the coal undergoing carbonization. Its principal uses are for reheating the coke ovens and if the ovens are sited close to the blast furnace for providing furnace heat. It consists principally of hydrogen and carbon monoxide and has a typical calorific value of 500 Btu/ft^3 (18.7 MJ/m^3).

coking coals

Coals of low to medium volatile matter, 23 to 30 per cent, with a low *sulphur* and *ash* content. These yield *coke* suitable for the metallurgical industries. Today coking coals for the iron and steel industries account for a fair proportion of general coal production.

combined cycle generation

The large scale generation of electric power by a combination of a *gas turbine* cycle and a *steam turbine* cycle.

In conventional thermal power stations fuel is burnt in a boiler to provide high pressure steam for expansion in a steam turbine which is linked to a generator. The highest efficiency achievable is approximately 38 per cent, measured as the percentage ratio of electricity output to heat content of fuel.

In a combined cycle generation system, the fuel (usually *fuel oil*) is first converted to gaseous form by partial combustion. The raw fuel gas produced is cleaned and fully combusted in a gas turbine. The high temperature (about 600°C) exhaust gas of the gas turbine is used to raise steam in an exhaust boiler and the steam is then expanded in a steam turbine. Power is extracted both from the gas turbine and from the steam turbine.

The primary gasification step is not required if the fuel is already available in gaseous form and coke oven gas and natural gas can be used directly.

If electricity generation is the major requirement of the plant, a condensing steam turbine is used. Efficiencies of 38-44 per cent are obtainable. If some process heat is required a *back-pressure turbine* or a *pass-out turbine* can be substituted..

Advantages of the combined cycle system are fast start-up, low cooling-water consumption, lower pollution by nitrogen oxides and, if several gas turbines are used with a single steam turbine, the ability to maintain high efficiency at part load. Operational experience with this type of plant is limited, but installed capital costs seem to be higher than those of conventional plants.

The development of gas turbines able to operate at well above 1100°-1200°C would further improve the overall efficiency of a combined cycle system – figures of around 50 per cent have been suggested as feasible if the required advances in blade materials technology can be made.

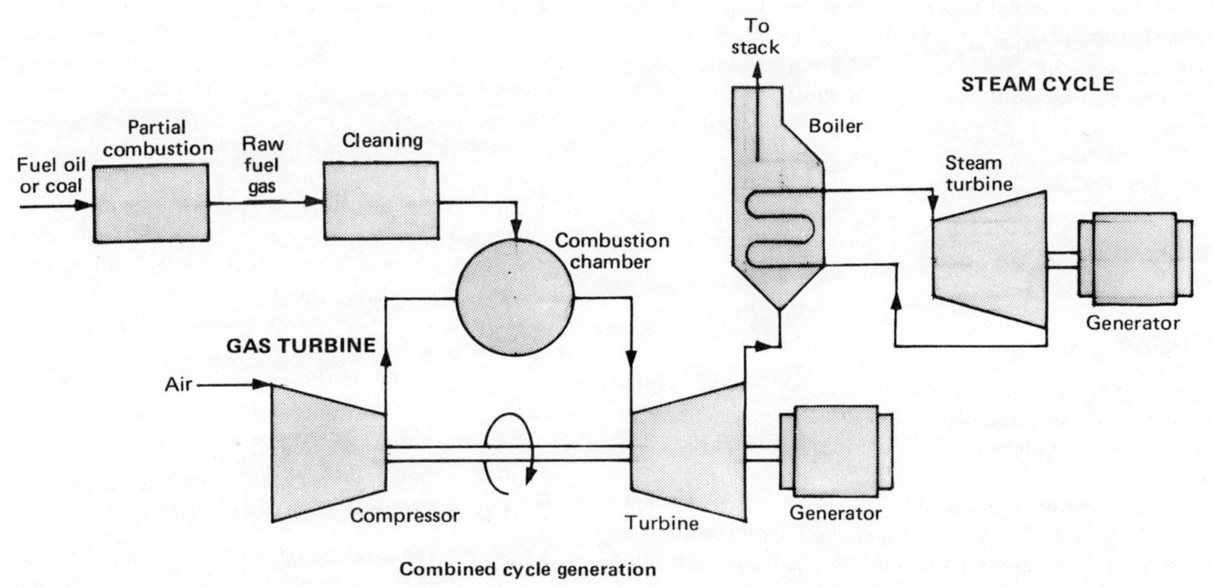

Combined cycle generation

combined heat and power (CHP)

The combination of heat and power generation within the same plant. This substantially reduces overall fuel requirements by exploiting otherwise wasted heat from conventional electricity generation to provide low-grade heating for industrial and domestic use.

Conventional power generation through *fossil fuels* obtains a conversion efficiency of approximately 30-35 per cent. For electricity generation this means that for every 1 MJ of electricity produced 2 MJ of energy in the form of heat is simply wasted. Rechannelling this energy to heat water can fulfil the process heating requirements of local industry and the domestic sector.

The fuel savings through CHP increase the conversion efficiencies to approximately 70 per cent but against this should be balanced the energy cost in building new installations and plant.

At present CHP stations operate in at least 25 countries either as *total energy* schemes or as part of a *district heating* scheme. District heating schemes may employ a CHP plant in conjunction with *heat-only boilers.*

Problems arise in the use of CHP plant due to the mismatch between heat demand and the electricity demand at any time. Where the CHP plant operates as part of a larger integrated electricity supply system these problems may lead to difficulties in integrating the electricity supply since a deficit in either heat or electricity supply may occur. This problem may be mitigated by use of an ITOC (Intermediate Take-off Condenser) system where the relative quantities of heat and power produced may be varied so as to follow more closely the demand for both heat and power.

The conventional CHP process employs a central power station plant producing steam for power and heating requirements. *Diesel*, gas or dual-fired engine driven generators may also be used with the waste heat supplementing existing boiler plant. Similarly, gas turbine driven generators may be used with the exhaust heat supplementing existing boiler plant. (Whereas gas turbine driven generators operate at a low conversion efficiency – 18-20 per cent – the exhaust heat can be utilized very efficiently, which brings the overall conversion efficiency back to the region of 80 per cent.)

CHP incinerators burning waste materials, such as *domestic refuse*, provide double energy saving in that fuel is efficiently utilized and conventional fuels are spared for alternative uses.

combustion

The rapid chemical combination of oxygen with the combustible elements of a fuel. Heat, light and flame may all be produced in an exothermic reaction and new substances are formed.

The three major combustible chemical elements are *carbon, hydrogen* and *sulphur* (sulphur has little value as a source of heat but causes problems of corrosion and pollution in connection with some of the major fuel sources).

The formulae for the complete combustion of carbon and hydrogen with oxygen can be expressed as:

$$C + O_2 = CO_2 + 34 \text{ MJ/kg of carbon}$$
$$2H_2 + O_2 = 2H_2O + 12.8 \text{ MJ/m}^3 \text{ of hydrogen}$$

combustion turbine

Alternative name for *gas turbine.*

compressed air storage

The storage of energy in the form of compressed air which is used for driving turbogenerators and producing electricity. The use of compressed air in this way has been suggested as a method of storing energy derived from non-continuous sources such as tidal and wind. The energy could be released later at low tide, for instance, or in periods when the wind speed is inadequate, so that it would be possible to have a continuous energy flow even though the source is fluctuating. It is suggested that underground caverns, such as those created by salt mining which often are near to proposed tidal sites, would be suitable storage chambers.

See also *energy storage; tidal energy; wind energy.*

condensate

A *natural gas liquid* consisting of a mixture of *hydrocarbons* (C_5 and upwards) occurring alongside lighter fractions in high pressure and temperature gas reservoirs. Alternatively known as natural gasoline it is used as a blending component in the petroleum industry. It condenses out of the gaseous mixture as the result of pressure and temperature drop on extraction from the reservoir.

conservation of energy

A general natural principle which asserts that energy cannot be created or destroyed. An apparent exception – the conversion of mass to energy in nuclear reactions – is explained as being due to the interchangeability of mass and energy.

Although energy is conserved in all non-nuclear processes, the way that fuels are used involves apparent energy 'losses', so-called because only a certain amount of the energy contained in a fuel is usefully employed. The remainder serves no use and goes into the surrounding environment. Efforts for greater *energy conservation* are principally concerned with the minimization of this 'loss', and rate of loss.

The principle of the conservation of energy is generally known as the First Law of Thermodynamics.

See also *energy, mass equivalent of.*

Consol synthetic fuel (CSF) process

A process for converting *coal* to a liquid fuel suitable as a refinery feedstock in which the coal is dissolved in an oil solvent at 400°C. The solvent itself acts as a *hydrogen* donor to the coal and on recycling is hydrogenated catalytically in order to prepare it for a subsequent batch of coal feedstock.

See also *coal liquifaction.*

contouring rafts

Also known as Cockerell's contouring rafts. These devices for harnessing *wave energy* consist of a series or bank of

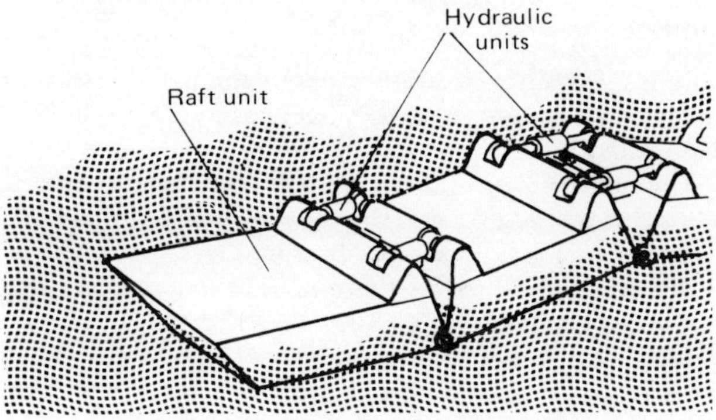

Schematic view of Cockerell contouring raft-units.
Many such rafts would be used in series.

rafts designed to convert the energy in the movement of the raft (caused by the rise and fall in wave movement) into high pressure in a fluid. Pumping is provided by hydraulic motors or pumps which connect the individual raft units and which are activated by the motion in the bank of rafts riding the waves.

Tests in the UK on rafts of a 1/50 scale have shown promising results and 1/10 scale tests are scheduled. The pressurized fluid could be stored within the body of the raft or pumped into separate storage vessels. Considerable further research and development is needed before the possibility of commercially viable full scale systems may be evaluated.

See *wave energy* for examples of other collection devices and the general problems and advantages of wave energy as an alternative source of energy.

controlled flash evaporation process (CFE)

A proposed open cycle ocean thermal energy conversion unit, used for extracting useful energy from the ocean. In this system the warm water from the surface of the sea is first processed in an evaporator chamber. The sea water passes through orifice spacer blocks and because of the pressure drop in the evaporator column, forms vapour or steam cavities. The core stream of vapour is separated from the outer chute of water which is returned through the base of the evaporator to the ocean. The steam passes to an expansion chamber coupled to a power generator and thence to a water cooled condenser. The CFE process represents a considerable advance on the solar sea thermal power plant originally designed by Claude, eliminating problems due to dissolved gases (employing a deaerating unit) and corrosion problems.

controlled thermonuclear reactor (CTR)

See *fusion reactor.*

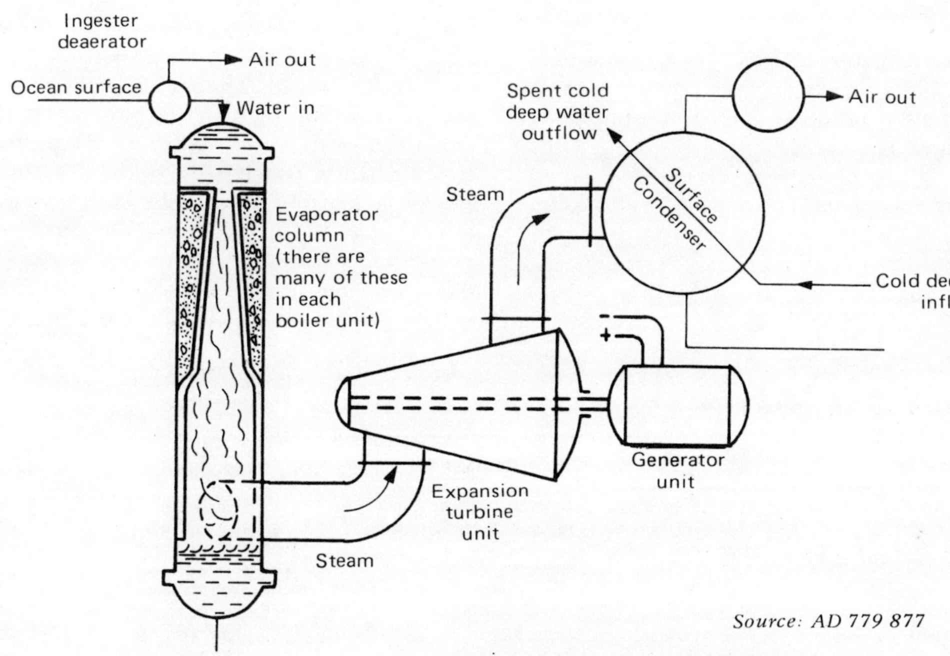

Source: AD 779 877

Schematic representation of controlled flash evaporation process plant

control rod

A device used to control the rate and the power distribution of the *fission chain reaction* in a *nuclear reactor*. It is commonly a rod of neutron-absorbing material, eg *cadmium* or *boron*, usually in an alloy steel.

Fine control rods (regulating rods) and coarse control rods (shim rods) are inserted into the core of the nuclear reactor. To start up the chain reaction the rods are slowly withdrawn so the *neutron* density increases. When *criticality* is achieved, usually when the rods are near the centre of their travel path, they are repositioned and the rate of reactions stabilized. To shut down the reactor the rods are reinserted into the core. This control of *reactivity* is possible only because about 1 in 150 of the neutrons feeding the chain reaction is 'delayed', ie it emerges later than the instant of fission because it is emitted by one of the fission products. Without delayed neutrons the speed of the chain reaction is such that it would be far more difficult to control.

conversion

1. Direct: a method in which one form of energy is directly converted into another, eg *chemical energy* into electrical energy in a *galvanic cell.*
2. Indirect: a method in which one form of energy is converted into another through one or more intermediate forms, eg in a coal fired power station chemical energy in the coal is converted to *heat energy* in steam which is converted to *mechanical energy* in the turbine and to electrical energy by the generator.

conversion efficiency

In any conversion process there is no loss of energy and the sum of all energy outputs equals the sum of all energy inputs. However, energy may be degraded to unusable forms, eg the waste heat from electricity generation. The conversion efficiency is the proportion of *useful energy* output to total energy input, and is normally stated as a percentage:

$$\frac{\text{useful energy output}}{\text{total energy input}} \times 100$$

See also *combined heat and power.*

conversion factor

1. Another name for *breeding ratio.*
2. The amount of heat yielded by a ton of oil as compared with a ton of coal.

See also *petroleum.*

conversion ratio

Another term for *breeding ratio.*

coolant

A fluid (gas, water or molten metal) circulated through the *nuclear reactor core* to remove heat generated by the *fission chain reaction.*

The amount of heat given off per second per unit volume of a reactor core (the 'power density' of a reactor) varies: from 1.1 kilowatt per litre in the *Magnox reactor* to 49 kilowatts per litre in the *boiling water reactor;* 102 in the *pressurized water reactor (PWR);* and, in the *liquid metal-cooled fast breeder reactor (LMFBR),* 646 kilowatts per litre. In reactors with low power densities the coolant may be a gas, *carbon dioxide* or *helium*, which has many advantages but relatively poor thermal conductivity, though this is improved if the gas is pressurized. In a reactor with higher power density a coolant with prime heat removal abilities is vital; the PWRs use water, the LMFBRs molten metal (*sodium* or sodium-potassium alloy).

Failure of the coolant in a nuclear reactor leads to overheating which can in turn lead to melt-down of the reactor core: the engineering of the cooling systems of reactors must be to the highest standards.

Some reactors have an open-ended cooling system, pumping water from a river through the reactor core and back to the river, or ordinary air from the atmosphere and back again. This means that the radioactivity picked up by the coolant in the reactor pollutes the neighbourhood around it. Alternatively, reactors can have one or two closed circuits which recycle the coolant. Closed circuits can be pressurized which improves the heat transfer properties of some coolants; dangerous or expensive coolants, such as molten sodium, helium or heavy water can be used in them; and they can be constructed to check the leakage of radioactive substances out of the reactor.

coolants used in representative nuclear reactor types

Type of reactor	Peak power density	Coolant	Coolant pressure	Coolant outlet temperature
MAGNOX (Dungeness A)	1.1 kW/l	carbon dioxide gas	19 atmospheres	245°C
AGR (Hinkley Point B)	4.5 kW/l	carbon dioxide gas	40 atmospheres	634°C
HTGR (Fort St Vrain)	6.3 kW/l	helium gas	46.4 atmospheres	485°C
PWR (Zion 1)	102 kW/l	light water	150 atmospheres	318°C
BWR (Browns Ferry 1)	49 kW/l	light water	68 atmospheres	285°C
CANDU (Pickering)	16.2 kW/l	heavy water	85 atmospheres	293°C
SGHWR (Winfrith)	11.2 kW/l	light water	63.5 atmospheres	282°C
LMFBR (Phénix)	646 kW/l	liquid sodium	1 atmosphere	562°C

cooling ponds

Deep tanks of water which serve to cool and shield highly radioactive, spent fuel removed from a *nuclear reactor*. Cooling ponds give time for the *fission products* with short half-lives in the irradiated fuel to decay and therefore reduce its level of radioactivity before reprocessing.

copper

A chemical element, copper is a reddish-brown, ductile metal. It is prepared from sulphide, oxide and carbonate ores by smelting, followed by electrolytic refining. Secondary sources of the metal, in the form of reclaimed scrap, are also important.

symbol	Cu
atomic number	29
atomic weight	63.55
melting point	1084.5°C
boiling point	2580°C
density	8933 kg m^{-3} (8.93 g/cm^3)

It is an excellent conductor of heat and electricity and finds many applications of these roles. The electrical industry is a large consumer of copper, using the pure metal as an electrical conductor in generators and motors and in transmission and distribution equipment. *Aluminium* is, however, preferred for overhead lines because of its lower weight and cost. The metal and its alloys such as brass are also used to a large extent for heat-transfer apparatus, such as *heat exchangers*, though the high cost of copper often dictates the use of alternative, less expensive materials. In such applications the corrosion resistance of copper is often an advantage.

copper sulphate

A blue salt of copper, chemical formula $CuSO_4$, which is soluble in water. It is used in electroplating in some types of *electric cell* (eg *Daniell cell*) as electrolyte and in *solar ponds* where it prevents growth of *algae* and improves heat absorption.

correlation index

A system of *hydrocarbon* classification produced by the US Bureau of Mines. *Hydrocarbon fractions* are divided into basic types, paraffinic, paraffinic-naphthenic, naphthenic; the higher the naphthenic proportion, the higher the index; crude oils from different sources can be compared by referring to their correlation indices.

cracking

A process by which hydrocarbons, usually obtained from *petroleum*, are decomposed by thermal or catalytic means to produce lower-boiling *fractions* suitable for use as *chemical feedstock* or *gasoline*.

critical assembly

An assembly of materials just sufficient to sustain a *fission chain reaction*.

criticality

In a *nuclear reactor*, the situation of a self-sustaining *fission chain reaction* being established, when at least one neutron and no more than one from each *fission* is achieving another fission, that is, when the reproduction factor is exactly 1.

critical mass

The smallest weight and configuration of fissile fuel that can support a self-sustaining *fission chain reaction*. It depends on the probability of *fission* in the material used, which is related to the nature of the *nuclei* under bombardment and the nature and energy of the bombarding particles. It also depends on the shape the material is in; it must be large in proportion to its surface to prevent most effectively too many high velocity *neutrons* escaping before causing further fissions. The most efficient shape is a sphere which has the minimum ratio of surface area to volume. It must also be above a certain minimum size; below it there will be too much surface through which neutrons may escape.

critical size

The minimum size of a *nuclear reactor core* of given geometry and combination of fuels that will make possible a self-sustaining *fission chain reaction*.

critical temperature

The term is most commonly used to describe the maximum temperature at which gas (or vapour) can be liquified through the application of pressure alone. Examples are for hydrogen −240°C, nitrogen −147°C, carbon dioxide 31°C.

See also *gas liquifaction*.

crude oil

Also called 'crude'. Oil from an underground petroleum reservoir that has been freed from *associated gas* but prior to any further processing.

Crude oil from different reservoirs has a varying appearance and composition which is the basis of its classification. Paraffinic crude, for instance, is rich in *paraffins* and typical of much of that obtained in the Middle East and northern parts of the USA.

See also *crude oil analysis; petroleum*.

crude oil analysis

The evaluation of *crude oil* to assess its main characteristics – physical and chemical. These include its *API gravity*, sulphur content, viscosity, *pour point*, percentage of C_4 and lighter fraction, salt content, acidity, smoke point, *char* factor, *hydrocarbon* content and yields of the main distillates; there are many others. Analysis is important in determining the type of processing operation used to convert different crudes into their products. It is possible to 'fingerprint' a crude accurately so that its exact source can be determined.

See also *petroleum.*

cryogenics

The production, maintenance and utilization of very low temperatures. It should be noted that cooling processes themselves require an investment of energy, and this has to be considered in assessing viability for future application. There are a number of ways in which cryogenic techniques are used or expected to be used in the future energy economy. These include liquifaction of gases and superconductivity.

Liquifaction of gases. (Gas liquifaction.) The principal application from an energy point of view is in the liquifaction of *hydrogen* gas, which requires the attainment of very low temperatures and considerable gas pressures. The development of suitable cryogenic containers has been pursued to enable hydrogen to be transported on a large scale, to provide a portable and flexible fuel. Another important application is in the liquifaction of *natural gas.*

Superconductivity. Superconductivity depends on the fact that electrical conductors lose their resistivity when the temperature is taken below a critical level. It is then possible to obtain very high magnetic fields without the use of impossibly large electromagnets. This effect is being used in the various experiments being conducted to obtain controlled *fusion* reactions such as in the Princeton experiment. Another experimental application has been in the use of supercooled underground transmission lines, the object of which is to remove losses due to conductor resistance, and in electrical generator windings where superconductivity has been experimentally exploited. To date, the cost benefits of these devices have not been demonstrated to provide viable systems.

cryogenic upgrading

The use of low temperatures to separate out impurities and unwanted products from a gaseous mixture.

CSF

See *Consol synthetic fuel process.*

Curie (Ci)

A unit of *radioactivity* reflecting the number of *nuclei* which break up each second in a particular sample. One curie is defined as 3.7×10^{10} disintegrations per second. This is approximately equivalent to the activity of one gramme of radium-226. (For finer measurements a picocurie is used which is 10^{-12} of a curie.)

cut

Another name for *petroleum fraction.*

cycloalkanes

See under *alicyclic hydrocarbons.*

cycloalkenes

See under *alicyclic hydrocarbons.*

cyclohexane

A *cycloalkane* produced from *benzene.* It is used principally as a solvent, in the manufacture of nylon and as an additive in blending motor spirit with a high octane rating.

See also *Appendix 4; Table 2.*

cylindrical reflectors

See *solar collectors; solar energy.*

D

dams

Barriers erected across rivers (or valleys) to create artificial lakes and thus a head of water which can be used to drive a hydroelectric generating unit (dams also have use for flood control purposes, to make rivers navigable, or for water storage). Their creation can have beneficial and harmful *environmental effects.*

Basic types of dam are timber, earthfill, rockfill and masonry of which gravity and arch dams are the two main sub-types.

See also *arch dams; hydroelectric power.*

Daniell cell

A form of primary *electric cell.* A zinc rod *electrode* is placed in a porous pot which is filled with dilute sulphuric acid. The porous pot is contained in a copper pot containing *copper sulphate* solution; the copper pot acts as the positive electrode. The *emf* of the cell is 1.1 volts.

dead oil

Petroleum deposits containing no *associated gas* and having no inbuilt pressure mechanism for driving the oil to the surface.

deasphalting

Removing the asphaltic components from residual stock by *solvent refining* for manufacturing lubricating oils. Liquid *propane* is commonly used as the solvent for precipitating out the asphalt.

decay

Radioactive disintegration of a *nucleus.*

See also *radioactivity.*

decay chain

A disintegration path, in which a succession of radioisotopes are formed. Eventually the chain ends when a stable *isotope* of an element is reached, eg *uranium-238* goes through a chain of disintegrations, by emitting *α-particles* and *β-particles*, ending up as the stable lead-206.

decay daughter

The product of a radioactive disintegration of a radioisotope.

decomposition

The breakdown of chemical compounds into simpler molecules. Decomposition is important in nature for dead animals and plants are broken down by microorganisms into simpler chemicals; in this way organic compounds are circulated. *Fossil fuels* are products of decomposition and other changes over millions of years.

See also *anaerobic fermentation; biological energy sources; waste materials energy.*

dehydrocyclization

A platinum-catalyzed *reforming* process in which straight chain *alkanes* are converted first into *naphthenes* and then into *aromatics.* For example, n-heptane may be converted into methylcyclohexane, thence to toluene. Hydrogen gas is produced as a result of this process which may then be used for other refinery processes requiring hydrogen such as *hydrocracking.*

dehydrogenation

Removing hydrogen from a chemical compound. An example is the cyclization process which occurs when *straight-run gasoline* is thermally reformed.

dehydrosulphurization

Another name for *hydrodesulphurization.*

delayed coking

A *thermal cracking* process in which vacuum or atmospheric residues from low-sulphur crude are heated to 500°C, the resulting *cracking* ultimately yielding a porous mass of electrode grade coke suitable for aluminium production. Lower boiling materials vaporize, are passed out of the coking drum and fractionated to yield gas, gasoline, light gas oil and heavy gas oil.

delayed neutrons

Neutrons emitted from atomic nuclei just following the moment of nuclear *fission.* The *fission chain reaction* would be much faster and extremely difficult to control if it were not for the phenomenon of the delayed neutron emitted by fission products, which accounts for less than 1 per cent of the neutrons lost from nuclei as the result of fission.

See also *prompt neutrons.*

delineation well

An exploration well used to determine the extent or boundary of an oil reservoir.

See also *appraisal well.*

delivered energy

Energy which is input to a process or system and is measured at the point of crossing the process or system boundary is termed delivered energy. For example, the system might be a factory and one of the components of the delivered energy to the factory would be electricity measured, by meter, at the point where the power lines enter the factory.

depleted uranium

Uranium which contains less than 0.72 per cent of uranium-235, a necessary result of *uranium enrichment.*

depletion policy

A complex subject concerned with the 'best' (optimal) rate at which non-renewable resources should be extracted and consumed. The technically optimal rate may differ from the broader, economically optimal rate of use and may, in turn, differ from a narrowly commercial or profit maximizing rate. Decisions are particularly sensitive in economies where there are sizeable but limited non-renewable resources (such as the North Sea oil fields).

DERV

UK abbreviation for diesel engine road vehicle. DERV fuel is used in compression ignition engines that turn at high speed.

desalination

The removal of salt from water to make it potable, less corrosive and of a suitable quality for industrial and domestic use.

The supply of fresh water is a great problem in arid areas

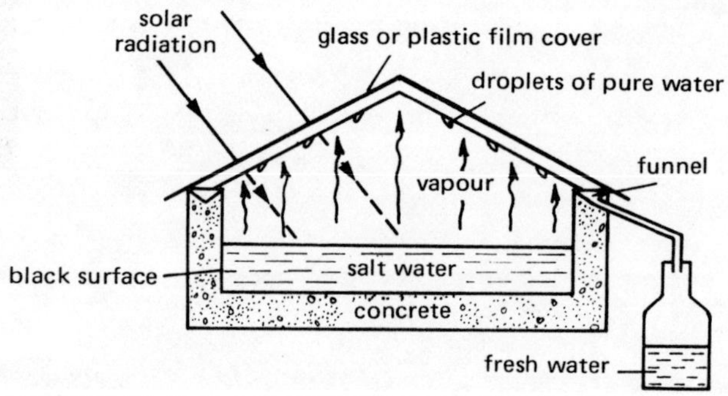

desalination (roof type of solar still.) About 90 per cent of the incident solar radiation is transmitted into the still where the sun's rays are absorbed and the water is heated.

and desalination plants typically rely on *solar energy.* Sea water is evaporated, desalinated water condenses on a suitable surface, and is collected. The fuel costs of this method are zero, but capital and land requirements are high. Other methods of desalination, such as multiple effect *distillation* and flash distillation, although requiring a considerable fuel input, use cheaper, more compact plant. Efforts to reduce the energy cost of desalination are being directed towards new processes such as reverse osmosis but distillation is still the only practical method.

Desalination plants are being installed in a number of oil-producing Middle East states where demand for fresh water is increasing rapidly through greater industrial activity and the higher standard of living.

desalting

The removal of inorganic salts (especially chlorides) from *crude oils.* Salt may be deposited in the heater tubes during *distillation* processes causing corrosion problems.

desulphurization

The removal of *sulphur* from coal, crude petroleum, petroleum fractions, or natural gas carried out prior to further refining or use. Desulphurization is essential in order to overcome problems associated with corrosion, catalyst poisoning and *air pollution* and is widely employed in the petroleum refining industry as most crude oils contain sulphur as an impurity. Natural gas may contain hydrogen sulphide as an impurity; such gas is referred to as 'acid' or 'sour' and the sulphide impurity is removed by washing with aqueous alkali. Most coals contain sulphur impurities which usually reappear as sulphur dioxide in the atmosphere, though *fluidized-bed* processes offer the opportunity of reduction in air pollution by combining the sulphur chemically in the combustion chamber.

See also *acid gas; dehydrosulphurization.*

deuteride

A chemical compound containing *deuterium* combined with one other metallic element. One form, lithium deuteride, is a component of the *hydrogen bomb* and is a possible fuel for *fusion reactors.*

deuterium

A heavy isotope of *hydrogen* with an *atomic weight* of 2. It is a constituent of *heavy water (deuterium oxide).* Deuterium is a probable fuel for *fusion reactors* and is a constituent of thermonuclear explosive devices. In the form of heavy water it is used as a *moderator* and *coolant* in *fission reactors* (eg *CANDU*). It is found as a naturally occurring constituent of water, in the proportion of one part in 6700. The total resources of the element are therefore vast if the water in the oceans is included.

Like hydrogen, it is an odourless gas, though being twice as heavy its density is double that of hydrogen (0.18 gram/litre as opposed to 0.09 gram/litre). Its boiling point is marginally higher, too;-249.5°C as opposed to -252.7°C for hydrogen. Deuterium is made by the reduction of *deuterium oxide* with pure calcium metal.

deuterium oxide

The deuterium analogue of water, known as *heavy water*, which has its principal use as a *moderator* and *coolant* in a number of designs of *nuclear reactors.*

Properties	
Appearance:	that of water
Boiling point:	101.6°C (214.9°F) – water 100°C (212°F)
Density (maximum):	1.106 kg per litre at 11°C (1.000 kg per litre at 4°C for water)
Melting point:	3.79°C (38.8°F) – water 0.00°C (32°F)

Production. The main sources are Canada and the USA, France, Norway, India and possibly Czechoslovakia. Although not as difficult to extract from the raw materials as uranium-235, it is still expensive. In 1982 Canada intends a production of 4600 tonnes per annum. A *CANDU* takes just under one tonne of heavy water per megawatt(e) and the steam-generating heavy water reactor *SGHWR* about a quarter as much. This is a once-only requirement: losses of moderator are small.

Demand. The main requirement up to the end of the century is likely to be from *nuclear reactors* fuelled with natural uranium and perhaps the SGHWR along with breeder reactors using the thorium-uranium-233 cycle; the demand from nuclear *fusion reactors* (assuming they come to fruition) will not be significant until well into the 21st century. Fusion reactors will be run on deuterium gas or compounds other than the oxide, but the oxide will be the source.

Isolation. Deuterium oxide can be isolated from naturally-occurring water by a number of processes, including *distillation, electrolysis* and chemical exchange, all of which make use of a small difference in chemical or physical properties between hydrogen compounds and their deuterium analogues. Although all of the processes mentioned have been used, the most important commercial process today is the Girdler-sulphide process. This is a chemical-exchange method and exploits the temperature-dependence of the distribution of deuterium atoms between hydrogen sulphide and water.

deviation drilling

Alternative name for *directional drilling.*

dewaxing

The extraction of waxes from lubricating oil *cuts* during their purification. Waxes are used in candles and for electrical insulation.

diesel engine

Often used as an alternative name for compression ignition engine. Fuel in such an engine is ignited by the heat of compression in contrast to a spark ignition engine in which the fuel is ignited by an electrical spark.

diesel fuel

A *petroleum fraction* used as fuel in compression ignition engines. It is obtained from crude fractions in the boiling point range 250°-350°C, that is *middle distillates.* High-revving engines need special quality diesel fuels such as *DERV.*

diesel index

A figure which indicates the ignition quality of diesel fuel. *Paraffins* have high diesel indices, and aromatics low indices. The diesel index is usually calculated from laboratory measurements of specific gravity and aromatic content. A figure of 60 or above indicates a good diesel fuel; one of 40 or below indicates an unsatisfactory fuel.

See also *cetane number.*

diesel oil

Alternative name for *diesel fuel.*

diffusion process

The most widely used process for *uranium enrichment*, this method operates on the principle that lighter gaseous atoms and molecules will diffuse through a thin porous membrane more quickly than heavier ones. Thus *uranium hexafluoride ('hex')* vapour may be progressively enriched by such a process repeated over and over again, leaving behind some hex containing *depleted uranium.* Because of its military importance, many of the details of the gaseous diffusion process remain secret, although it is known that the energy requirements for operating such a plant are enormous. It is hoped that the *centrifuge process* may achieve the same efficiency of separation but at a lower energy cost.

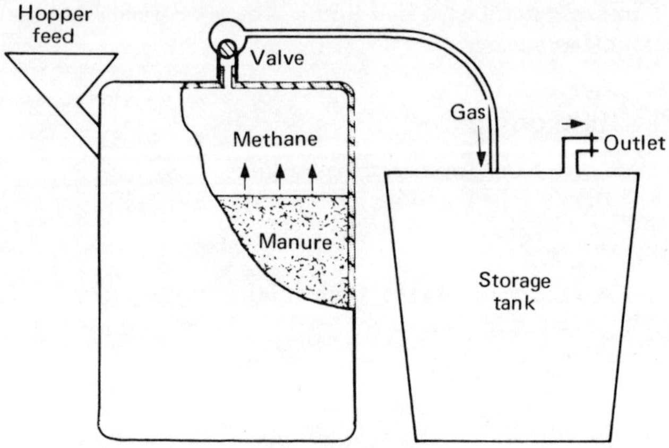

Digester for use on a farm

digester

A device in which biological wastes can be 'digested' or fermented anaerobically to produce *methane* gas as fuel. Digesters are in use on farms where waste materials such as manure are readily available. Shredded material is placed in a large digester tank where *anaerobic fermentation* breaks down the waste with the release of methane. The methane is collected in a storage vessel and used to provide power for farm processes. This has been common practice at *sewage* farms for some years.

diode generator

A *thermionic generator* in which a heated *cathode* emits *electrons* to form an electric current, collected by the *anode.* On a small scale (as in satellites) the energy that is used to dislodge the electrons could be provided by *solar energy* or by emanation from *radioactive isotopes.* On a larger scale, the heating of the cathode could be carried out by wrapping it around uranium fuel, or by heating the cathode using exhaust gases.

direct current (dc)

A unidirectional flow of electric current. Typical sources of direct currents are *electric cells*, rectified power units, and *direct current generators.* The term can apply to fluctuating currents provided they are unidirectional.

direct current generator

An *electric generator* that provides a unidirectional (dc) electric current. The ordinary generator produces an *alternating current* (ac). The dc generator is a modified version of an alternating generator, with commutators providing rectification to produce a direct current.

directional drilling

A technique of drilling oil or gas wells that starts with a vertical bore which is then diverted by as much as 60 per cent from the vertical. This enables a wide area of the field to be exploited from a single platform.

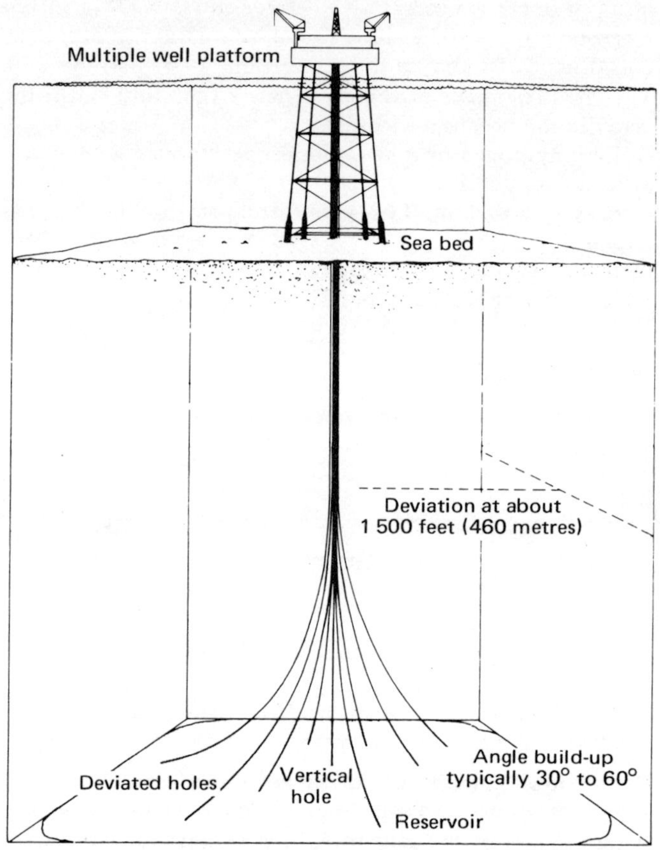

Directional drilling

dirty oil

Alternative name for *black oil.*

distillate fuel

A general term used to describe fuels obtained by the fractional distillation of crude petroleum.

Fuel	Boiling range °C
Gases	<0
LPG	−30–+30
Gasolines	30–200°
Kerosines	150–250°
Gas Oils	200–350°

See also *heavy distillate; light distillate; middle distillate; petroleum fractions.*

distillation

A process used to separate liquid mixtures by boiling followed by condensation of the vapours produced. The more volatile components, boiling at a lower temperature, are collected first. Improved separation of complex mixtures is obtained by fractional distillation, in which the vapour of high-boiling components is condensed by contact with lower-boiling components which are thereby evaporated. This technique has been largely developed for petroleum refining, where the preliminary stage of crude oil processing involves the use of large *fractionating towers* to achieve an initial separation into *petroleum fractions*.

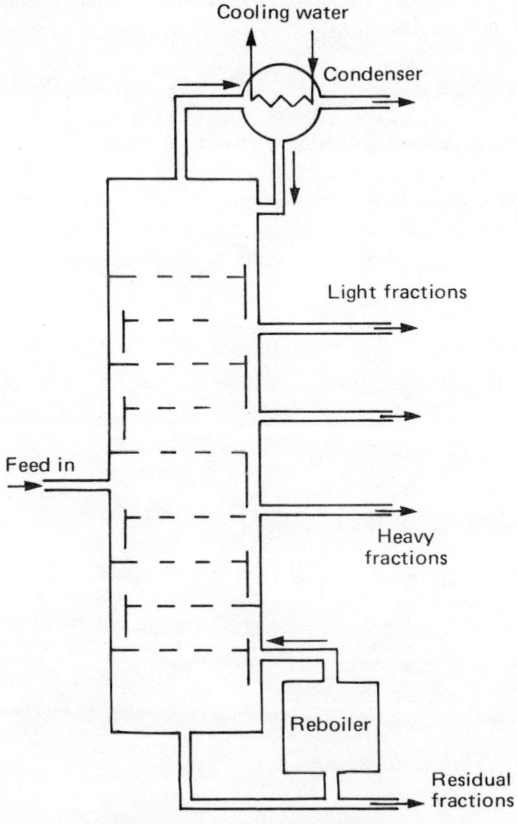

Schematic representation of fractional distillation process

Coal tar is produced by the carbonization or 'destructive distillation' of coal in horizontal retorts and is separated into its constituents by a distillation process known as 'topping'.

Distillation may be carried out at pressures below atmospheric — to lower boiling points and reduce thermal decomposition — a process known as *vacuum distillation*. High pressure distillation is used to separate the components of air for the supply of industrial gases.

distribution losses

See *transmission losses*.

district heating

As its name implies, heating a large area of commercial, industrial, or domestic premises by a network of pipes employing hot water or steam as *energy carriers*. In parts of the world, such as Iceland, the heat comes directly from geothermal sources, but usually the energy is produced by a centralized unit such as a *heat-only boiler* or as part of a *combined heat and power* scheme. District heating is relatively common in Scandinavia and Eastern Europe and as the economics of using fossil fuels more efficiently become increasingly advantageous such schemes will become more commonplace elsewhere, particularly in areas of high housing density and on new town sites.

diurnal variations

In the energy context, this refers to changes in demand that take place during a 24-hour period. It is particularly important for any centralized system to be able to handle these demand variations which can be quite marked particularly on cold winter days. For this reason, gas distribution systems contain large elements of diurnal storage and electricity supply utilities have *stand-by* and *peak-load* plant available to meet anticipated increases in demand.

divergence

In a *nuclear reactor*, the situation where the rate of the *fission chain reaction* is increasing, ie when more than one *neutron* from each fission is causing a further fission (the reproduction factor is more than 1).

domestic heating

The heating of domestic premises and of water, usually by solid fuel, gas, oil, or electricity. With the increase in price of conventional fuels, heating (or partial heating) by *solar energy* is becoming more attractive though there are problems to be overcome including high capital costs, overheating in summer, and mismatch between availability and need. There is great scope for *energy conservation* in the home as well as greater efficiency in end use, both of which could do much to increase standards of comfort and conserve fossil fuels.

See also *solar collectors*.

domestic heating oil

See *burning oil*.

domestic refuse

See under *waste materials energy*.

dose level

The amount of radiation to which the population is exposed, usually expressed in 'rems' or 'millirems'.

See also *background radiation; dose limit; maximum permissible dose; rem.*

dose limit

A threshhold level of radiation, set by the International Commission of Radiological Protection (ICRP), above which the general public should not be exposed. The whole body limit for the general population is 0.5 rem per year. For workers in the radiological and nuclear industry, different limits apply.

See also *maximum permissible dose.*

doubling time

The time required for a *breeder reactor* to double its inventory of fissile material by conversion of *fertile material.*

dry cell

A kind of *electric cell* in which the *electrolyte* is in the form of a paste rather than a liquid. The term is usually used to describe the dry version of a *Leclanché cell*, and this is commonly used in *batteries* for torches and transistor radios.

dry gas

The lighter extracts from *natural gas* that have a very low proportion of liquid hydrocarbons (ie less than 3.6g/1000m^3 of gas).

dry hole

Another name for *dry well.*

dry steam coal

A high rank *bituminous coal*, low in water and sulphur content, suitable for raising steam in industrial and power station boilers.

dry well

Also called dry hole or duster. A well that is unlikely to produce oil or gas in sufficient quantities to warrant its commercial development.

dung

A renewable source of energy, used especially in the third world for heating purposes.

See also *waste materials energy.*

duster

Another name for *dry well.*

E

economies of scale

Financial savings that are the result of an increased scale of production. These are particularly marked in the energy supply industries and arise through several effects: 1. Overheads: fixed cost overheads, such as sinking a mine shaft or manning a power station represent a smaller cost component as the scale of production increases. 2. Efficiency: energy losses can generally be reduced for larger conversion units. Some of the historic rise in electricity generating efficiency can be attributed to this effect. 3.'The two-thirds rule': the cost of vessels, reactors, pipes, etc, increases as the square of the linear size, whilst the volume enclosed increases as the cube. The cost of plant therefore varies as the two-thirds power of the capacity, so that doubling the plant size increases the capacity by a factor of 8, while the capital costs increase by a factor of 4. (Another way of saying this is that the capital cost per unit throughput is inversely proportional to the linear size.) This rule is, of course, very approximate. 4. Distribution effects: a large-scale energy supply network with an interconnected transmission system may need a smaller capacity reserve to cope with demand fluctuations, or shortfall of supply. 5. Other effects: these include labour specialization in larger units of production and research co-ordination in larger establishments.

It is probable that, like other economic activities, energy supply and conversion processes have an optimal scale of production, beyond which various factors contribute to higher costs. These diseconomies of scale include: 1. Breakdowns: the loss of, for example, one large generating set may cause serious difficulties to an electricity supply system. There is evidence to suggest that such equipment is more prone to failure than smaller plant. If this is the case then the optimal size for this sort of machinery has probably been exceeded. 2. Transport/transmission costs: larger units of energy production may entail either higher freight costs, or greater transmission losses, depending on whether they are sited close to centres of demand or to sources of supply. 3. Demand fluctuations: an energy supply system made up of a few large units of production may be less able to cope with fluctuations of demand than a similar system made up of smaller units. Although the scale of the distribution system gives greater potential of control, the unit size may hinder its exercise.

economizer

A particular type of *heat exchanger* used to recover heat from the flue gases of a *boiler* by transfer to the boiler feedwater. Economizers increase thermal efficiency and are used on most steam-raising boilers for electric power production.

Edison battery

See *nickel-iron accumulator*.

efficiency

The ratio of the total output of *useful energy* from a process to the total energy input (usually expressed as a percentage).

The definition of what is a useful energy output is open to question and will be completely dependent on the nature of the process in question. Thus a gas-fired water heater produces the useful energy in the form of hot water and the energy in the hot flue gases is generally lost. If, however, this heat could be used to good effect, the efficiency of the the heater would necessarily rise.

In power stations, electricity is produced at 35 per cent efficiency, which simply means that every 100 units of fuel input provide 35 units of electrical output. Large amounts of energy, much of which is theoretically recoverable, are for all practical purposes 'lost' but if this *low grade heat* could be harnessed usefully the overall efficiency of the conversion process could be said to have increased.

electrical conductor

A substance that permits *electric current* to flow through it. *Copper* and *aluminium* are good conductors. Because of its lightness the latter is used for high-voltage transmission lines.

See also *electricity transmission*.

electric cell

A device in which *chemical energy* is converted directly into electrical energy. In its simplest form, in the primary 'galvanic' cell, two dissimilar metallic *electrodes* are immersed in an *electrolyte* and, as the result of interaction between *ions* in the electrolyte and the electrode materials, a charge transfer occurs that produces a potential difference between the electrolyte and each electrode. When an external load is connected between the electrodes an *electric current* can therefore flow.

The ordinary galvanic cell described above is non-rechargeable because the chemical reactions which occur while the cell is on-load are irreversible. Such a cell is called a primary cell. The *emf* provided by such cells depends on the nature of the electrodes and electrolytes. Two or more cells connected in series will provide a total emf that is the sum of the individual emfs. Such an arrangement is properly called an electric battery, although the term is often used to describe a single cell as well.

Rechargeable electric cells are called secondary cells, accumulators or storage batteries. In these, a charging current causes chemical reactions between the electrolyte and the electrode, so the electrical energy input is stored in the form of chemical energy. But the process is reversible and the stored energy is reconverted into electrical energy when the cell is put on-load.

Another form of electric cell is the fuel cell, in which the electrochemical reaction that provides the electric power is fed by continuous fuel replenishment to the electrodes.

Practical forms of electric cell. 1. *Primary cells.* Most manufactured cells or batteries do not use liquid electrolytes, a paste form being employed. It is only in specialized applications, such as in precise electrical calculations and measurement work that liquid electrolyte cells, such as the Weston cell and the Clark cell are used, eg the original *Leclanché cell* is now manufactured in the form of a dry cell. Other examples of primary cells are the *alkaline primary cell, mercury cell, silver oxide cell.* 2. *Secondary cells.* The most important is the *lead-acid accumulator* familiar as the source of stored electric power in the motor car. Other examples of secondary cells are *nickel-iron accumulator, nickel-cadmium cell, sodium-sulphur battery.*

One of the most important aspects of the secondary cell is the ability to store electricity and for this reason the capacity of, say, a lead-acid accumulator will often be quoted as 'x ampère-hours', ie it will give, say, a current of 1 amp for x hours (or 5 amps for x/5 hours) before needing recharging. Weight and volume are often important factors, too.

See also *energy density.*

electric current

The rate at which *electricity* flows through an *electrical conductor*, usually measured in ampères (often abbreviated to amps).

electric generator

A device which converts mechanical energy into electrical energy. There are two basic types of generator, the alternator and the dynamo which produce *alternating current* and *direct current* respectively.

Both generate on the same scientific principle that an *electric current* is induced in a coil when it cuts the lines of force in a magnetic field. In a modern power station, electricity generation is effected by coupling the generator rotor, a large, cylindrical electromagnet, to the *turbine.* When the turbine and rotor rotate, alternating current is induced in the coils of the generator stator and fed to the station transformer where the voltage is raised to a level suitable for transmission. Direct current may be produced in the same way, rectification being achieved by use of a commutator.

Electric current is usually generated in three phases by employing three pairs of coils in the stator windings in planes lying at an angle of 120° to one another. Current is conducted away from the generator by three live wires and one neutral.

See also *three-phase supply.*

electricity

An energy form resulting from the flow of charged particles, such as *electrons* or *ions*, in a conducting medium; *electric current* is usually thought of as a flow of electrons through an *electrical conductor.* It may be compared with the flow of liquid through a pipe, in which case electric current is analagous to the rate of flow and the potential difference (voltage) required to maintain that flow is analagous to the difference in pressure at the two ends of the pipe.

Electricity is a particularly versatile energy form, being capable of providing heat, light and motive power at high efficiency. Its principal disadvantages are the large conversion losses which occur in generation and the inability to store it on a large scale without great inconvenience and expense.

See also *electric cell; electricity generation; energy storage; pumped storage schemes.*

electricity generation

The generation of *electricity* (a *secondary fuel*) from *primary fuels* such as coal, oil, hydroelectric sources or nuclear power. This is usually carried out at centralized power stations, but some industrial concerns generate both steam and electricity for their own use. Such *combined heat and power* schemes are also operated by some public utilities but, in general, the large, modern power stations are sited close to coalfields and oil refineries, away from centres of population, and are thus unsuitable for heat transmission.

The first stage in producing electricity from coal or oil at a power station is the combustion of the primary fuel in the boiler. The resulting heat boils water which is circulated through the boiler tubes at high pressure, converting it to high pressure steam. (In nuclear power stations, heat is generated in the *nuclear reactor core* and is conducted away for steam raising.) The steam is used to drive turbines which in turn drive the *generator* at high speed, thus producing electricity. (In hydroelectric schemes, the water is used to drive the turbines directly.) The efficiency of a modern power station is of the order of 35 per cent and it is not foreseen that any great improvement will be possible, limited as the *efficiency* is by thermodynamic considerations.

See also *energy conversion.*

electricity transmission

The carrying of electricity over long distances by overground power lines or underground cables. Large coal- and oil-fired power stations are usually situated close to the source of supply of their primary fuel (the coalfield or oil refinery), often some distance from areas of high demand. There is a requirement, then, for large-scale distribution of electricity and this is usually integrated into a national or regional transmission system. There are *transmission losses* associated with the distribution of electricity but these are kept to a minimum by using very high voltages, eg the 'Supergrid' system employed by the UK Central Electricity Generating Board operates at 400,000 volts. A light electrical conductor is favoured for overhead lines and aluminium is therefore widely used.

electric power

Colloquially, the changing of electrical energy to mechanical energy as in an electric locomotive, factory lathe, or domestic food mixer. Its strict definition, however, is the rate at which electrical energy can provide heat or motive power, and is measured in watts, or more commonly *kilowatts.* Most appliances have a *power rating*, eg a light bulb is approximately 100 watts, an electric fire 2 kilowatts and this is a measure of the amount of energy used by these devices when in operation. Following up the example of the electric fire, in one hour this would consume 2 *kilowatt-hours* (kWh) of electrical energy.

electrode

A conductor through which *electric current* passes in to or out of a conducting medium. The medium can be ionized gas, a vacuum in an electronic tube, an ionized liquid (electrolyte) or certain kinds of solids (eg semiconductors).

See also *diode generator; electric cell; electrolysis; fuel cell; magnetohydrodynamic power generation.*

electrolysis

The decomposition of an ionic compound caused by the passage of an electric current through it. When a potential difference is applied between a pair of *electrodes* immersed in the compound, or a solution of it, the positively and negatively charged ions migrate respectively to the negative and positive electrodes.

A possible application of electrolysis in future energy provision is in the harnessing of intermittent sources such as *wave energy* which may be capable of providing electrical output when there is no specific demand for it. This electricity could be used in the electrolysis of water which would release *hydrogen* for use as a fuel.

electrolyte

A chemical compound which, either in solution, or in the fused state, conducts electricity by virtue of the migration of charged constituents (ions).

See also *electrolysis; primary cell.*

electromagnetic radiation

A form of energy characterized by its ability to travel

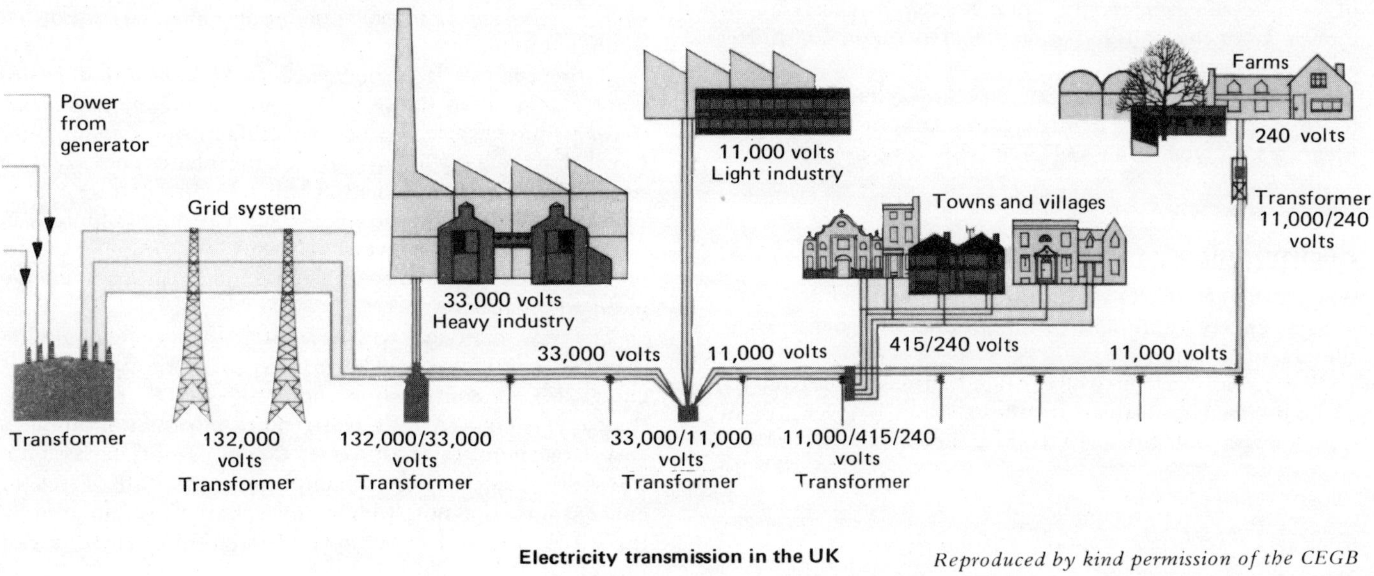

Electricity transmission in the UK *Reproduced by kind permission of the CEGB*

through a vacuum at the speed of light; sometimes simply referred to as 'radiation'.

Electromagnetic radiation can be described in two complementary ways: as particles (photons) whose energy determines the properties of the radiation; and as waves having electrical and magnetic components whose wavelength determines their properties.

Electromagnetic radiation is observed in nature as a continuously varying spectrum which, for convenience, is subdivided into bands according to the nature of the radiation. In order of increasing wavelength (decreasing photon energies) these are gamma rays, X-rays, ultraviolet light, visible light, infra-red radiation, microwave radiation and radio waves. Some sources, such as the *sun*, emit radiation over a wide range of wavelengths; others, such as lasers, over a very narrow band indeed.

Electromagnetic radiation differs from other forms, such as alpha radiation, beta radiation and neutron radiation, in that photons have zero charge and zero mass. Alpha particles (helium nuclei), beta particles (electrons) and neutrons all have measurable mass. Alpha and beta particles are also charged.

electromotive force (emf)

The difference in electric potential produced between the output terminals of generators of electrical energy, such as electric cells. It is the emf in a circuit that drives the electric current around it.

electron

A negatively charged particle which is much lighter than the other basic constituents of the atom, the *proton* and the *neutron*. Protons and neutrons constitute the atomic *nucleus*, the electrons orbiting the central part of the atom. When an electron is removed from the vicinity of an atom, the atom is said to be positively charged; conversely, a gain of an electron renders the atom negatively charged. The movement of electrons in an *electrical conductor* constitutes an *electric current*.

Radioactive isotopes release electrons from the nucleus by the breakdown of a neutron into a proton and electron. Electrons ejected from nuclei in this way are known as *β-particles*.

electron volt

When a charged object moves through a potential difference it gains energy equivalent to the product of its charge and the potential difference.

When one electron is accelerated through one volt the amount of energy gained by the electron is given by one electron volt, equal to 1.6×10^{-19} joules.

emf

Abbreviation for *electromotive force*.

energy

A scientific concept by which the basic processes of nature, whether physical, chemical or biological, are rationalized. Often defined as the 'capacity for doing work', energy cannot be directly observed or measured. Rather it must be studied by the effects it produces on matter. Concepts of energy flow and *energy conversion* are central to almost all sciences and energy occurs in many forms which all fall into one of two categories: *kinetic energy* and *potential energy*. Kinetic energy, the energy that a body possesses by virtue of its motion, is exemplified by *thermal energy* and *mechanical energy* of moving or rotating masses. Potential energy is that due to the state of a system and *chemical energy*, *nuclear energy* and electrical potential, or *emf*, are all of this type.

Two fundamental natural laws are observed to govern energy flows. The law of *conservation of energy*, also known as the 'First Law of Thermodynamics', states that energy cannot be created or destroyed. The 'Second Law of Thermodynamics' is a formal statement of the natural limitations that control the interconversion of energy forms.

Fuels were originally employed to obtain heat (thermal energy) for keeping warm and for cooking, but more recently have also been employed to provide power (mechanical energy) for industry and transport. Like food and shelter, energy is a basic requirement for the continuing functioning of society, and indeed recent events have demonstrated that without secure energy supplies industrial societies face considerable problems.

energy analysis

The systematic tracing of *energy* flows through an economic system or sub-system in such a way as to attribute an *energy requirement* to each commodity or service produced by that system.

If the analysis is carried back to fuel resources, ie primary energy, then the result is a *gross energy requirement*, (GER). For example, a process manufacturing a commodity C will have energy inputs to run the plant. These will be supplied by the fuel industries which in turn extract energy from primary energy sources. This chain of supply will necessarily involve conversion losses, and these must be taken into account when calculating the primary energy input to the process.

The raw materials will have their own energy requirements for production. The process plant too will require energy for its construction and maintenance. It is obvious that any energy analysis will require information about a very large number of processes unless simplifying assumptions are made. These assumptions constitute a set of conventions: the most widely used set are those adopted by the 1974 IFIAS (International Federation of Institutes of

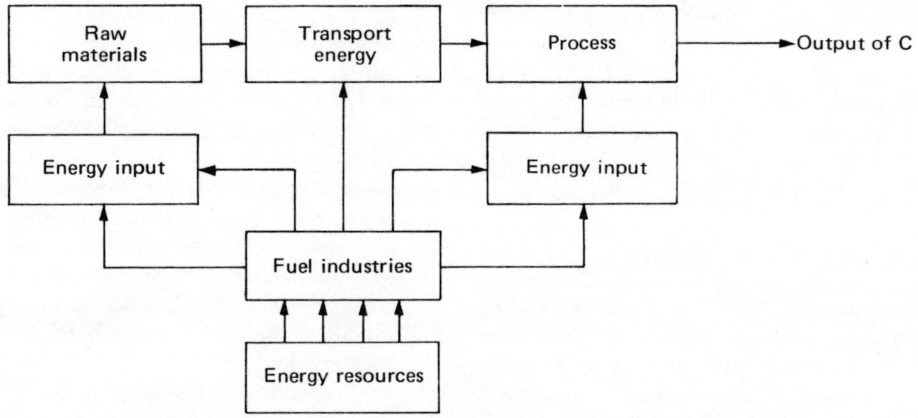

Energy analysis

Advanced Studies) conference at Stockholm and are taken as IFIAS conventions.

An analysis that is concerned only with the direct and immediate energy inputs to a process leads to the calculation of the so-called *process energy requirement* (PER).

energy audit

The systematic documentation of energy flows within an establishment, such as a factory, that has the aim of determining where energy losses occur, or where significant energy savings can be achieved. The energy audit enables energy conservation methods to be tested and as such is an important feature of energy management. Energy audits are usually carried out at regular intervals.

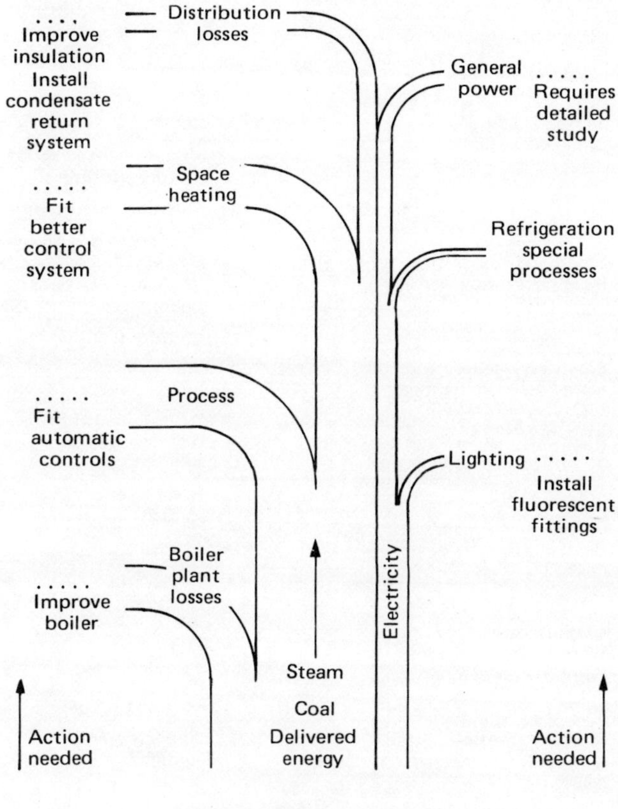

Result of a simple energy audit for a small factory

energy carrier

A term generally used to describe an agent by which energy is transferred from sources of production and supply to points of usage. Both *primary* and *secondary fuels* may be used as energy carriers, typical examples being solid fuels, petroleum, natural gas and electricity. The term also covers non-fuel materials such as liquids and gases which have had heat energy transferred to them by the combustion of a fuel. Examples here include steam and hot water in *district heating* schemes and carbon dioxide in a gas-cooled *nuclear reactor*. In the latter case gas is used as an energy carrier to take the heat away from the 'production source' (the *nuclear reactor core*) to be transferred to the next carrier (water), which in turn is converted into steam which drives the turbo-generators to produce electricity. Thus the energy originally taken from its primary source (uranium) has been conducted down a chain of carriers, finally appearing at its point of use as electricity.

Energy carriers little used today that have been suggested for future use include hydrogen, methanol and microwaves.

energy coefficient

The ratio between the change in a nation's primary energy consumption and the change in Gross Domestic Product for a particular year. Although the measure of energy coefficient may fluctuate markedly from year to year, the longer term average is comparatively stable. For instance, in the United Kingdom the average value of the energy coefficient over the last 25 years is approximately 0.7 which indicates that a 1 per cent increase in Gross Domestic Product requires a concomitant 0.7 per cent increase in primary energy use. In less developed countries, the energy coefficient ratio tends to be higher, indicating that in comparison more primary energy is needed for the same percentage increase in Gross Domestic Product.

energy conservation

The saving of a non-renewable energy resource for use at a future date, without any decrease in present utility. As a result of recent international events and changed circumstances which have resulted in unprecedented rises in the price of energy, conservation is being taken increasingly seriously by industry, governments and international agencies. It is generally acknowledged that, despite their political origins, the OPEC price increases reflect a genuine economic scarcity of a valuable resource (oil) which, along with other fossil fuels, needs to be used as carefully as possible.

It is hoped that conservation will give nuclear energy and the renewable sources a longer period to be able to establish themselves, so that economic and social dislocation due to fuel shortages may be avoided. There is vast scope for saving energy that is at present wasted in the home, in industry and in transport. Ways in which energy may be saved include: 1. Domestic: loft insulation, hot water tank and pipe lagging, cavity wall insulation, draughtstripping, double glazing, more efficient heating appliances, regular maintenance of appliances. 2. Industry: adoption of more energy-efficient processes, regular maintenance and checking of equipment, less wastage of low grade heat (eg power stations), more energy-consciousness and better housekeeping. 3. Transport: smaller vehicles, lighter vehicles, more efficient engines and transmission.

The implementation of many of these measures depends crucially on the economics of the required investment and, as the price of energy increases, more of these schemes will become economic and those at present viable will have shorter payback times.

energy conversion

The transformation of energy from one form to another.

Overall process	Conversion stage	Energy conversion From: Input energy form	To: Output energy form	Efficiency best present practice	Maximum possible efficiency
Electricity generation from fossil fuels	1	Chemical energy in coal or oil	Heat energy of generated steam	85–95	100
	2	Heat energy of steam	Mechanical energy of rotating turbine	40–48	63*
	3	Mechanical energy of rotating turbine	Electrical energy	95–98	100
	Overall	Chemical energy in coal or oil	Electrical energy	32–42	63*
Electricity generation from hydro-power	1	Mechanical energy in falling water	Mechanical energy in rotating turbine	85–90	100
	2	Mechanical energy in rotating turbine	Electrical energy	95–98	100
Electricity generation from fossil or chemical fuels using fuel cell	1	Chemical energy in fuel	Electrical energy	50	80–100**
Electricity generation by photo-voltaic cell	1	Radiant energy in sunlight	Electricity	20***	93
Solar heating	1	Radiant energy in sunlight	Heat at ~100°C	30–60	100
Electric resistance heating	1	Electricity	Heat at any temperature	100	100
Gas heating	1	Chemical energy in gas	Heat up to ~100°C	80–85	100
Mechanical energy from fuels via internal combustion engine	1	Chemical energy in fuel	Heat energy in combustion gases	95	100
	2	Heat energy in combustion gases	Mechanical energy in rotating parts of engine	32†–40††	75–85
Photosynthesis	1	Radiant energy in sunlight	Chemical energy in plant tissues	2–3	95

** for steam at 560°C; heat rejection at 35°C*
*** measured by electrical output (for reversible operation)/enthalpy change of cell reaction*
**** partly due to fact that solar cells cannot use whole frequency spectrum of sunlight*
† Otto (gasoline) cycle
†† Stirling cycle

Some examples of energy conversion

Energy conversion is fundamental to any energy system, whether natural or man-made. Photosynthetic plants convert the *radiant energy* from the sun into the *chemical energy* which keeps the plant alive and ensures growth and propagation. The chemical energy contained in plants is used by the animals that eat them to provide heat energy and motive power. Fossil fuels, containing chemical energy in the form of hydrocarbons, etc are converted to heat, and hence to electrical energy by the use of steam turbines.

All energy conversion processes are subject to two fundamental natural principles: 1. The First Law of Thermodynamics which is a formal statement of the fact that energy has never been observed to be created or destroyed. (The apparent exception is nuclear reactions, where the conversion of mass into energy is allowed for by the equivalence of mass and energy.) This means that the total energy inputs, of all forms, to a system must equal the total outputs of all forms of energy. 2. The Second Law of Thermodynamics which states that it is impossible to construct a *heat engine* whose only effects are the production of *work* and the exchange of *heat* with a single reservoir. In effect this means that any heat engine must necessarily reject some heat to a second reservoir. The important result of this is that any engine producing power from heat, such as a *steam turbine* has a maximum possible efficiency which is determined solely by the temperature of the heat input (the steam) and the temperature of the reservoir to which rejected heat is discharged, invariably the atmosphere or hydrosphere. Thus a steam turbine operating between 560°C and 35°C has a maximum possible efficiency of 63 per cent, and in best current practice, efficiencies close to 48 per cent can be obtained. The table lists some common energy conversion processes together with their efficiencies both practical and theoretical.

energy cost

Alternative term for *energy requirement*, but also a general term for energy benefits or liabilities accruing to a decision.

energy crops

Crops which are or could be grown wholly or partly for their energy content. The process of *photosynthesis* enables plants to capture the sun's radiant energy and store it as *carbohydrates.* Plants can therefore be seen as solar energy converters and those plants that produce significant amounts of carbohydrates are potential energy crops.

Trees, for example, contain a high proportion of *cellulose* as wood, and have long been grown for their energy content as well as for construction purposes. Other plants produce large amounts of *starch.* Both cellulose and starch can be fermented anaerobically to yield alcohols, such as *ethanol* and *methanol*, or *methane.* Char and oil can be produced by *pyrolysis.* Cellulose can also be burnt directly.

Among the plant types being investigated as possible energy crops are tropical grasses, cereals, sugar cane, algae and seaweed. Sugar cane waste (bagasse) is currently used to provide process heat in the sugar refining industry.

See also *anaerobic fermentation; biological energy sources.*

energy density

A term used to describe the amount of energy contained in a unit weight, or unit volume of an *energy carrier* or *energy storage* medium. It is most often used when reference is made to storage cells. For instance, it has been said that electric-powered road vehicles will not come into their own until high energy density batteries can be readily and economically produced. The *lead-acid accumulator* has a theoretical energy density of only 600 MJ/tonne (200 MJ/tonne achieved) whereas the prototype *sodium-sulphur battery* has a theoretical energy density of 2800 MJ/tonne with 800 MJ/tonne achieved.

energy economics

A rather loose and ill-defined expression used to describe the basis on which an investment in an energy-producing or energy-saving scheme is judged to be viable or not. It is often very difficult to gauge whether a large energy investment will prove to be economic when it comes to fruition as the profitability of a scheme may depend on a number of uncertain factors such as the level of economic activity, price of OPEC crude oil, unforeseen technological breakthroughs, or lack of demand for that particular commodity. Prominent examples of areas where unfavourable energy economics are partly responsible for slow expansion are *coal gasification* and *coal liquifaction* processes in the United States and *solar energy* technologies. By modifying the methods of accounting used to judge the economics of a particular investment, eg changing the discount rate and thus the pay-back period, an unattractive scheme can be made to seem financially attractive, even though the energy production, or savings, has remained unchanged. An example of this is suggested investment in *combined heat and power schemes* for domestic premises.

energy farming

See *energy crops.*

energy intensity

In *energy analysis* the energy input to an industry or industry sector per unit financial output. Energy intensity is a measure of the importance of fuel costs to the economic factors of production for each industry. Commonly used units are gigajoule per pound value (GJ/£); kilowatt hours per dollar value (kWh/$).

energy management

Ensuring that energy is used efficiently in the industrial, commercial and domestic fields. The soaring cost of energy in the last decade has provided the necessary stimulus for organizations to look carefully at the level of their energy consumption and take steps to reduce their energy expenditure. Many have achieved this through, for example, improved *thermal insulation* of buildings, process plant and heating equipment. Insulation reduces heat loss from buildings and in warmer countries also reduces the rate at which air conditioning plant has to operate.

See also *energy analysis; energy audit; energy conservation; heat recovery; integrated environmental design.*

energy, mass equivalent of

Energy produced as a result of a mass defect, as in a *fission* reaction or *fusion* reaction. A loss of mass, m, produces energy, E, in accordance with Einstein's equation $E=mc^2$, where c is the velocity of light.

energy pay-back period

A term, borrowed from economics, used to indicate the time taken for an energy-saving or energy-producing development to repay the energy investment incurred as a result of the materials and fuels consumed in its deployment. Such periods, usually measured in months or years, are always shorter than their corresponding financial pay-back periods, since the energy costs are only one component of the full development costs, the energy-benefits being measured at their energy value only.

See also *energy analysis; energy economics.*

energy policy

A much discussed subject concerned with the optimum strategies for exploiting the variety of energy resources at a community's, or a nation's, disposal. The problem has been particularly well highlighted since 1973 and it is a widely held view that comprehensive energy policies need to be adopted if a smooth transition is to take place into a future where fossil fuels, and petroleum in particular, become increasingly scarce and expensive. Others contend that the necessary warnings will naturally arise from signals in the energy market-place, even though there are long lead times and huge amounts of capital investment involved. Energy policy has to take into account a variety of matters such as environmental effects, the desirability of nuclear power, public acceptability of new energy schemes, energy conservation, etc and it is thus very difficult to formulate in such a way as to gain wide acceptance from all individuals, institutions and organizations.

energy ratio

A term used in *energy analysis* which gives the ratio between the energy output of a process and the amount of energy input to that process. An example of its application is in the study of energy flows in agricultural production processes. Because of the increased mechanization of agriculture and application of energy-intensive fertilizers, the energy ratios associated with modern production systems are markedly lower than those prevailing in less sophisticated agricultural economies where the principal energy inputs are traditionally human and animal labour.

In economics, the term energy ratio may also be taken to mean the ratio between a nation's primary energy consumption and its Gross Domestic Product.

See also *energy coefficient.*

Crop	Approximate Energy Ratios
Potatoes (UK)	1.0
Wheat (UK)	2.0
Soybeans (USA)	2.0
Corn (USA)	3.0
Rice (SE Asia)	25–50

energy requirement

The total amount of energy taken from some stock or resource that is needed to supply a commodity or service, as measured under an agreed set of conventions. (Such a set of conventions constitutes the framework of *energy analysis*.)

Energy requirements are usually measured in units of energy per physical unit, such as gigajoules per tonne (GJ/te) or kilowatt hours per pound (kWh/lb).

See also *gross energy requirement; process energy requirement.*

energy sources

The sources from which heat, light and power are obtained. They include the *fossil fuels* – *petroleum, coal* and *natural gas* – nuclear *fission*, and the so-called alternative energy sources, some of which are already in use – *hydroelectric power, solar energy, wind energy, tidal energy, wave energy, geothermal energy* and *biological energy sources.* Some low-grade fossil fuels – *oil shales* and *tar sands* – are on the verge of commercial development. Possible future sources of energy include *ocean thermal energy conversion*, and nuclear *fusion.*

Although manpower and animal power are still important, especially in some agrarian societies, increasing mechanization, heating and lighting requirements and transport needs have led to a rapid rise in world energy consumption. The sharp rise in oil prices of 1973 has led to a greater awareness of the finite nature of the fossil fuels and has provided a stimulus for research into new sources of energy and for *energy conservation.*

The industrial revolution was based on plentiful supplies of cheap coal. In the 20th century oil and natural gas have been used increasingly especially since the end of World War II. By the end of 1976 coal accounted for only 30 per cent of total energy consumption, oil just under 45 per cent, and natural gas nearly 18 per cent. Nuclear energy supplied only 2.2 per cent of world energy consumption.

energy storage

The storage of fuels at times when supply exceeds demand, for occasions when demand outstrips supply. Obvious examples include stockpiles of coal at power stations and tanks of refined petroleum products at oil refineries, though the concept of energy storage should not be confused with stocks of primary fuels required for the smooth continuous running of deliveries of fuel and energy. Natural gas stores are replenished during the summer months to deal with winter demand which may well exceed supplies readily producible from natural sources. Electricity suffers from the disadvantage that it is particularly difficult and expensive to store, this only being presently achievable by means of *batteries* and *pumped storage schemes.* Hence, any generating system requires a number of *peak-load* power stations to cope with increases in demand.

As conveniently storable fuels such as petroleum and natural gas begin to deplete, storage problems are likely to become more acute and thus research at present is aimed towards designing and building effective heat stores. A crude example of such a store is the concrete block used in electric storage radiators.

See also *heat storage*.

energy, units of

The various measures used to define the quantity of energy contained in a fuel. Many of these units have their origins in the physical sciences, eg the *Btu,* the *calorie,* the *erg,* the *joule,* the *kilowatt-hour* and the *therm.* The latter two units are often used in specific connection with electricity and gas respectively, though they also have wider applications, eg the unit 'therms per tonne' is often used as a measure of the calorific value of a petroleum product or solid fuel. Rates of energy flow are often quoted in arbitrary and quite haphazard units, eg barrels per day (petroleum), million tonnes per year (coal and petroleum) or million cubic feet per day (gas).

enriched uranium

A form of *uranium* that contains more than 0.72 per cent of *uranium-235.* It is made by one of the *uranium enrichment* processes and finds a specific use in *nuclear reactors* which require specially enriched fuel.

environmental effects

Widespread or local changes in the surroundings, whether harmful or otherwise. All forms of energy use have associated environmental effects and these range from trivial to potentially destructive on a large scale. Two of the most important, *air pollution* and *water pollution*, are discussed elsewhere. They are exemplified, along with others of a more general nature, in the table. A less obvious, but long-term potentially harmful environmental effect is that due to discharges of waste heat.

See also *thermal pollution.*

erg

A unit of *energy* or *work.* It is defined as the work done by a force of 1 dyne moving through 1 cm in the direction of the force. This unit is the product of the old 'cgs' system of physical measurement and is unrealistically small for any practical applications. Ten million (10^7) ergs are equivalent to one *joule.*

ethane

A colourless, odourless gas, chemical formula C_2H_6; a member of the *alkane* series of hydrocarbons. It is a minor constituent of *natural gas* and finds its major use as a *feedstock* for ethylene production.

ethanol

Also known as ethyl alcohol and popularly called alcohol.

Environmental effect	Nature	Source	Example
Air pollution	Chemical	Fossil fuel combustion	SO_2 from coal-burning power station
	Radioactive	Nuclear fission reactor stock gases	Kr-85 I-131
Water pollution	Chemical	Oil transport	'Torrey Canyon' 'Amoco Cadiz'
	Radioactive	Nuclear fuel reprocessing	Cs-137
Thermal pollution	Increase in local temperatures of water or atmosphere	Any large scale energy use	Cooling water discharge from larger power station
Seismic effects	Earthquakes	Weight of water in lakes behind large dams	Large hydroelectric power stations
Micro-climatic effects	Alternation of local weather patterns	Thermal updraughts	Power station cooling towers
		Increase in local water evaporation	Large hydroelectric power projects
Landscape damage and amenity loss	Spoil tips	Coal mining	Parts of Northern England
	Severe land disturbance	Strip (open-cast) mining	Parts of Eastern USA

(Absolute alcohol is ethanol free from water and any impurities.) A colourless liquid which is miscible with water and most organic solvents, it has a melting point of −114°C, boiling point of 78.4°C and chemical formula C_2H_5OH. Industrially it is made from water and ethylene at high temperatures in the presence of a catalyst and a small proportion is manufactured by fermentation of natural sugars. Its chief industrial uses are as a solvent and a precursor for a wide range of organic chemicals. It is also used in some parts of the world as a blending agent in motor spirit.

See also *anaerobic fermentation; biological energy sources.*

ethylene

Also called ethene. A colourless gas and member of the alkene group of hydrocarbons. Its chemical formula is C_2H_4, melting point −169°C and boiling point −103.7°C. Ethylene is produced by the *cracking* of *petroleum* and is a most important *petrochemical feedstock*.

ethyne

An alternative name for *acetylene*.

expansion turbine

In common usage, any form of *turbine*, excluding *steam turbines* and *gas turbines*, that extracts mechanical power from a gas stream by allowing the gas to expand in a controlled fashion. 'Gas' here is taken to include vapour and the alternative terms, 'vapour expansion turbine' or 'vapour cycle turbine', are in use.

Expansion turbines may be of the axial-flow or radial-flow type. Axial flow expansion turbines are used for power recovery from high temperature waste heat sources. Radial flow types are more suited to power production from lower temperature waste heat sources and operate at high rotational speeds.

Working fluids for expansion turbines include air and many of the fluorocarbons used in refrigeration systems, such as 'Freons'.

As well as being useful for power recovery, expansion turbines can be employed for refrigeration and find application in cryogenic operations for air separation and petroleum refining plants.

external combustion engine

A form of *heat engine* in which heat is transferred to a working medium, invariably a gas, by the combustion of a fuel in a separate chamber. Heat transfer is achieved either by conduction through the cylinder walls, by transfer of a fluid, or by heat exchange with the working medium.

Of the various types of external combustion engine, the *steam turbine*, closed cycle *gas turbine* and the Stirling engine are the most important today. Historically the steam engine, in which steam is used to drive a reciprocating expansion engine, has also been important. The Stirling engine is also a reciprocating engine, using air or other gases in a closed cycle. This particular type of engine is useful as a portable power source able to operate from any fuel and has been developed for military use, with hydrogen as the working medium.

An important advantage of all external combustion engines over the majority of *internal combustion engines* is the continuous nature of the combustion process. This enables air-fuel ratios to be precisely controlled, a difficult problem in many internal combustion engines, where fuel must be injected and burnt in a very brief period. Not all IC engines suffer from this: the open cycle gas turbine uses a continuous combustion process. The ability to control the burning process, especially the use of lean mixtures, together with the elimination of ignition problems associated with reciprocating piston engines, has led to continuing development of external combustion engines as a means of reducing the air pollution associated with vehicle emissions. In particular, the use of such engines could reduce emissions of nitrogen oxides, carbon monoxide and unburnt hydrocarbons, and eliminate the problem of particulate lead caused by the use of *anti-knock components.* The practical development of such engines, however, is thwarted by numerous difficulties, of which the problem of capital cost is especially acute. In the case of the Stirling engine, the poor power-to-weight ratio and acceleration response inhibit its application to vehicles.

F

fast breeder reactors

Nuclear reactors using a *fission chain reaction* dependent on fast *neutrons* to produce more *fissile fuel* than they consume, that is, to achieve a *breeding ratio* of more than 1 (*burner reactors* have a breeding ratio of less than 1). To 'breed', the reactor must be designed so that each fission produces enough neutrons both to cause further fissions (to keep the chain reaction going) and for absorption by *fertile nuclei* which then transmute into nuclei of new fissile fuel. The reactor is designed to create a high neutron density in the core, a condition under which sufficient fissions will occur to sustain the chain reaction and occasion a net increase in the quantity of fissile material present. For this reason, and because the breeder principle depends upon the presence of fast neutrons, no *moderator* is used in the breeder reactor.

The core is of highly enriched fissile fuel (plutonium and uranium dioxides) pouring out fast neutrons into a blanket of fertile material surrounding it. The *coolant* must be extremely efficient as the rate of heat production is very high. Liquid metal is used in the *liquid metal-cooled fast breeder reactor (LMFBR)* – either *sodium* or sodium-potassium alloy – and the *gas-cooled fast breeder reactor (GCFBR)* under development will use *helium*.

The ratio between the amount of fissile fuel produced in a breeder reactor to that consumed is known as the *breeding ratio*, or conversion factor, and the *doubling time* refers to the length of time taken for a breeder reactor to double its inventory of fissile material.

The breeding of new fissile fuel from fertile material is an attractive feature of breeder reactors, given the limited nature of economically exploitable uranium reserves. However, there are many important technical problems to overcome before a full-scale commercial reactor, providing sufficient power for a conventional turbogenerator unit, can be constructed, notably the reprocessing of spent fuel from *burner reactors* to provide plutonium fuel and safety aspects of the cooling circuit.

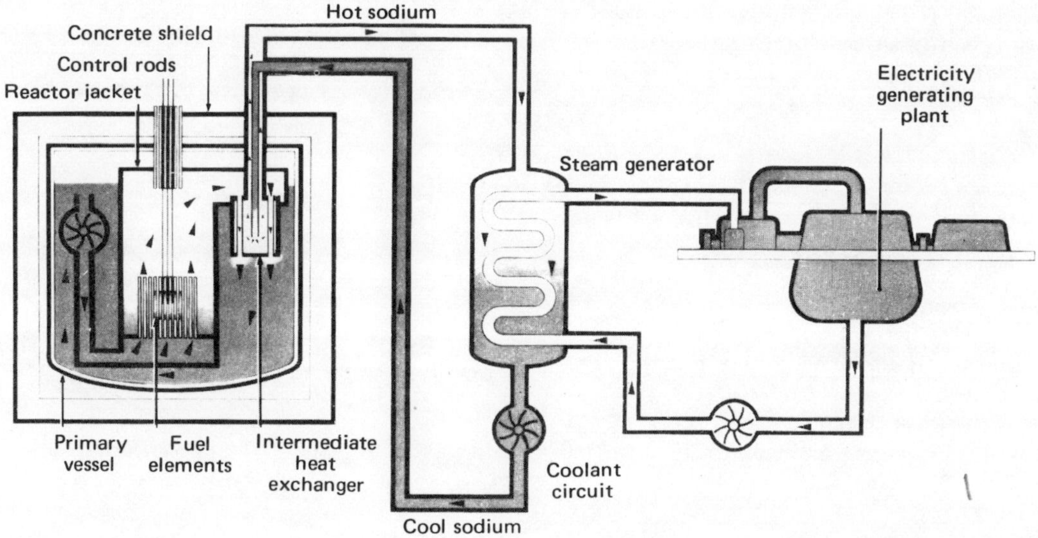

Reproduced by kind permission of the UKAEA

Fast breeder reactor

fast neutrons

Neutrons travelling at, or close to, the speed at which they were ejected from the fissioning nucleus. They are of special importance in the *fast breeder reactor.*

See also *thermal neutrons.*

fault

A phenomenon originating deep in the earth's crust which gives rise to a longitudinal crack through layers of rock which can set up a trap for oil or gas pockets. Such accumulations are known as fault trap reservoirs.

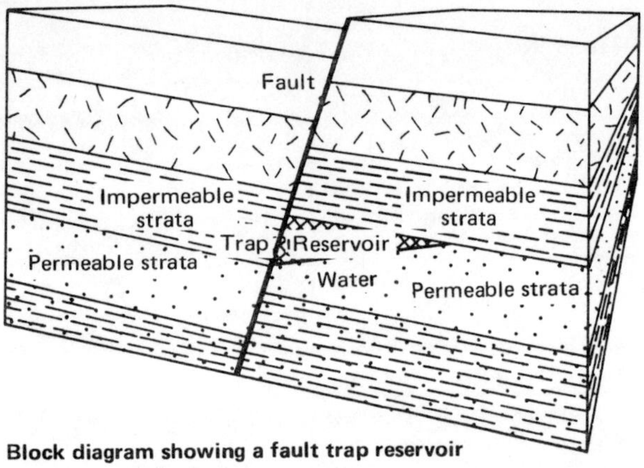

Block diagram showing a fault trap reservoir

feedstock

See *chemical* and *petrochemical feedstock.*

fermentation

The breakdown of complex organic compounds by micro-organisms or enzymes usually with the release of gas and heat energy. Alcoholic fermentation is the breakdown of carbohydrates with the formation of alcohol.

See also *anaerobic fermentation; digesters; methane; sewage.*

fertile fuel

In *nuclear reactors* the fertile fuel or material is that which is capable of being converted into *fissile fuel.*

Nuclear fission fuels are classified as fertile or fissile by the behaviour of their *nuclei* in fission reactions. The fertile fuels are those whose nuclei, if they react with a neutron, will be likely to capture it and undergo transmutation into the nuclei of fissile fuels.

Uranium-238 is the most commonly used fertile fuel. It decays on absorption of a *neutron* and loss of a *beta-particle* into *plutonium-239.* This happens in all reactors, even 'thermal' or 'burner' reactors. They all use a mixture of fertile and fissile material in the composition of their fuel and, although the *conversion ratio* is less than 1, plutonium-239 is produced and may undergo fission, making a significant contribution to the energy output, particularly in the *CANDU* reactor. When the fuel is reprocessed outside the reactor, the plutonium-239 is recovered and used for weapons material.

Thorium-232 is also fertile, being transmuted by neutron absorption into *uranium-233.* Thorium-232 is being used in the fuel of some designs of *high-temperature gas-cooled reactors*, HTGR.

The conversion of fertile to fissile fuel is the principle on which the *fast breeder reactor* operates. This is a reactor designed with a conversion ratio greater than 1 which over a period of years may produce more fissile fuel than it consumes.

fertile material

See *fertile fuel.*

fertile nuclei

The nuclei of *uranium-238* and *thorium-232* are called fertile in *nuclear reactor* terminology because they can be made, by *neutron* bombardment, to decay radioactively to produce the *fissile fuels plutonium-239* and *uranium-233.*

firedamp

A combustible gas, mainly a mixture of hydrocarbons occurring naturally in coal. The main constituent of firedamp is methane which evolves during the coalification process as water and carbon dioxide are eliminated. It is a serious problem in underground mines since it explodes on contact with air.

Fischer-Tropsch process

A *coal gasification* and *coal liquifaction* process in which coal may be converted to a wide range of hydrocarbons and organic chemicals, depending on reaction conditions. A *synthesis gas* is produced by heating coal in a mixture of steam and oxygen and on passing this through a catalytic

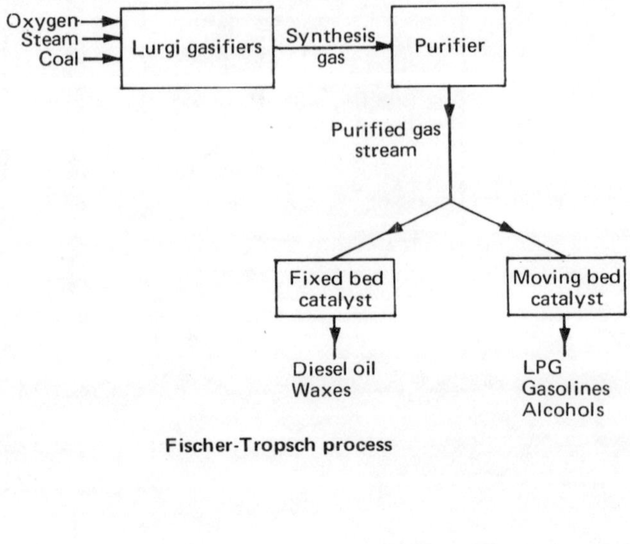

Fischer-Tropsch process

bed containing an iron or cobalt catalyst, a range of hydrocarbons is obtained including petrol, *diesel oil*, and *fuel oil*. Separation of the various products is achieved by *fractionation*. The types of product obtained as a result of this synthesis are crucially dependent on pressure, temperature, and choice of catalyst, and the major practical application of this process is at the Sasolburg plant in South Africa which has used Fischer-Tropsch synthesis to manufacture liquid fuels for some years. See *Sasol* process.

fissile atoms

See *fissile nuclei*.

fissile fuel

In nuclear reactors, the fissile or fuel material is that which will undergo *fission* on absorption of a neutron.

Only a few isotopes of the heavy elements are fissile in this sense. Only one, *uranium-235*, exists in nature, where 1 in 140 atoms of natural purified uranium is uranium-235, the rest being *uranium-238*. This proportion can be increased in *uranium enrichment* plants. The other two are 'bred' in the fission process in nuclear reactors, *uranium-233* out of *thorium-232* and *plutonium-239* out of uranium-238.

See also *fission chain reaction*.

fissile material

See *fissile fuel*.

fissile nuclei

Nuclei of *fissile fuels*, which are capable of undergoing nuclear *fission*.

fission

The splitting of a heavy atomic *nucleus*, either spontaneously or as the consequence of the absorption of a *neutron*, into two or more lighter, often radioactive nuclei accompanied by 2 or 3 free neutrons (fission products). These fly apart with enormous force because in the *fission* process mass is converted into energy.

The amount of energy released in the fission process may be calculated by taking the difference between the mass of the original fissile material and the total mass of all the *fission products*, the energy released being given by Einstein's relationship:

$$E = mc^2$$

where E = energy released
m = mass difference
c = velocity of light.

Thus large amounts of energy are released on fission and indeed the fission of *uranium-235* yields 2.5 million times as much energy as the combustion of the same amount of carbon. Sustaining and controlling the *fission chain reaction* make it possible to harness this 'nuclear energy' which has direct applications in powering large ships and in electricity generation.

fission chain reaction

A process by which *neutrons* liberated by the *fission* of a heavy nucleus may strike and split other *fissile nuclei* liberating more neutrons which in turn may cause more fissions.

Such a chain reaction is sustained if from each fission at least one neutron achieves another fission. If each set of fissions is reproducing itself exactly, the chain reaction is said to be 'critical', with a reproduction factor of 1. If the reproduction factor is even slightly more than 1, the chain reaction will be 'divergent' and if not controlled will accelerate into explosion. If the reproduction factor is less than 1, the chain reaction is 'sub-critical' and will die out.

To reach and maintain the situation of 'criticality' certain pre-conditions must be met. First, enough natural uranium must be found and purified – any impurities will absorb neutrons. There must be enough purified uranium to make a lump of 'critical mass': a high velocity neutron fresh from fission may shoot straight out of the lump of material it is in and so be lost; to cut down on this leakage the lump has to be large in proportion to its surface – a sphere is the most efficient shape as it has the minimum surface area to volume ratio.

In a lump of natural purified uranium, one atom in 140 is *uranium-235*, the rest being *uranium-238*, a more stable isotope. Under neutron bombardment, the probability is that uranium-238 nuclei will absorb neutrons and be transmuted into *plutonium-239* rather than undergo fission. They may fission if hit by a neutron of more than a certain energy, but neutrons from this fission are not energetic enough to split other uranium-238 nuclei, so there is no chain reaction. Uranium-235 is much more fissile. It will split if hit by a neutron of any speed, even a thermal neutron, that is, one with reduced energy as a result of previous collisions. On average, only just over 2 new neutrons are produced from each fission caused by a thermal neutron, but for the purpose of sustaining a chain reaction in uranium-235 this is enough. If extra neutrons are wanted (to breed plutonium-239 from uranium-238 in a *breeder reactor*, for example) then the chain reaction must be run on fast neutrons because they produce more new neutrons per fission, although they have a lower probability of causing fission. Simply to sustain a chain reaction thermal neutrons are most effective. They have the highest probability of colliding with a nucleus because they remain in its neighbourhood for a longer period of time.

Thus the chances of getting 'criticality' when using uranium as fuel will be increased if the proportion of uranium-235 to uranium-238 is increased (*uranium enrichment*) and if the very fast 'fission-fresh' neutrons are slowed down by the addition of a *moderator* – material which will not absorb neutrons colliding with it but will reflect them back into the fissile material with their velocity reduced by collision.

Control of the chain reaction – keeping the reproduction factor at 1 and shaping the pattern of power distribution – is maintained by the use of *control rods* made of a material which strongly absorbs neutrons. The control rods are initially withdrawn from the reactor core to encourage the fission chain reaction to proceed and become self-sustaining and are gradually reinserted to control the reaction, or to shut it down altogether when the need arises.

The control device is effective because a very small proportion of the ejected neutrons are 'delayed', ie they emerge later than the instant of fission because they originate from one of the atoms produced by a fission. If this were not so the speed of the chain reaction is such that it could not be controlled.

The fission produces heat and a fission chain reaction cannot proceed for long without overheating. This can be prevented by pumping heat-absorbing fluid around the fissile material. Liquid (water, molten metal) or gas (carbon dioxide, helium, air) can be used as *coolant.*

The object of keeping a fission chain reaction going can be simply to produce fuel for nuclear weapons, plutonium-239, for example, in which case the heat generated is a waste product. But the heat can be used, like heat from burning coal or oil, in particular to raise steam to drive electrical generators. In this case, the fission chain reaction is the basic process of nuclear power stations.

An arrangement of materials and hardware capable of sustaining a fission chain reaction is called a *nuclear reactor.*

fission energy

The energy released as the result of nuclear *fission.*

fission products

The products of the nuclear *fission* process. These include nuclei lighter than those of the original fissile nucleus and free *neutrons.*

fission reactor

A *nuclear reactor* that exploits the phenomenon of nuclear *fission* to produce energy.

fixed-bed catalyst

A *catalyst* which remains stationary in the reaction vessel or chamber. Such an arrangement is used in some petroleum refinery operations and in *coal liquifaction.*

flaring

The practice of venting and burning unwanted gas or petroleum products associated with oil deposits. Once a common practice, it is becoming less so nowadays as the natural gas is considered to be a more economic and valuable by-product than it used to be. Small amounts of waste *refinery gas* are flared but this is only usually carried out when it cannot be used for providing heat for refinery processes.

flash point

The lowest temperature at which the application of a flame to a combustible, volatile substance will cause the vapours to ignite or 'flash'.

flat-plate collector

Also known as 'solar panel', a form of *solar collector* in which a flat surface is arranged to absorb *radiant energy* from sunlight. Heat is extracted from the surface by means of a thermal transfer medium, usually water or air. Unlike a *focusing collector*, such a collector has no concentrating effect and the output temperature is therefore limited, but it is able to collect diffuse radiation, a feature particularly important in temperate climates.

This type of collector makes use of the greenhouse effect whereby a surface exposed to solar radiation becomes hotter if a transparent medium, such as glass, is arranged to be close to, but not touching, the surface. Short wave radiation from the sun passes through the glass but the longer wave radiation emitted by the heated absorber plate is largely reflected by the glass. In this way the use of one or more layers of glass enables higher temperatures to be attained. Some transparent plastic materials can also be used.

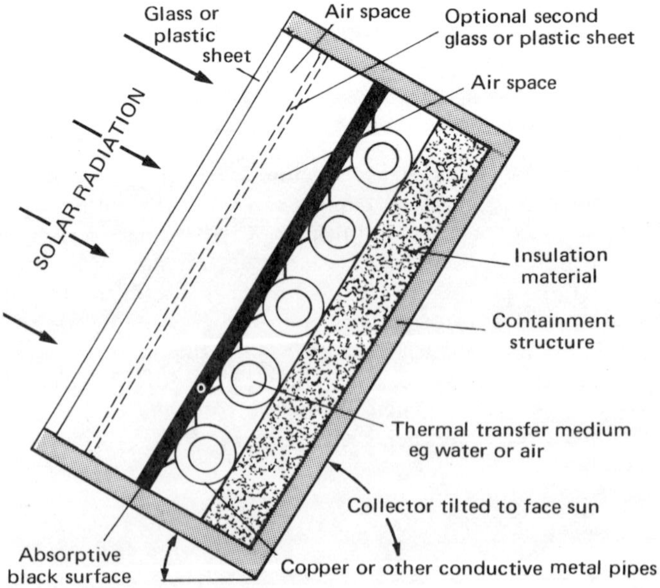

Flat-plate collector

Since a large proportion of the energy from the sun arrives as relatively short wavelength visible light and near

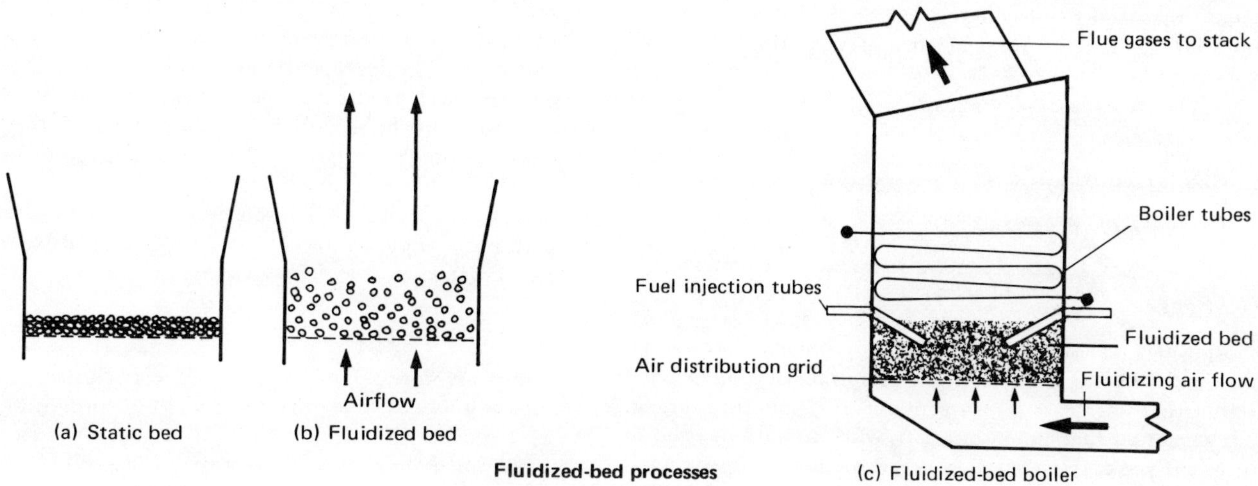

(a) Static bed (b) Fluidized bed (c) Fluidized-bed boiler

Fluidized-bed processes

infrared, a surface capable of absorbing this is required. If this surface can be tailored so as to be a poor emitter of long wave radiation (that emitted by a body at temperatures up to several hundred degrees C), the radiative losses from the collector will be further reduced and high temperatures can be attained. Such selective coatings are used on some production collectors.

The use of double or triple glass layers enables higher temperatures to be attained, but reflection losses from the air/glass surfaces are increased. Optically coated glass transmits more light by reducing these losses.

Evacuation of the space between the absorbing surface and the glass improves the collector efficiency by reducing losses due to thermal conduction through the air.

Although the flat-plate collector is basically a simple device, the use of selective surfaces and coated glass increases costs considerably and higher temperatures are attainable only at considerable capital expense. However, the use of cheaper, single-glazed collectors to preheat water for a small number of high-temperature collectors may enable temperatures of up to 200°C to be obtained economically.

The efficiency of a collector is measured as the ratio of the output heat contained in the coolant to the input of solar radiation over the collector area. Losses occur due to absorption by dirt or self-shading reflection from the glass, re-radiation through the glass, convection in the air gap and conduction through the support structure. Since all these increase as the temperature of collection rises, flat-plate collectors exhibit a falling efficiency with rising temperature characteristic. The maximum temperature of the collector is that at which the losses equal the gains, and the efficiency is zero. (This is also known as the 'idling temperature'.)

fluidized-bed

See *fluidized-bed processes.*

fluidized-bed combustor

See *fluidized-bed processes.*

fluidized-bed processes

Those which rely on the tendency of finely divided solids to float in a low velocity gas or liquid stream and behave as a fluid. As the upward velocity through a stationary bed of particles is increased a point is reached at which the drag on the particles is sufficient to support their weight and the bed becomes fluidized. The technique has been used for some years in certain petroleum refinery operations, where a fluidized *catalyst* enables intimate contact between reactants and catalyst. The use of fluidized-beds for coal combustion offers significant advantages over present pulverized fuel (PF) (and the older stoker-grate) furnaces, and considerable development has been done in France, West Germany and the UK.

The good heat transfer characteristics of the fluidized-bed allow more compact units and lower operating temperature (leading to lower stack-gas emissions of nitrogen oxides). In addition the tendency for agglomerated ash to sink to the bottom of the bed means that low-rank coals can be used effectively; and the incorporation of limestone or dolomite into the bed reduces the air pollution associated with the use of high-sulphur coals by chemically absorbing much of the sulphur dioxide. A pressurized fluidized-bed is being developed for the gasification of coal in a stream of air, or steam and air, to produce a clean, low-Btu fuel gas suitable for local distribution or for use in gas turbines. A pilot plant in the USA has demonstrated the feasibility of using fluidized-beds to generate electricity and heat from municipal solid waste. The shredded waste is delivered through an air separator, where metals and glass are removed, to a fluidized-bed combustor, and the hot combustion gases are expanded after cleaning through a gas turbine, which also provides the fluidizing air. A waste heat boiler provides steam. Disadvantages of the fluidized-bed include the need to provide power for fluidization, and the fact that agglomeration problems prevent fluidization of some materials.

See also *catalytic cracking*.

fluidized-bed reactor

See *fluidized-bed processes*.

flywheel

A rotating mass which is able to store mechanical energy and which finds its major application in the rotating internal combustion engine as a means of ensuring smooth running. It is suggested that much larger flywheels could be used to store off-peak electricity, generated during low-demand periods. Energy would be extracted from the flywheel to help meet *peak load* requirements.

focusing collector

A form of *solar collector* in which curved mirrors or other devices are used to concentrate the direct *radiant energy* from the sun onto something capable of absorbing it. This would usually be a blackened tube, cooled by water or some other heat transfer medium.

Such collectors must usually be made to track the sun, since they generally require accurate alignment of incoming radiation with an optical axis to achieve maximum efficiency. They are very sensitive to fouling of mirror surfaces by water or dust and cannot operate well in cloudy conditions when the direct component of the radiation is minimal.

Focusing collectors are, however, capable of high output temperatures and although present designs are costly to make it may be possible to reduce costs in the future if designs suitable for large-scale production can be developed.

Although generally conceived for use in desert or semi-desert areas with high insolation and low precipitation, it may be possible to develop certain types for use in temperate climates.

forest resources

These are forest lands and the tree resources growing on them. Between one quarter and one third of the earth's land surface is covered in forests. Forest distribution is determined by climatic factors such as light intensity, temperature and water supply. These in turn depend on latitude, altitude and rainfall. Competition for the use of land also determines to an increasing extent where forests are found. Accessible areas are just those likely to be used for other purposes.

The productivity of a forest is influenced by the availability of nutrients – dependent on the type of soil – and a forest will itself influence the soil content over a period of time through leaf fall, the amount of light its canopy allows through to the floor and the consequent effect on plant and animal life on top of and within the soil.

There are many ways of classifying forests but they can broadly be divided into the following types: cool coniferous

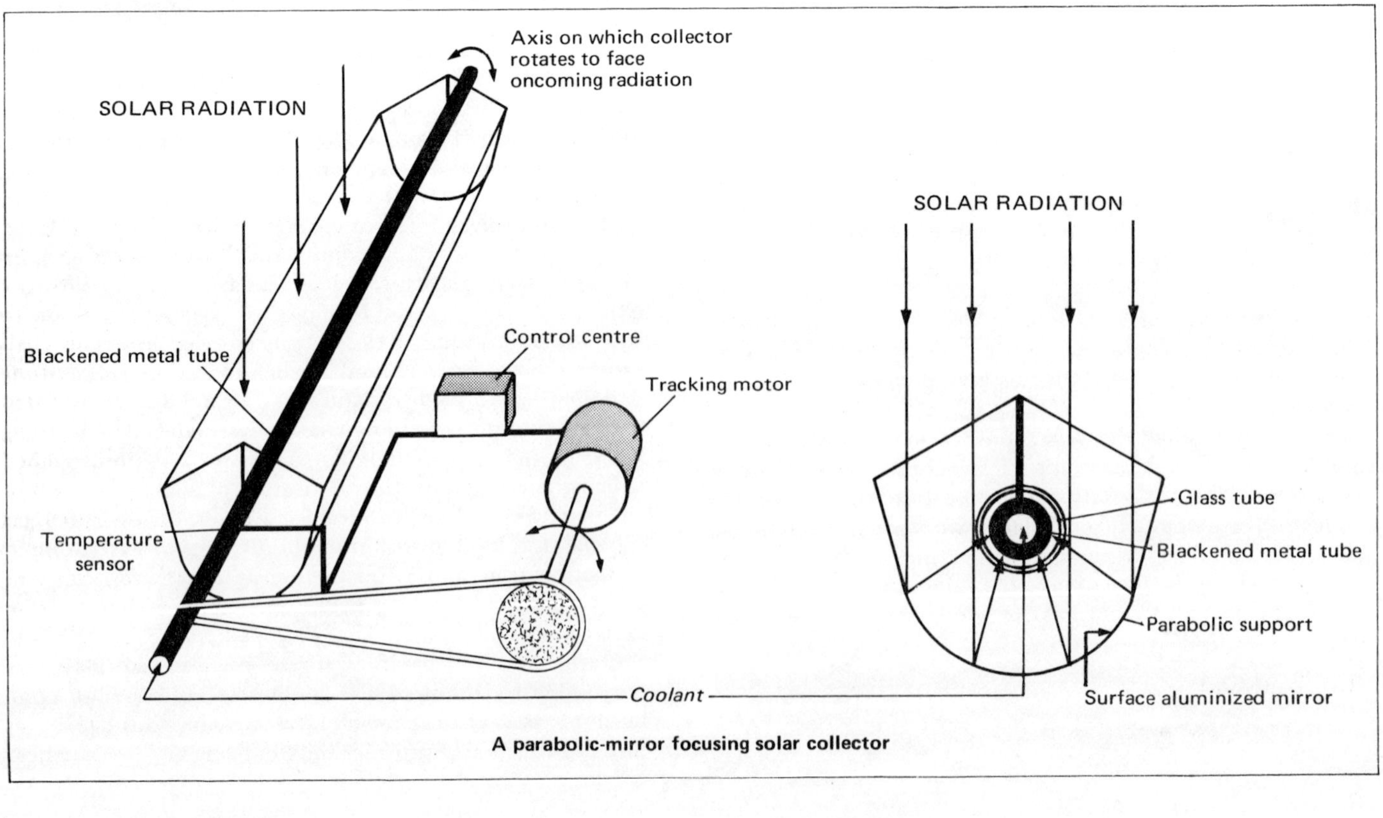

A parabolic-mirror focusing solar collector

forest, temperate mixed forest, warm temperate moist forest, equatorial rain forest, tropical moist deciduous forest and the sparser dry forest of the savannas. The forests of the USSR, Europe, North America and Japan have predominantly coniferous trees though there is much mixed woodland too. The equatorial rain forests contain only broad-leaved species and unlike the coniferous and mixed temperate forest zones are little developed.

The average increment per annum per square kilometre of forest is about 350 tonnes of wood but there is variation from region to region due to the differing lengths of the growing seasons and the varying light intensity. It is highest in the tropics with about 1950 tonnes per square kilometre. Production is less than 35 tonnes per square kilometre north of the Arctic circle even in favourable regions and the figure is similar where rainfall is less than 25 cm per annum.

The growth of the world's forests is calculated to add 13×10^9 tonnes dry weight per annum of wood and about one eighth of this is used, almost equally split between fuel and non-fuel uses. The proportion is heavily weighted in favour of fuel in the under-developed parts of the world, however. The energy available in 13×10^9 tonnes of dry wood is equivalent to about 7×10^9 tonnes of coal or about two thirds of the present world energy requirements. Of course not all of this energy is available for use as fuel.

It is possible to increase our forest resources each year through skilful management: unlike our fossil fuel reserves they are (theoretically) inexhaustible. Not only can forest sizes be increased but tree growth adds to the total cellulose energy resources each year up to the time a tree is felled. In this respect cellulose in the form of trees would appear to have longer term prospects than fossil fuels as an energy source. However, there is a net reduction in the world's forest each year – the appetites of the paper and construction industries, furniture and related trades, the needs of agriculture, and increasing urbanization, all contribute to this resource depletion. Unless a conscious effort is made to develop cellulose fuel resources in preference to others it seems unlikely that cellulose will become a more important part of the world's overall energy supplies.

Undoubtedly forest productivity could be increased, but the rate of this increase is unlikely to keep pace with the increasing demand for energy. The cost of wood fuel is low but this would rise markedly, especially in terms of energy expenditure with the application of more energy intensive methods. This has been the experience of agriculture in the developed countries, for while productivity per acre has gone up considerably, the energy content of the crops produced has been only about twice the energy input needed to grow them, and sometimes the ratio of input to output has been much lower.

See also *cellulose; photosynthetic energy; wood;* and *Maps 19 and 20.*

fossil fuels

Collective term for all fuels derived through fossilization processes from past living organisms. Examples are *peat, coal, petroleum (crude oil)* and *natural gas*, and related materials such as *tar sands* and *oil shales.* Total world economically recoverable reserves of fossil fuels are calculated at around 500 billion tonnes oil equivalent. Fossil fuel consumption in 1976 totalled 5,963 million tonnes oil equivalent. Because fossil fuels are finite resources there is an increasing pressure to develop alternative sources such as *geothermal energy; nuclear energy; solar energy; tidal energy; wave energy; wind energy.*

See also *Map 14.*

fraction

One of a number of parts of the distilled material produced during the fractional *distillation* of a mixture of organic materials (eg *petroleum*) each of which has a different boiling point.

See also *petroleum fraction.*

fractionating tower

A vertical *distillation* column used to separate petroleum into 'fractions' according to their boiling points. (Such devices are also used in the petrochemical industry.)

Separation is effected by the repeated vaporization and condensation of the mixture between the hot lower end,

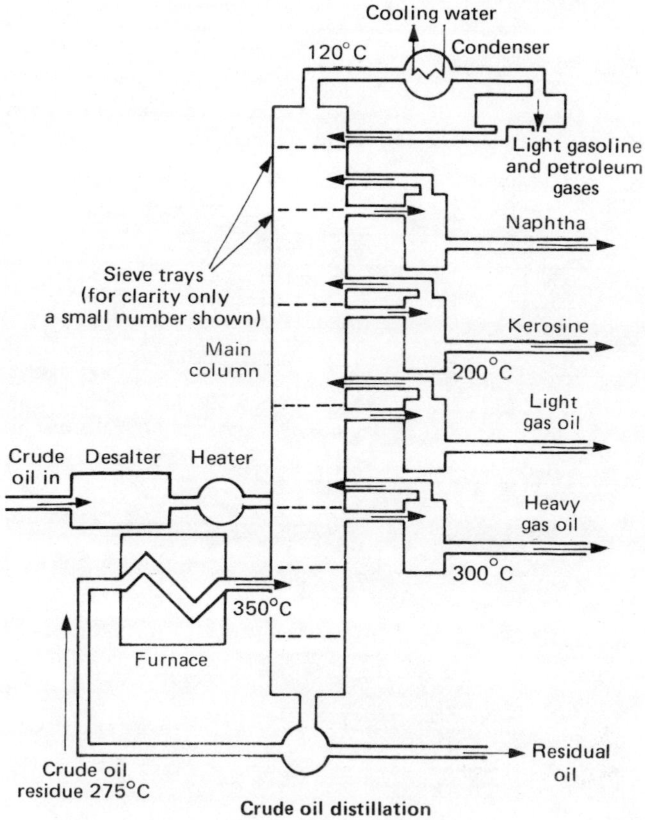

Diagram of crude oil distillation process in a fractionating tower

where heated residual oil is recirculated, and the cold top of the tower. This temperature gradient allows fractions to be drawn off at a height corresponding to their boiling point – lower boiling fractions are drawn off at the top while the high boiling, heavy fractions are taken from the foot of the tower. Preheated crude oil is injected into the column at a point between the middle and the lower end. To ensure good separation, the rising vapour must be in close contact with falling liquid, and a commonly used design employs 'sieve-plates' at vertical intervals of two feet or so, at which the vapour bubbles through liquid (see diagram). Alternative designs use bubble-caps over the holes to prevent liquid falling through the holes in the absence of upward vapour pressure or dispense with plates altogether in favour of packed shapes contoured to ensure maximum vapour-liquid contact.

Crude petroleum fractions generally need to be further processed to obtain products with the required qualities for use as fuels. Such processing may include further distillation.

See also *petroleum fraction; petroleum refining.*

fractionation

The separation of a complex chemical mixture, for example, *petroleum*, into its individual components, or fractions. This separation is usually carried out by means of *distillation.*

See also *fractionating tower.*

Francis turbine

A kind of *hydraulic turbine* used in hydroelectric schemes to drive generators and generate electricity. The Francis

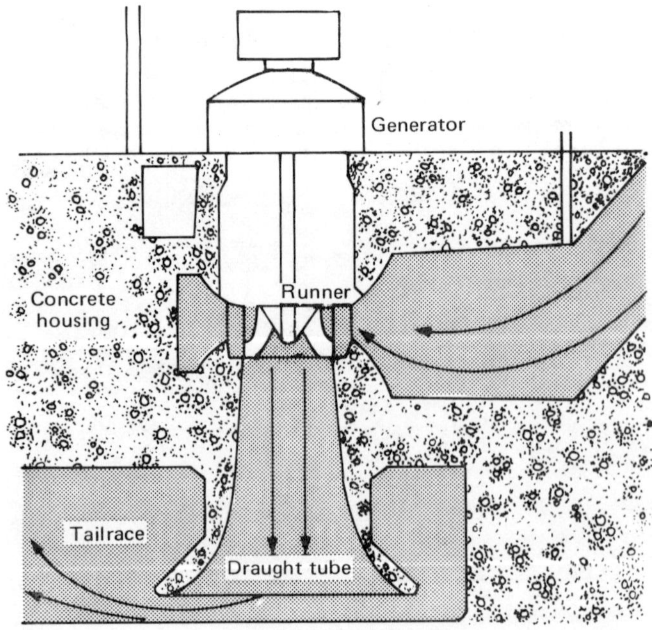

Cross-section through a Francis type turbine installation

turbine is a reaction turbine, running full of water and submerged and is used for medium heads of between 50 metres and 500 metres.

fuel

A substance capable of undergoing change by a chemical process, or by radioactive means, with a simultaneous release of energy. The most important fuels are the *fossil fuels*, which provide most of our energy needs though *nuclear energy* is beginning to make an important, albeit small, contribution. Not all *primary energy* forms involve the degradation of fuels, these renewable forms of energy including *wind, tidal, geothermal, solar* and *wave energy.*

fuel assembly

Another name for *fuel element.*

fuel cell

An *electric cell* in which chemical energy is converted directly into electrical energy, as a result of a continuous supply of fuel being fed to *electrodes* in contact with an *electrolyte.* Fuel cells have considerable potential as sources of electrical energy that are effectively portable, silent and do not cause pollution. They can be fed with fuels such as *hydrogen, ammonia,* hydrazine, and various *hydrocarbons.*

In the simplest, original form, devised by Bacon, porous electrodes submerged in an alkaline solution are supplied separately with hydrogen and oxygen gases. As a result of this the following reactions occur:

at the anode: $4OH^- + 2H_2 = 4H_2O + 4e^-$
at the cathode: $4e^- + 2H_2O + O_2 = 4OH^-$

The surplus of electrons at the anode and the deficit at the cathode gives rise to a current flow in the external circuit.

In the original cell the electrodes were porous nickel sheets, arranged in parallel, with the inner surfaces containing fine holes, widening to larger holes in the outer facing surfaces. The cell was filled with hot concentrated potassium hydroxide solution. The surfaces containing the small holes were thus wetted with the hydroxide solution, which presented an interface of large surface area for the gases that were pumped to the outer surfaces of the electrodes. The holes were small enough to prevent the penetration of the gases through the electrodes. The oxidation product, water, can be a nuisance because it can build up to such levels as to effectively dilute the electrolyte, but in Bacon's cell this was overcome by keeping the electrolyte temperature high. The cell voltage produced in the oxygen-hydrogen cell is 1.18 volts, and in practice batteries of such cells are used. A single cell is capable of providing a capacity of 200-500 watts, but considerably larger units are being developed.

Practical developments and uses. A number of different electrolytes have been employed in fuel cells, a prime consideration being the ability of the electrolyte to deal with impure, rather than pure hydrogen. The three main classes of electrolyte developed are aqueous, fused salts, and very high temperature electrolytes in which oxygen ions are active. A major economic factor is the need for cheap *catalysts* to be employed in the electrodes (the catalyst effectively increases surface activity). The higher temperature systems can employ cheaper catalysts and use impure hydrogen as a fuel. In the lower temperature aqueous electrolyte cells, the unwanted accumulation of water described above can be overcome by sophisticated electrode design. In such cells gas fuel is injected at high pressure and this necessitates strong and bulky containers.

Applications. The major use of the fuel cell thus far has been in the space programme where cryogenically stored

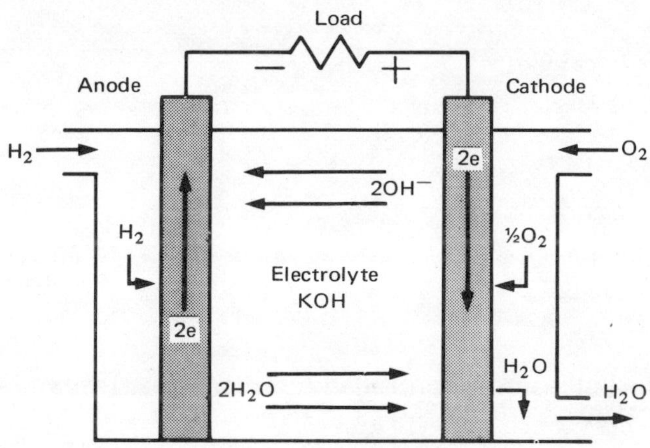

Hydrogen-oxygen fuel cell

fuel feeds the cell. The water produced as the result of the oxidation of the hydrogen in the cell is used as drinking water. It has been hoped for some years that fuel cell development would become sufficiently advanced to allow it to be used as the power source in motor vehicles, but the required technical advances have not been made. Considerable research and development continues on the outstanding problems associated with electrode and electrolyte design.

fuel element

A unit of nuclear fuel plus its cladding which can be inserted or removed individually from the core of a nuclear reactor. A fuel element may be made up of more than a hundred *fuel pins.* Materials used for cladding are stainless steel, zirconium alloy (zircaloy), aluminium and in *Magnox* reactors, a special magnesium alloy. The elements are designed to allow gas or liquid to be pumped over their surfaces to extract the heat produced by the *fission chain reaction.* This heat is then conveyed directly in steam or via heat exchangers to produce steam for driving turbo-generators. The fuel is commonly pelleted uranium oxide. Numerous fuel elements are enclosed in the reactor core, a typical PWR having 50,000 to 60,000 fuel pins. There is a limit to the amount of time a fuel element can be left in the reactor; this is determined by the rate at which neutron-absorbing fission products build up. The fuel elements, when only partially spent, must be removed by remote control in view of the intense radioactivity present. They may be stored temporarily in a *cooling pond* and then taken to a radiochemical separation plant where the fissile and fertile materials are recovered from the radioactive waste.

See also *nuclear fuel cycle.*

fuel enrichment

See *nuclear fuel cycle.*

fuel fabrication

See *nuclear fuel cycle.*

fuel gas

A gaseous fuel that can be distributed by pipeline to the point of consumption. Fuel gases include *natural gas, liquified petroleum gas* (propane, butane, and pentane which are gaseous at normal temperatures but which may be liquified), refinery gases (produced at petroleum refineries and mainly used for refinery process heat) *coal gas*

calorific values of fuel gases

gas	Btu/ft^3	MJ/m^3
natural gas	1000	37.4
propane	2450	91.6
refinery gases	1600	59.8
coke-oven gas	500	18.7
blast furnace gas	90	3.4
producer gas	125	4.7

and various low and medium Btu gases. Examples of the latter group are *producer gas, blast furnace gas* and *coke-oven gas.*

fuel oil

A collective term covering a range of heavy *petroleum fractions* which, in the course of the normal refining process, may include both distillate and residue. It is used for steam-raising purposes in both large industrial boilers and at power stations.

fuel pin

A thin-walled metal tube (called cladding) in which nuclear fuel, for example pellets of uranium oxide, is packed. The tube – usually of stainless steel or zircaloy – may be several metres long and quite narrow, about one centimetre in diameter in *pressurized water reactors* (PWRs) and less than six millimetres in diameter in the prototype fast reactor (PFR) at Dounreay. Sets of fuel pins constitute *fuel elements.* The cladding tubes or cans may be provided with fins to facilitate heat transfer and the metal used must be such as to resist an environment that is abnormally corrosive due to a combination of pressure, temperature, chemical and radiation hazards. Aluminium and a special magnesium alloy (*Magnox*) are used in some reactors. The cross-section of the pins and the thickness of the metal are kept to a minimum to prevent excessive neutron absorption, to increase thermal conductivity from the poorly conducting fuel material, and to provide a larger surface area for heat transfer to the cooling medium, be it a gas or a liquid.

fuel reprocessing

See *nuclear fuel cycle.*

fuel rod

A form in which fuel is used in a *nuclear reactor*. This usually consists of uranium metal or the oxide sealed within a *cladding*, such as *Magnox* or *zircaloy.*

fusion

A reaction in which two or more light atomic particles are brought together so that their nuclei fuse. This reaction can only occur at very high temperature, because at such temperatures the atoms move at sufficiently high speeds to overcome the mutual repulsion of their positively charged nuclei.

The mass of the resulting single nucleus is less than that of the original nuclei, and a large amount of energy is released. It is the fusion of hydrogen nuclei to form helium that provides the fuel for the self-sustaining fusion process in the *sun*. A fusion process also occurs in the hydrogen bomb, a 'conventional' uranium (fission) bomb providing the necessary high temperature for the fusion reaction to take place.

A number of experimental *fusion reactors* are being constructed in the hope of producing a controlled fusion reaction on earth. The successful development of such reactors could provide almost limitless energy, using readily available fuels.

Conditions for a fusion reaction. To obtain useful energy it is necessary to attain a high enough rate of reaction, which has to yield sufficient energy to more than compensate for any losses in a reactor. In all the systems the nuclei carry electric charges of the same sign so that they repel one another, the force of repulsion being very strong when the nuclei are within reacting distance. Also the force of repulsion varies as the product of the nuclear charges, which is one of the reasons why the lightest elements are favoured as fuels, (they have smaller nuclear charges). For a nuclear fusion reaction to take place it is essential that the nuclei make a mutually close approach.

The higher the velocity of mutual approach, the closer together a given pair of nuclei can come. But increasing energy input leads to greater losses, and the solution is to increase the frequency of collisions between reacting particles. The only way of getting repeated collisions, with no energy lost from the system as a whole, is to raise the temperature of a suitable mixture of atoms to a very high value, about 10^8 degrees Kelvin, ten times the temperature of the interior of the sun.

The proportion of high energy particles increases rapidly with the temperature and the chance of a successful nuclear reaction also increases with temperature. Given a high enough temperature and pressure, the rate of production of energy will exceed the rate of loss, but this temperature has to be maintained for a time sufficient to allow adequate generation of energy, and the extremely high temperature plasma has to be confined. A device for realizing these conditions is called a thermo-nuclear reactor or *fusion reactor.*

fusion energy

Energy released during the process of *nuclear fusion.*

See also *fusion reactor.*

fusion fuels

Materials likely to be used in *fusion reactors* to produce energy by the *nuclear fusion* process. The most probable substances are *deuterium, lithium* and *tritium.*

fusion reactor (controlled thermonuclear reactor, CTR)

A reactor (in the early stages of development) that it is hoped will use the *nuclear fusion* process to produce energy in a controlled fashion.

In fusion processes nuclei of light atoms (such as *deuterium*) are energized so that they move into collision at very high velocities. The effective equivalent temperatures of the mass of gas containing the nuclei (the *plasma*) has to be 40 million degrees Kelvin, if fusion between nuclei is to occur. To bring this about a large amount of energy must be invested. For a fusion reaction to be sustained the rate of fusion of nuclei must be such that the energy released in the fusion reaction must exceed the energy input to the plasma.

Generating high temperatures. Several methods are under development, which in some cases are combined with the technique of containment of the plasma. Heavy electric currents may be passed through the plasma for a short time and this is done by discharging a bank of very low inductance capacitors through the system. The simplest arrangement is an electric arc struck between the electrodes situated in a vessel filled with the reaction mixture (gaseous deuterium and tritium) under low pressure. A current of millions of amps is passed and this heats up the plasma formed in the arc, but this arc current is large enough to set up a magnetic field, which acts so as to compress the plasma, as in the theta pinch experiments. Compression of the plasma increases the particle concentration and the

fusion reactor

density of the electric current, which raises the temperature. Such a system is known as linear pinch, and the self-compression of an electric arc by its own magnetic field is known as *pinch effect*. It is found, however, that energy and plasma are able to leak from the ends of the arc. Lateral movement of the plasma will be resisted by the magnetic bottle but not longitudinal movement.

Two methods are used to counter this leakage. In the first method the magnetic field, which is arranged to lie in the direction of the axis of the arc, is also arranged to have its highest intensity near the ends of the arc. This 'magnetic mirror' tends to contain the plasma, but leaks still happen. In the second method a toroidal discharge is used. A toroid has no ends and this configuration therefore eliminates leakage but instability of the discharge in various forms is still a major problem. It has not been possible to confine a plasma for much more than a tenth of a second. Several different kinds of toroidal discharge apparatus exist, eg the stellarator (used at Princeton) and tokamak (in the USSR, at the Kurchatov Institute of Technology).

Adiabatic compression using magnetic fields has also been tried. By squeezing plasma very rapidly by a powerful magnetic field, the temperature will go up before there is much chance of losing energy (hence the term adiabatic, no exchange of heat), but again the plasma is able to escape.

Energizing and containing the reaction. The maintenance of very high temperatures and the containment of the fuel are opposing requirements. The act of heating a mixture will tend to make it fly apart. There are two main containment methods which are being used in the nuclear fusion experiments that are being carried out.

In the first, the magnetic bottle reactor, containment is by a magnetic field and takes advantage of the ionization into positively and negatively charged particles of the reaction mixture (plasma) inevitably brought about by the high temperature of operation. Magnetic rather than electrostatic fields have the advantage that they act equally

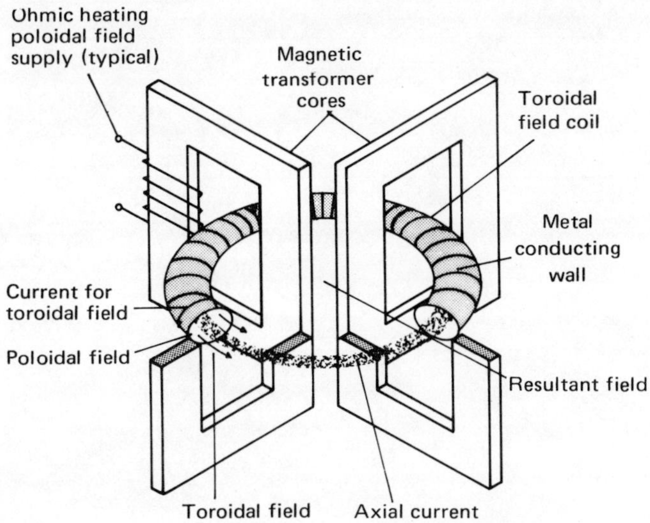

Basic elements of Tokamak fusion reactor design

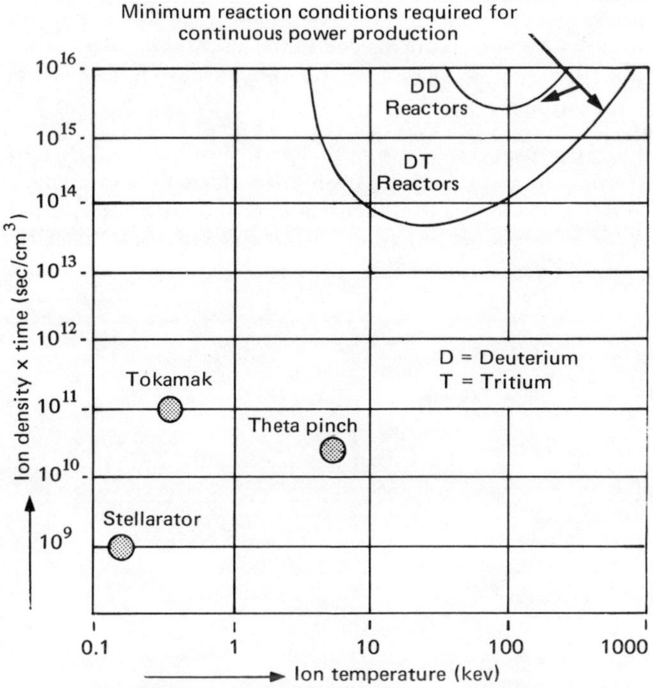

Diagram showing reaction conditions achieved in various experimental fusion reactor designs, compared with minimum reaction conditions required for practical power production

effectively on positive and negative ions. Magnetic containment, in principle, can withstand any temperature but a limit is set by the strength of the magnetic field required to contain a high particle concentration at a high temperature. Magnet technology will be aimed at producing magnetic fields of the desired performance at the lowest cost; superconducting magnets are likely to be used.

Many configurations of magnetic field are being tried. In addition to externally applied fields, there may be magnetic fields set up by currents arranged in the plasma and these fields, in particular, may be pulsed. In this way, the containment of plasma may be combined with the generation of high temperatures. The problems of handling such plasma are complex and not fully understood.

In the second, the inertial containment — laser fusion reactor, the plasma is heated up so rapidly that it is able to react appreciably before it disperses. The method is to apply a concentric pulse of extremely intense light from an exceptionally high power laser to a pellet of lithium deuteride. The intensity of the light is such as to compress the reaction mixture (which increases the particle concentration) and heat it to the required temperature. Since the particle density is very high it is only necessary to contain the plasma for a very short time.

With a laser it is possible to attain a light intensity comparable with that produced by the core of a star. The light is in the form of pulses each lasting 10^{-12} seconds to 10^{-9} seconds. The light from the laser is directed on to a fuel pellet. By directing a number of laser beams concentrically upon it the fuel pellet is very considerably com-fusion reactor

pressed and this enables the reaction to take place, but even so very high powers are needed in the laser (about 10^{14} watts for 10^{-9} seconds). The laser for this purpose has yet to be built but seems feasible.

Future prospects. Thermo-nuclear fusion has been demonstrated on a small scale, both with magnetic containment and with laser inertial containment. In each case, much more power was used to drive the magnetic field, heat the plasma or operate the laser than was liberated by the fusion reaction.

The next objective is to attain a break-even point between input and output energy. Calculations suggest that very large pieces of equipment will be needed to demonstrate this and a number of nations are starting to build these. These include the giant Princeton-based experimental apparatus, the large-scale Soviet experiments, and the UK-based Joint European Torus (JET) project.

G

gallium arsenide solar cell (GaAs solar cell)

An electric cell for the direct conversion of sunlight to electric power using single crystal gallium arsenide. This type of cell has a high theoretical efficiency (27 per cent) and practical values range from 13 per cent to 19 per cent. The cells perform well at high temperatures and have a high open circuit voltage. However, their cost is high, particularly because of the need for single crystals and because the materials are inherently expensive. For this reason it does not seem likely that GaAs cells will displace *silicon cells* or *cadmium sulphide ceramic cells* as likely candidates for the large scale generation of electricity from solar energy.

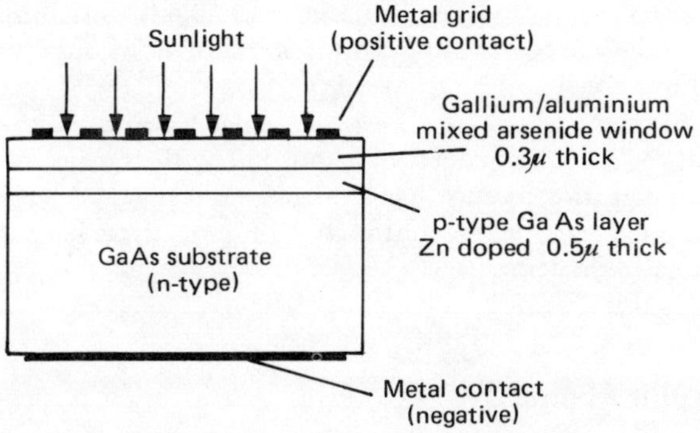

Schematic representation of GaAs solar cell

galvanic cell

Another term for *electric cell* named after its originator.

gamma radiation

Ionizing radiation consisting of *gamma rays* emitted from atomic nuclei as the result of radioactive decay. Emission of gamma radiation from a nucleus often accompanies the loss of an *alpha particle* or *beta particle.* The level of penetration of gamma radiation is very high and the gamma rays with the highest energy are able to travel through over a metre thickness of concrete.

gamma ray (γ ray)

Very high energy, short wave electromagnetic radiation emitted by nuclei of radioactive materials. Gamma rays are similar in form to X-rays, but their wavelength is generally shorter.

gas cap drive

A natural drive mechanism in which pressure of gas in the gas cap displaces oil downwards in the direction of the producing well. It occurs in a limited or closed reservoir in which there is no effective *water drive* mechanism. The reservoir pressure drops as oil is extracted from the reservoir so that the gas cap expands into the pore spaces previously occupied by the oil. *Solution gas drive* will also contribute to the drive mechanism: the larger the gas cap in relation to the oil zone the smaller the amount of solution gas drive.

gas cap gas

Also called *associated gas.* Excess natural gas which accumulates in a reservoir on top of oil already saturated with gas. Gas cap gas provides an important drive mechanism.

See also *gas cap drive.*

gas centrifuge process

See *centrifruge process.*

gas compressor

A device used to compress gas and in association with cooling to liquify it. The process is used to separate *propane, butane* and other hydrocarbons from *methane* in natural gas so that the methane can be piped to the consumer and also to produce *liquified petroleum gas (LPG)* and *liquified natural gas (LNG).* Large gas compressor stations are also used to pump natural gas over long distance through pipeline networks.

gas-cooled fast breeder reactor (GCFBR)

A type of fast breeder reactor (under development) in

which gas – probably *helium* – will be used as coolant. It is proposed that this reactor would use a mixed oxide fuel, probably including *thorium-232* which, under neutron bombardment, changes to fissile *uranium-233*. There would seem to be considerable doubts about the safety of a GCFBR despite the experience of operating gas-cooled burner reactors.

gas-cooled reactors

Nuclear reactors that use gases for cooling the nuclear reactor core. Examples are the *Magnox reactors, advanced gas-cooled reactors (AGRs)*, using carbon dioxide gas coolant, and the experimental *high temperature gas-cooled reactors (HTGRs)* using *helium*.

Gases have poor heat transfer properties and the power consumed in circulating the gas is substantial. The *power density* reached in the core of gas-cooled reactors is designed to be lower than that attained in reactors with liquid coolants; because of low thermal capacity gases are less efficient coolants than liquids.

gaseous diffusion process

See *diffusion process*.

gas generator

Part of a *gas turbine* that supplies high pressure gas flow.

gas lift

The injection of gas under pressure into the oil column in the well bore to help raise oil to the surface. The method is commonly used in the Middle East when the natural pressure has become insufficient for the well to flow unaided.

gas liquifaction

The conversion of a gas to the liquid state. Natural gas consists mostly of *methane* with lesser amounts of *ethane, propane, butane* and heavier hydrocarbons known collectively as the *pentane-plus fraction*. Where possible, methane will be supplied direct to the consumer by pipeline after removal of the impurities and other hydrocarbons. The latter are usually liquified by a process involving compression and cooling, and individual constituents can be separated by fractionation. The liquid propane or butane or a mixture of the two is referred to as *liquified petroleum gas (LPG)*. This can be purchased in portable steel containers as calor gas for use as a fuel for heating, cooking and lighting. The residual gas – methane, ethane and small amounts of propane – may be liquified and transported by tanker if it cannot be piped to its destination. This mixture is known as *liquified natural gas*.

gas oil

A *petroleum fraction* occurring in the *middle distillate* range and with a boiling point of between 200°C and 350°C. It owes its name to its original role which was to enrich *water gas* as part of *town gas* manufacture. Its chief use today is as a fuel for diesel engines (*DERV, marine diesel oil*) and for larger domestic and industrial heating boilers. It may also be used as a feedstock in refinery cracking processes when it is converted into *gasoline*.

gas-oil contact or level

The depth at which the interface between gas and oil accumulations occurs in a reservoir.

gas/oil ratio

A term referring to the amount of gas produced along with the oil from a producing well. It is usually expressed as cubic feet of gas per barrel of oil and an average value is 1500 cubic feet per barrel.

gasoline

1. In the United States, a popular term for *'motor spirit'* or *'petrol'*.
2. A *light distillate* falling in the boiling range 30°-200°C. Gasolines are colourless, volatile and highly inflammable and find their chief applications as the major constituents of motor spirit and *aviation spirit*.

For many years the *kerosine* range comprised the oil products most in general demand, but with the rise of the motor car the picture has changed radically and refinery techniques are being continually developed to keep pace with gasoline demand.

See also *cracking; petroleum refining*.

gasoline engine

See *spark ignition engine*.

gas reinjection

See *reinjection*.

gas separation

The removal of dissolved gases from crude petroleum. This takes place before the oil is refined by distillation and is usually carried out in a number of stages, depending upon the nature and concentration of the dissolved gases.

gas storage

See *natural gas storage*.

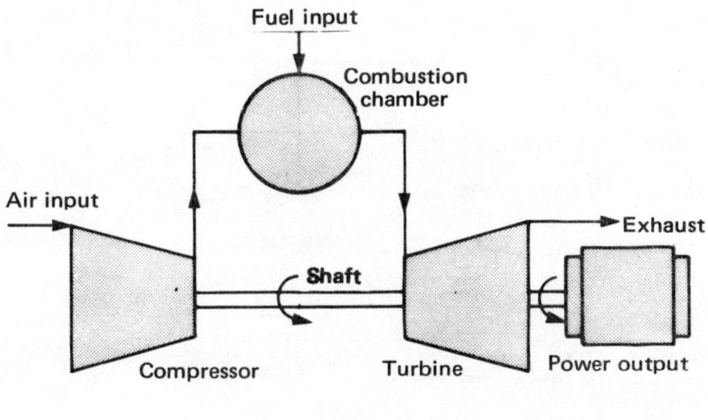

Simple open cycle gas turbine (schematic)

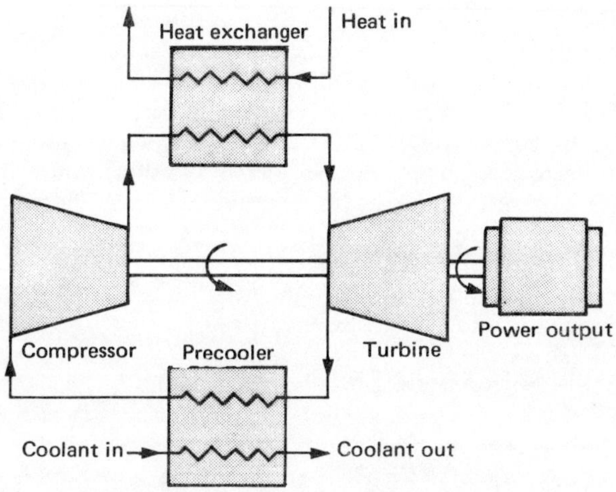

Simple closed cycle gas turbine (schematic)

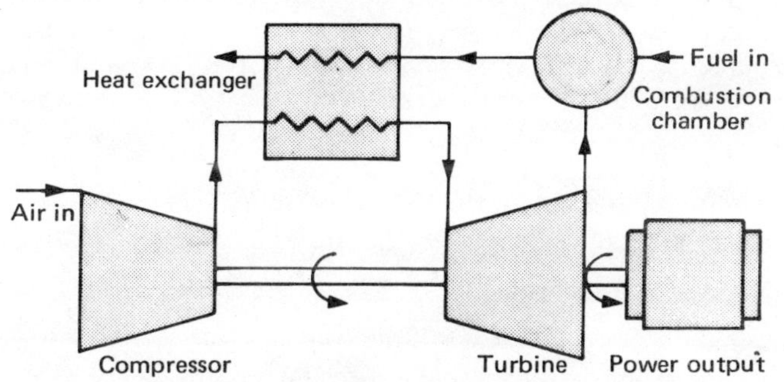

Open cycle gas turbine system for burning solid fuel

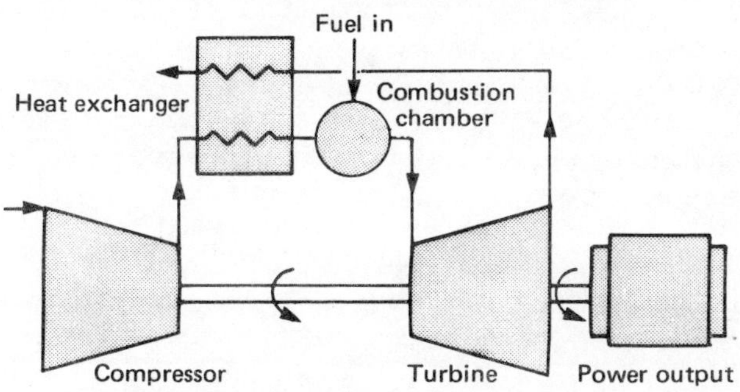

Open cycle gas turbine with heat exchanger (schematic)

gas turbine

A rotary *heat engine* in which a gas is heated by the combustion of a fuel in a pressurized chamber and expanded to produce mechanical power. Compression of the gas is achieved in a separate compressor stage mechanically linked to the power shaft.

In the more common open cycle gas turbine the fuel is burnt in compressed air, expanded through the power stage and exhausted to the atmosphere. Fresh air is drawn in to the compressor. This type is also known as the internal combustion turbine since heat is transferred to the working medium, air, by combustion of the fuel directly in that air.

In the closed cycle gas turbine heat is transferred, via a heat exchanger, to a self-contained volume of gas which is constantly recirculated through the turbine. This type has similarities with the *steam turbine* in that the products of combustion do not pass through the turbine. The condenser of the steam turbine is replaced in the gas turbine by a 'precooler' which cools the gas before it enters the compressor.

The gas turbine cycle can be combined with the steam turbine cycle in the so-called *'combined cycle generation'* system to generate electrical power.

The use of a heat exchanger to transfer heat from the exhaust gas to the air entering the combustion chamber can improve the overall thermal efficiency. This may not always be a large improvement, however, due to resistive losses as the gases pass through the heat exchanger passages.

In certain cases it may be required to burn a fuel, such as pulverized coal, whose combustion products contain erosive or corrosive components. The use of a heat exchanger to supply heat to the compressed air enables such fuels to be safely utilized in an open cycle, but this is achieved at the expense of loss of efficiency, due partly to the pressure and temperature drops in the heat exchanger.

When the gas turbine is required to operate at steady power output, as in the case of gas turbogenerators for peak-load electricity generation, the compressor may be driven directly by the turbine shaft. Such 'single-shaft' machines are less flexible, and less efficient at part load, than the more complicated twin-shaft types. In these a high-pressure turbine drives a compressor on one shaft and acts as a 'gas generator' for a mechanically independent 'power turbine'. Twin shaft turbines are used for road, rail and marine power units where flexibility and part-load efficiency are important.

The efficiency of the gas turbine as a heat engine can be improved by inter-stage cooling of the compressor and by the use of a multi-stage power turbine, with reheating between stages. In practice, however, the inherent compactness and simplicity of the basic gas turbine cycle are important considerations, and the more complicated cycles cannot often be justified.

The compact nature of the gas turbine is a desirable feature for aircraft propulsion, and in various forms this type of engine powers most of the aircraft in commercial service today. In the turbojet engine the shaft power does little more than drive the intake compressor and the propulsive effort derives from the expansion of the exhaust gases through a jet. (Hence the term 'jet engine'.) In the turboprop and turbofan engines some shaft power is delivered to a propellor or a fan (to provide a separate jet of cool air), respectively.

The short start-up times of gas turbines (typically two minutes) has led to their widespread use as *peak load* generators in electricity networks, particularly in western Europe. Fuel costs are high, partly because of the lower efficiency of the gas turbine cycle compared with the steam turbine, and partly because such sets usually run on premium fuels, such as natural gas or gas oil.

The inherent reliability of the gas turbine makes it a good choice for industrial power generation, where exhaust heat can provide useful process steam. Industrial gas turbines are often developed from aircraft units and modified to ensure longer running periods between overhauls.

Closed cycle gas turbines have been suggested for use in nuclear power stations, where the self-contained nature of the working gas means that it can be circulated through the core directly without leading to emission of radioactivity. Helium has been proposed for this purpose. As well as having desirable nuclear properties, helium is a better working gas than air from the point of view of gas turbine cycle efficiency. The gas is also a better heat conductor and leads to lower resistive losses than air.

One of the major limitations to the efficiency achievable by the gas turbine is due to the constraints on inlet temperatures imposed by the materials used in the turbine blades. Since higher inlet temperatures, as with any heat engine, lead to higher efficiencies, the development of blade materials able to retain strength at high temperature would improve the prospects for more widespread use of the gas turbine.

gas turbine generator

An electrical power production device consisting of a generator linked to a *gas turbine.* A reduction gearbox is generally employed between to reduce the high speed of the gas turbine to a value suitable for the generator. Sometimes abbreviated to gas turbogenerator, this type of generating plant is widely employed to meet peak electricity demand due to its short start-up time and relatively low capital cost.

gas-water contact or level

In petroleum terminology, this refers to the depth at which the interface between gas and water occurs in a reservoir.

GCFBR

Initials for *gas-cooled fast breeder reactor.*

generator, electric

See *electric generator.*

geochemical survey

A method of searching for concealed fossil fuel or mineral deposits which employs chemical analysis of rock materials, both surface samples and cores.

geophysical survey

A method of prospecting in which the principles of physics and geology are applied to subterranean rock strata. Geophysical techniques include *magnetic surveys, gravimetric surveys* and *seismic surveys.* Magnetic and gravimetric surveys are used for reconnaissance purposes and seismic surveys to provide more detailed information. Radioactive and electrical measurements can also be made and the analysis of cores from boreholes also provide data which help in the location of *fossil fuel* and mineral resources.

geothermal energy

Energy available as heat emitted from within the earth's crust, usually as hot water or steam. It also refers to the energy in *hot rocks* in the earth's crust.

Emissions of hot water and steam occur in many parts of the world, but are most common in volcanic regions. Steam can be used directly to generate electricity by feeding it to a turbine, as at Larderello in Italy. Hot water is principally used for space heating – in houses, greenhouses, factories, sometimes in whole towns; and to provide industrial process heat. The technology needed to utilize the energy in hot, dry rocks is still being developed but they are potentially the largest source of geothermal energy.

On a world scale present geothermal energy use is relatively insignificant – geothermal electricity generating capacity is little more than 1000MW(e) – but further plants under construction are likely to more than double this by 1980. In addition, about 5000MW(t) are derived from geothermal hot water sources. The prospects for its development as a significant contributor to our energy requirements have changed greatly following the energy crisis of the early 1970s. More than 50 countries are now actively engaged in geothermal research related to energy needs.

Origin of geothermal heat. The interior of the earth is hot but between it and the surface is an insulating layer of crustal rocks. The heat is thought to originate through the breakdown of *radioactive elements.* Some may also result from rock movements and chemical changes taking place under intense pressure.

The crust is not a continuous shell as was once thought

but a group of rigid plates – so-called tectonic plates – which move separately rather like a group of rafts. Where the plates move against each other at the edges, the crust is fractured or stretched and the hot underlying materials move closer to the surface. In such areas the rate of increase of temperature with depth will be much greater than the average 30°C/km. A relatively shallow borehole therefore is needed to exploit the geothermal energy. If the underlying strata are sufficiently permeable and the temperature gradient exceeds 80°C/km then subterranean steam fields may be located and tapped commercially.

Hot water fields. These contain water at approximately 60°C to 120°C and are successfully used for domestic, agricultural and industrial heating purposes. Geothermal water provides nearly all the heating for Reykjavik, the capital of Iceland. Around 50 per cent of the population in Iceland relies on such water for space heating and there are plans to increase this to 70 per cent. The USSR makes extensive use of geothermal water for space heating. In several plants in Hungary, greenhouses covering an area of over 1.5 million m^2 are heated by geothermal water. There is also considerable development in France. In Paris several large apartment blocks are heated by geothermal water extracted from depths of less than 2000 metres, and in the Melun region some 2000 house units have been similarly heated since 1971. It is planned to heat up to half a million dwellings in this way by 1985.

The principal current use of geothermal water for space heating may be extended as technology improves. Turbines that can be driven by vapours from low-boiling point organic liquids are being tested. Certainly, in terms of cost, geothermal water compares favourably with other energy sources. Estimates from France suggest a saving of around 13 per cent compared with oil-fired installations.

Dry steam fields. Fields of dry steam are relatively easy to exploit. The first plant to operate using this energy source was at Larderello in Italy which started electricity generation in 1904. Its output is about 405MW. The Geysers in California, USA, is the largest known geothermal field. It started operation in 1960. Construction and operating costs for the plants at both fields have been lower than for conventionally fuelled plants. The major problem is that of pollution, for the steam contains impurities such as hydrogen sulphide, but the level is equivalent to the sulphur released in burning a low sulphur oil in a plant of similar size. Steam at 150°C to 300°C may be obtained by drilling wells to depths of between 500 and 3000 metres. It is usually piped direct to turbines. These are constructed of corrosion- and abrasion-resistant materials and several types are used. *Back-pressure steam turbines* are frequently used in the early phase of field development because they will cope with steam containing large quantities of non-condensable gases such as carbon dioxide. Such steam is often concentrated in the upper part of the reservoir. Reduction of the 'contaminants' enables the subsequent changeover to condensing turbines. In addition it is possible to predict reservoir life at an early stage in the field development programme.

It is not economic to pipe steam over large distances so each power plant handles the steam production of several wells in the immediate vicinity. Power conversion efficiency is of the order of 20 per cent. Much higher usage efficiencies are possible when geothermal energy is used directly as heat.

Hot brine (wet steam) fields. The exploitation of such fields for power generation has serious drawbacks: corrosion, pollution and the disposal of waste water. In New Zealand the salinity is low but generally the water contains large quantities of dissolved minerals. Wells in California's Imperial Valley release water and steam containing a total 25 per cent dissolved matter. The tendency is to dispose of condensates and waste water by reinjecting it into the producing formation. Chemical and thermal pollution of surface waters is reduced, the fluid supply of the reservoir is replenished and there is less danger of subsidence due to fluid withdrawal.

Hot dry rocks. The idea of exploiting the heat contained in dry rocks is a recent one. Such rock is present under the whole of the earth's land surface. But since the average geothermal gradient is 30°C/km rocks at a temperature of 300°C will be found on average at depths of 10,000 metres. With current technology drilling beyond 6000 metres is not routine; rarely is it economic beyond depths of 4000 metres. Obviously, therefore, only where the geothermal gradient is abnormally high will the hot rock formations be sufficiently near the surface for their exploitation to be considered. Nevertheless there are a number of areas where hot rocks occur from 2000 to 3000 metres down.

The limiting factor at the moment is inadequate technology. Rock is a poor conductor of heat so a heat conducting material must be injected into the drill hole and a large heat transfer area created underground so that the circulating fluid can extract sufficient heat and then be recovered from the hole. The thermal energy would be removed at the surface using a heat exchanger and the cooled fluid returned to the hole to extract further heat.

The use of nuclear explosives has been suggested to produce an adequate heat transfer surface. The idea would be to fragment a large volume of rock underground by the explosion and then pump cold water down the borehole and through the fractured rock. The water could be recovered through a second hole as hot water or steam. Obviously there are environmental problems associated with such proposals. An alternative method is to employ hydraulic techniques of rock fracturing such as are already used in the petroleum industry for well-completion.

Since the hot rock method of extracting geothermal energy has not been tried, cost projections are uncertain. Plant costs should be about the same as for other geothermal systems; corrosion and pollution problems should be lower. It may be possible to produce energy by this method at costs which are competitive with conventional or nuclear fuelled plants.

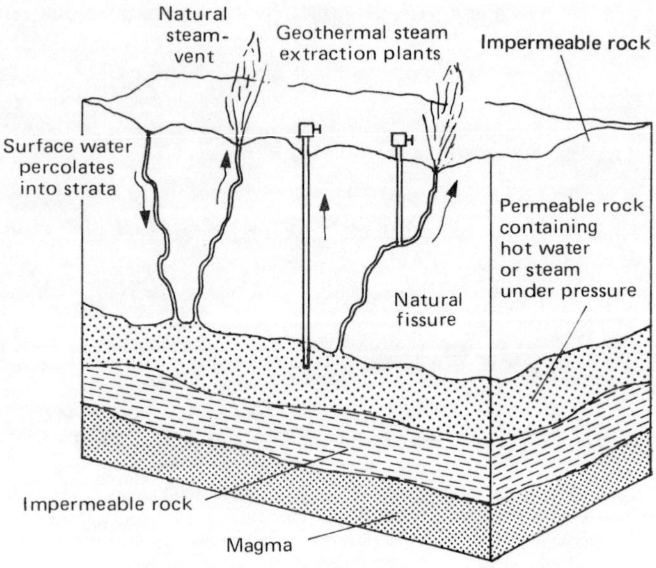

Schematic diagram of a wet geothermal reservoir

A recent development in the search for energy from hot rocks is that at Los Alamos in the Jemez mountains. Twin boreholes, linked by directional drilling, have been made to a depth of 3 km. A large rock volume has been fractured hydraulically at the bottom of the boreholes. Water is to be injected down one borehole and superheated steam will be recovered from the other. It is planned to link two electrical generating systems into the closed loop steam system which has been set up by previous drilling. A life of 20-25 years has been estimated for the geothermal well. Rock temperatures at the Los Alamos site are above 200°C at the depths drilled. When the technology allows drilling to 5 km or more much higher temperatures will be achieved – of the order of 350°C to 450°C.

The Los Alamos experiments will allow investigations of rock fracture techniques, an improvement in field life predictions, development of thermal flow measurement techniques, the testing of different turbine types and the analysis of the corrosive and possible 'plugging' effects of dissolved minerals.

See also *Map 21.*

geothermal gradient

The rate at which rock temperature rises as the depth below the surface increases. The average value for the earth's crust is 25°-30°C per km, but values of 80°C per km are found at the edges of the tectonic plates and such gradients are likely to be exploitable for extraction of *geothermal energy.*

GER

Abbreviation for *gross energy requirement.*

Girdler Sulphide process

Process used to manufacture *deuterium oxide* (heavy water), used as both coolant and moderator in *nuclear reactors*, for example the Canadian Deuterium Uranium (CANDU) reactor.

GOR

Abbreviation for *gas/oil ratio.*

graphite

A form of *carbon*, commonly called 'black lead' with a melting point of 3500°C. In graphite, the carbon atoms are tightly bonded in sheets which are only loosely held together in layers. The layers are free to slide, giving graphite its lubricating properties.

Graphite is found in nature but is usually made artificially by the action of heat on carbon-containing materials in a limited supply of air. Coke from *coal* or *petroleum* is mixed with *pitch*, heated mildly to drive off volatile material, then moulded into shape and baked at about

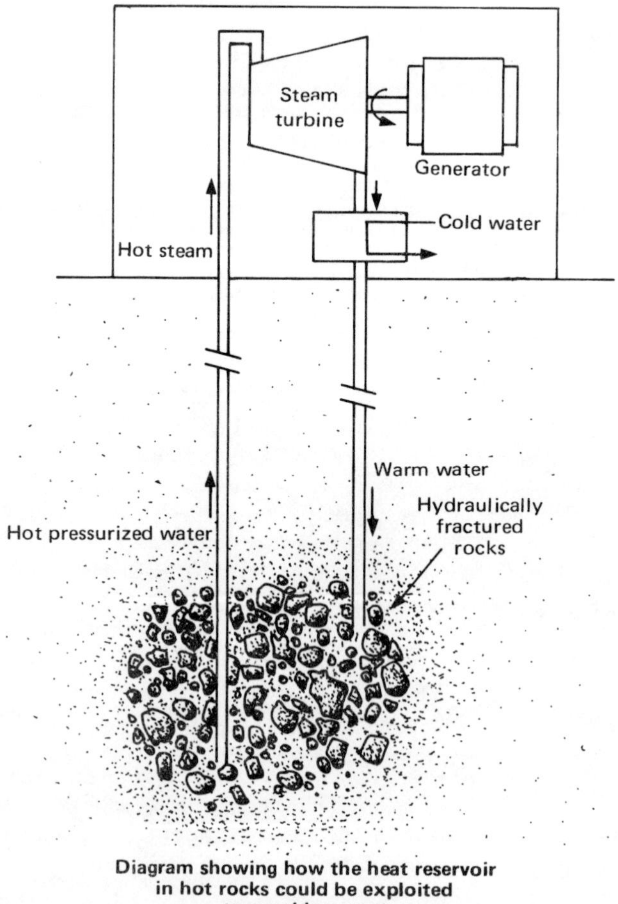

Diagram showing how the heat reservoir in hot rocks could be exploited to provide power

1000°C. The process is completed by passing an electric current through the partially graphitized material to sinter it at about 2800°C.

Graphite has excellent mechanical strength which increases at higher temperatures reaching a maximum at about 2500°C; it also has very good thermal and electrical conductivity. Graphite has a number of important uses: 1. It is one of the materials which can act as a *moderator* in thermal *nuclear reactors.* For this purpose graphite must be treated at 2500°C with fluorine in order to drive off any boron (which is a neutron absorber) in the form of volatile fluoride. 2. As a pyrolytic graphite which is impermeable to gases and liquids and which is used in some forms of electrical apparatus. 3. In the production of electrodes for electrical furnaces, in carbon arc lamps and in the *Leclanché cell.*

See also *ceramic reactor; high temperature gas-cooled reactor (HTGR).*

gravimetric survey

A technique employed by geologists for determining the arrangement of rock strata when searching for minerals and *fossil fuels.* It makes use of the variation in the earth's gravitational field due to differences in the density of subsurface rocks. Because base rocks generally have higher densities than overlying sedimentary rocks, when they are close to the surface unusually high readings are recorded. The converse also applies, and different sedimentary rocks give different readings. The various data obtained can be used to provide a broad idea of underlying terrain at the reconnaissance stage, but more accurate geophysical methods are necessary for precise identification of subsurface features.

gravity

Abbreviation for *API gravity* (US) or for *specific gravity* (UK).

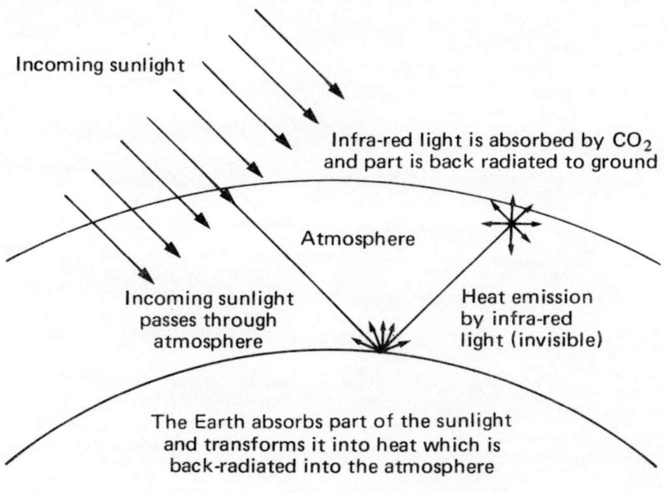

The greenhouse effect in the atmosphere

gravity drainage

The drainage downwards under the influence of gravity of oil in a gas cap reservoir where production has depended on gas drive. To capitalize on gravity drainage, holes are drilled at the edges of the reservoir to provide pressure drive so that production can continue after the *gas cap drive* or *solution gas drive* mechanisms have ceased to be effective.

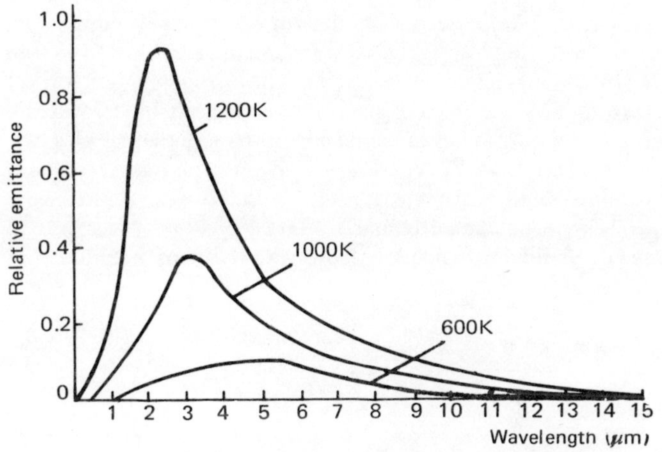

Emittance of a black body at different temperatures

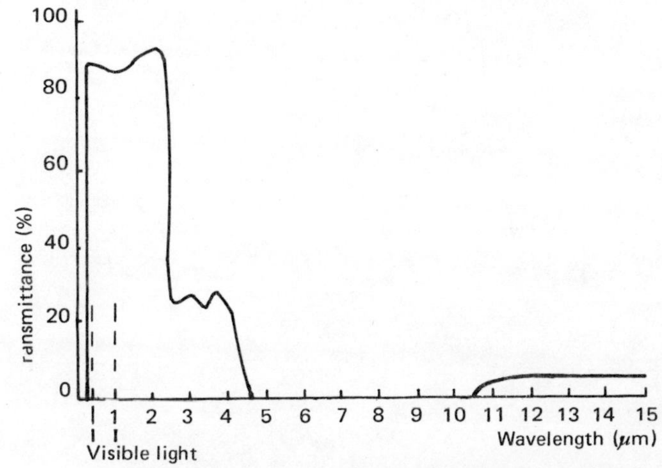

Transmittance of window glass as a function of wavelength

greenhouse effect

The phenomenon by which a greenhouse traps the sun's heat and so provides improved conditions (compared with open ground) for cultivating plants in cooler weather. The term is also used to describe the way in which the earth's atmosphere acts as an insulating blanket, keeping the earth at a temperature much higher than would otherwise be the case.

The transparent glass or plastic walls and roof of the greenhouse transmit the shorter wavelength, visible light and *infrared radiation* from the sun but not the longer wavelengths. Thus, during the hours of daylight infrared radiation passes into the greenhouse and warms its atmosphere and the ground on which the house stands. The transparent walls and roof of the greenhouse partly reflect and partly absorb the longer wavelength radiation. Subsequently the walls and roof re-emit the absorbed radiation into the house. Coating the glass with a non-heat-radiating film, which is transparent to sunlight, further maximizes the heating effect of radiation received.

In an analogous fashion the carbon dioxide and water vapour in the atmosphere transmit short wavelength solar radiation but reflect (by absorption and subsequent re-emission) the longer wavelength heat radiation from the warmed surface of the earth. The rising proportion of carbon dioxide in the atmosphere, caused by large scale fossil fuel burning, is causing concern to some climatologists, who assert that the increase in the greenhouse effect will produce higher average surface temperatures. If this were the case widespread climatic effects, possibly disastrous in scale, could be expected. However, it seems possible that the atmospheric dust associated with fossil fuel use, by causing increased solar absorption in the upper atmosphere, has counteracted the rise in carbon dioxide production. The situation is far from clear at the present.

See also *solar collector; solar energy.*

grid

Colloquial name for *electricity transmission* system.

gross energy requirement

The total amount of *primary energy* taken from resources that is needed to supply a commodity of service, as measured under an agreed set of conventions. Often abbreviated to GER, this is the most generally useful result of an *energy analysis.*

See also *energy requirement; process energy requirement.*

H

half-life

Term used to describe the characteristic average rate of decay of a particular radioactive isotope. It indicates the time in which half the nuclei of a sample of the isotope will decay.

For example, the half-life of strontium-90 is 28 years. This means that of 1000 strontium-90 nuclei, 500 will have decayed after 28 years. Of the remaining 500, 250 will have decayed in the next 28 years; of these 250, 125 will have decayed in the next 28 years and so on.

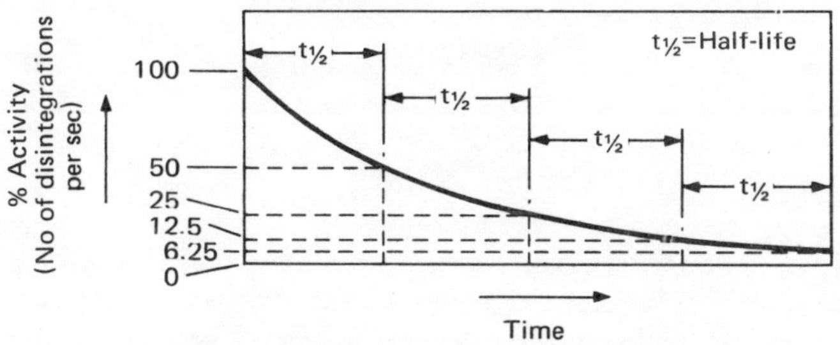

The relationship between the number of radioactive disintegrations and time, and half-life, of a radioactive material

The rate of disintegration of any radioactive nucleus may be defined in terms of its half-life. It is the long-lived nature of some constituents of radioactive waste which give rise to concern over storage of such materials for a long period of time, eg plutonium-239 which has a half-life of 24,000 years.

H-coal

A *coal liquifaction* process yielding a synthetic crude oil suitable for conventional refining. In the H-coal process a pulverized slurry of coal and recycled oil is subjected to high pressure (200 atmospheres) and high temperature (450°C) catalytic hydrogenation. The slurry fed in at the bottom of the reactor is kept in intimate contact with the pelletized metal catalyst by constant agitation and the solid, liquid, and gaseous products are removed from the top. Unreacted coal and ash are separated out and ammonia and hydrogen sulphide removed by scrubbing. Some of the oil produced is used in preparation of fresh slurry feed and the remainder can go for refining. Very little char or heavy tars appear to be produced, and it is hoped that the H-coal process may eventually become a commercially viable proposition for the production of light and middle petroleum distillate from coal.

hard coal

A collective term covering anthracite, bituminous and sub-bituminous coals. Brown coal and lignite are usually referred to as *soft coal*.

head of water

The *potential energy* possessed by a mass of water by virtue of its height above a fixed point of reference. The energy contained per unit mass of water is equal to the product of the acceleration due to gravity and the height:

$$\text{potential energy} = mgh$$

Thus 1 kg of water raised through 1 metre possesses a potential energy of 9.81 joules since g in the *SI system* of units is equal to 9.81 ms^{-2}.

When water is allowed to fall, the kinetic energy due to its motion is equal to the potential energy equivalent to the head through which it has fallen. This kinetic energy can be converted at high efficiency to electric power and in *hydroelectric power* schemes the head of water available is an important measure by which potential new sites are selected. The head of water will also help determine the type of turbine used.

heat

Energy that flows from one body to another as a result of a temperature difference between them. Although heat can be described in terms of *thermal energy*, ie the random motion of the atoms or molecules of a body, it can only be measured if allowed to flow. For this reason there is no way that the amount of heat in a body can be measured, since it is impossible to separate, even conceptually, the thermal energy content of a body from the potential energy of its constituent atoms and molecules.

The effect of heat addition to a body can produce any of a number of changes. These include temperature rise, changes in length or volume or changes of state, such as melting and boiling. Each of these changes is associated with a quantitative measurement that relates the magnitude of the effect to the heat flow. As regards temperature, a 'specific heat capacity' can be defined for every substance and this is a measure of the heat input required to raise the temperature of unit mass of the substance by unit temperature. Typical units are $Jkg^{-1}\ deg^{-1}C$, (SI); and Btu/lb/$^{\circ}F$, (British). Temperature can be regarded as a measure of the 'intensity' of heat. The term 'specific heat capacity' is often abbreviated to 'specific heat'.

For changes of length, (or volume), coefficients of linear (or cubic) expansion can be measured, and these are usually defined in terms of the fractional increase or decrease per unit rise in temperature. Most coefficients of expansion are positive, ie most substances expand when heated, but there are exceptions.

As far as phase changes are concerned, these are usually associated with relatively large heat changes. For example, the melting of ice at $0^{\circ}C$ to water at $0^{\circ}C$ requires a considerable input of heat, known as the *latent heat* of fusion. Similarly, a latent heat of vaporization can be measured for the process of boiling.

The conversion of heat to *mechanical energy*, or 'work', occupies a central role in the energy supply industry since it is by means of this conversion, in the various forms of *heat engine*, that the chemical energy of fossil fuels, as well as the nuclear energy of uranium, is converted to power for electricity production, industrial use and transport.

A fundamental and far-reaching result of the laws of energy conversion is that heat can only be exploited to produce power if a temperature difference exists between a source of heat and a heat sink. This means that sources of heat at or around the temperature of the surroundings are of no use for power production. Power production from heat relies on the existence of a thermal gradient, as in the case of *geothermal energy* and *ocean thermal energy conversion*.

heat engine

In general terms, any system whose effect is the conversion of heat into mechanical energy ('work'). Heat engines take on many disguises and exist in natural systems as well as man-made devices, but all have in common two essential features. These are respectively a source of heat, or a hot source, and a source of cooling, or cold 'sink'. (In thermodynamics the use of the term heat engine is restricted to closed systems, ie those exchanging only heat and work with their surroundings.)

The maximum amount of work that it is possible to obtain from a heat engine for unit heat input depends solely on the temperature of the hot source, T_1, and the temperature of the cold sink, T_2. If these two temperatures are measured on the absolute (Kelvin) scale, the relationship between the work-to-heat ratio, ie the efficiency of the engine, and T_1, T_2 is particularly simple:

$$\text{maximum efficiency} = \frac{W}{Q_1} = \frac{T_1 - T_2}{T_1}$$

In practice, no heat engine can attain this ideal efficiency. Nevertheless this theoretical or 'Carnot' efficiency is an important quantity since it gives the engineer a good idea of how far any particular design of engine departs from the ideal.

The ratio of practical efficiency to ideal efficiency (always less than unity) is sometimes known as the 'second law efficiency' or 'effectiveness'.

Examples of heat engines include the atmospheric system of the earth that produces the winds and waves and a vast number of man-made devices, including the *internal combustion engine*, the *gas turbine*, the *steam turbine*, rockets, jets and guns.

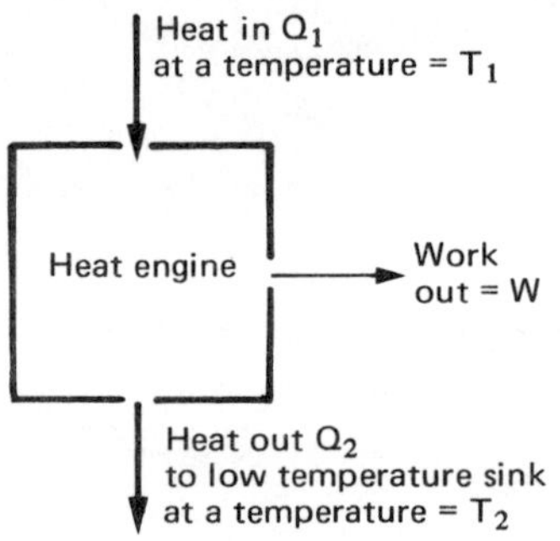

Ideal heat engine

A heat engine in reverse (ie a system capable of absorbing work, producing heat at a high temperature, and taking heat in at low temperature) is known as a *heat pump*. Like heat engines, heat pumps are limited by the Second Law of Thermodynamics. This limitation is an essential corollary of the maximum efficiency constraint of the heat engine.

Heat engines can be classified in several ways: 1. Open or closed cycles: open cycle heat engines take in a fresh charge of working fluid (which is therefore usually air) for every cycle; in the closed cycle type a constant inventory of working fluid is cycled continuously. Closed cycle variants have the option of using more expensive gases which may have desirable thermodynamic properties. This would be prohibitively expensive for the open cycle type. 2. External or internal combustion: *external combustion engines* rely on a source of heat outside the working chamber. In *internal combustion engines* a fuel gives up its heat of burning directly to the working fluid. 3. Mechanical configuration: both reciprocating and rotary forms of heat engines have been devised. As their names imply, reciprocating engines make use of a back-and-forth motion, which is usually converted to shaft power, whilst rotary types produce rotary motion directly. Within the category of rotary engines, axial and radial variants can be recognized, according to whether the flow of working fluid is along, or perpendicular to, the axis of rotation.

heat exchanger

A device which enables thermal transfer between two fluid streams at different temperatures without allowing physical contact between them.

The performance of a heat exchanger depends markedly on the area of contact it is able to create between the streams, so that convoluted designs, using pipes and fins, are common. There are many varieties of heat exchanger, but broadly speaking they fall into one of three categories: gas to gas; liquid to liquid; and gas to liquid. Heat exchangers may be direct, involving the simple placement of a conducting barrier of thin metal between the fluids, or indirect, involving the use of an intermediate heat transfer fluid that is circulated through a pair of heat exchangers in a continuous cycle.

For any particular design application of a heat exchanger, one of the most important operating parameters is the temperature differential under steady flow conditions. A low value indicates a good heat exchanger but can only be obtained by the use of large surface areas, and this leads to high costs. The design of equipment where heat exchangers play a central role, such as heat pumps and heat recovery systems, therefore involves a compromise between the energy savings resulting from better heat transfer and the capital cost of heat exchangers.

See also *economizer; heat pipe; heat wheel; recuperator.*

heat insulation

See *thermal insulation.*

heat limit

That level of thermal discharges to the environment beyond which it is suggested that adverse climatic or other effects may appear.

Although the magnitude of present heat discharges to the atmosphere and hydrosphere is not seriously questioned, considerable controversy arises about whether this represents a threat to the balance of climatic forces or not. It has been demonstrated that certain areas, such as major conurbations and concentrations of power plants, already show local climatic changes, and these effects seem to arise when the output of heat approaches a certain percentage, probably about one to ten per cent of the solar input over that area. It is not at all clear what the effects of total world discharges of heat reaching one per cent of the average solar input would be, and at present rates of growth even this level will not be reached until the next century. However, local concentrations of heat discharges, as is already clear, can have marked effects.

See also *energy conversion; thermal pollution.*

heat-only boilers

Large centralized boilers with a high heat output suitable for fuelling *district heating* schemes. The designation 'heat-only' denotes that there is no turbogenerator coupled to the boiler installation, as in *combined heat and power* schemes where electricity is produced for distribution in addition to heat. Thus the capital costs of installation are lower.

heat pipe

A device in which heat can be transferred by the evaporation and subsequent condensation of a liquid in a closed vessel. Unlike a *heat pump*, a heat pipe operates at substantially constant pressure throughout, and cannot produce a heat output at a higher temperature than the input. The device uses no power, however, and the condensed liquid is returned to the evaporation region by capillary action within a wick of absorbent material. The simplicity of the device results in low installation and maintenance costs. By suitable choice of working fluid and operating pressure, a heat pipe can be designed for use at almost any required temperature. The thermal conductance of a heat pipe can be high – as much as several hundred times that of copper

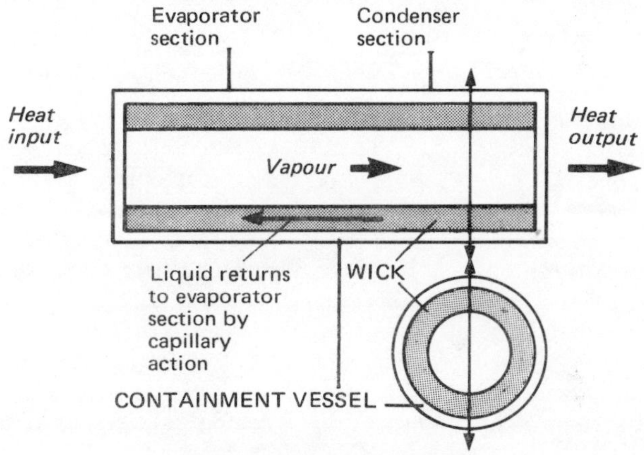

Basic elements of heat pipe

– and power handling capacity is high. Heat pipes find application in *waste heat recovery* in industrial processes.

heat pump

A device in which heat is transferred from one environment to another at a higher temperature, as a result of the input of mechanical work to the device. In this way heat can be made to flow against its natural inclination, rather as water can be made to flow uphill by the use of a water pump. A heat pump can be considered to be a *heat engine* operating in reverse, absorbing, instead of producing, mechanical energy.

A heat pump can be used either for heating or cooling. A refrigerator is a heat pump designed so as to absorb heat at a temperature well below ambient and reject heat at a temperature just above ambient. However, the term 'heat pump' is generally used to describe a device used for heating, ie one that gives out heat well above ambient tempera-

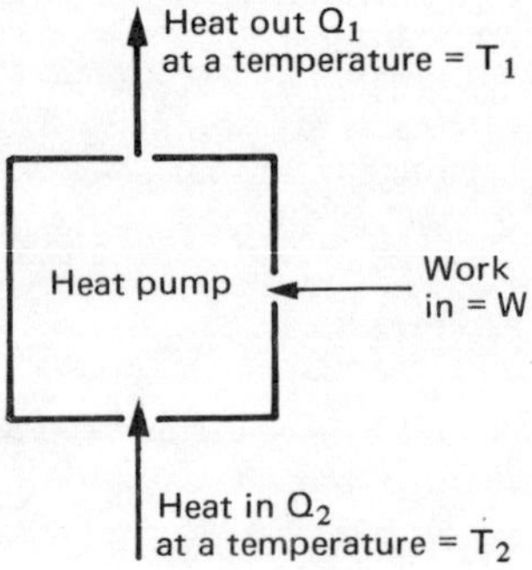

Ideal heat pump

ture while absorbing heat at a temperature just below ambient. Clearly the same device can function both as a cooling system in summer and a heating system in winter, and it is in this dual role for air conditioning that heat pumps have the greatest domestic and commercial application.

In winter, heat is absorbed from an exterior heat exchanger and transferred to the interior. Problems arise with icing of the exchanger surfaces, especially in humid climates, and means must be found for removing this since it hinders heat transfer.

In the summer the heat pump absorbs heat from the interior, rejecting heat to the exterior environment. By condensing water vapour from the circulated interior air dehumidification is also achieved.

Heat pumps are increasingly coming into use for industrial waste heat recovery, since upgrading a heat source to a higher temperature will render it more useful. This is particularly attractive economically when a cooling function can be combined with a simultaneous heating requirement.

In the conventional compressor-driven heat pump a refrigerant fluid (known as the 'working fluid') is continuously circulated through a closed cycle. (The refrigerant is most frequently one of a large range of fluorochlorocarbons, but ammonia and sulphur dioxide are also used.) The gaseous refrigerant is first compressed and then allowed to condense (in a 'condenser') by giving out heat to a cooler environment. The liquid refrigerant is then throttled to low pressure, by means of an expansion valve, and allowed to evaporate in another heat exchanger (the 'evaporator') drawing in heat. The vapour is then circulated to the compressor to begin another cycle.

The effectiveness of a heat pump as a refrigerator is determined by the cooling effect produced per unit of work done. The 'coefficient of performance', usually abbreviated to COP, is defined as the ratio:

$$\text{COP} = \frac{\text{heat absorbed by evaporator}}{\text{work input to compressor}}$$

If a heat pump is used for heating, the COP must be redefined in terms of the heat output, and to avoid confusion the term heat gain is used. The heat gain of a heat pump is defined as the ratio:

$$\text{heat gain} = \frac{\text{heat output from condenser}}{\text{work input to compressor}}$$

Since the heat output from the condenser is equal to the sum of the work input to the compressor and the heat absorbed by the evaporator, it follows that:

$$\text{heat gain} = \text{COP} + 1$$

For thermodynamic reasons, the heat gain of a heat pump cannot exceed the heat gain of an ideal 'Carnot' heat pump, which is a function only of the upper and lower temperatures:

$$\text{heat gain (ideal)} = \frac{T_1}{T_1 - T_2}$$

T_1 = upper, condensing temperature
T_2 = lower, evaporator temperature
(temperatures in degrees Kelvin)

For a lower temperature of 5°C (278K), and an upper temperature of 50°C (323K), the heat gain of an ideal heat pump can be calculated to be 7.18.

Heat gains of practical heat pumps, based on the Rankine vapour compression cycle, fall far short of the ideal, being between 2 and 3 for similar operating temperatures.

This is a useful heat gain, but since the compressor is usually driven by an electric motor, using power generated at an efficiency of about 30 per cent, there is little overall primary energy advantage, for heating purposes, compared with the direct use of fossil fuels. Development work is being carried out on heat pumps driven by fossil fuel-fired engines whose waste heat can supplement the output, leading to overall efficiencies of the order of 150 per cent.

In the 'absorption cycle' heat pump compression of the vapour is achieved by absorption in a liquid solution. The cycle is completed by distillation of the vapour from the solution, using an external source of heat. This type of heat pump has an inherently lower COP value, but is able to use a heat source as driving energy. For this reason they are used for *solar cooling* applications, and development work is proceeding on the use of such heat pumps for *heat storage*.

heat recovery

An important energy conservation method based on the utilization of otherwise wasted sources of heat, either to substitute for fuels or to reduce fuel consumption.

Heat can be recovered from gases, liquids or solids, and both sensible and *latent heat* must be considered. Since some waste heat is produced in almost every energy conversion the sources are apparently innumerable. However, many sources cannot economically be recovered – those at too low a temperature, contaminated by pollutants or impurities that prevent heat extraction, intermittently available or simply too insignificant.

Technologies important for heat recovery include *heat*

exchanger; heat pipe; heat pump; heat storage; heat wheel; integrated environmental design; waste heat boiler.

heat storage

The containment of excess useful thermal energy, ie heat available at a temperature above that of the environment, for use at times when demand for heat exceeds supply, or to act as a 'buffer store' between supply and demand. Most hot water or steam distribution systems for domestic and industrial purposes make use of tanks or pressure vessels for heat storage. Incorporating these buffer stores into the system reduces the required rated output of the boiler, and leads to more continuous operation, so reducing maintenance and increasing thermal efficiency.

Thermal buffer stores are also widely used in *district heating* schemes, where they enable fluctuating demand to be accommodated. District heating systems relying on *combined heat and power* (CHP) stations generally require a heat store to absorb wide disparities between demand for electric power and demand for heat. The alternative strategy of storing excess electricity when heat demand is high is usually more costly. The deployment of steam turbines in which the ratio of power to rejected heat can be varied eases the problem of matching, but does not eliminate it.

Heat storage assumes particular importance in the exploitation of unpredictable energy sources such as solar and wind energy, and in heat recovery from industrial processes. The full use of solar energy, in temperature climates, for heating is hindered by the low insolation in winter, when most heating is required. Conversely maximum solar flux occurs in summer when the heating load is reduced to hot water requirements alone. (The problem is less severe with wind power, but the high installed cost of aerogenerators largely precludes their use for direct space heating anyway.) Heat storage has been suggested as a means of using excess summer solar energy in the winter for space heating and hot water in houses. If, however, this was to be accomplished by the cheapest means, ie hot water tanks, the volumes required would be prohibitive for single dwellings. Present systems concentrate on supplying summer hot water only, with tank storage to deal with several cloudy days.

The development of economic inter-seasonal storage systems is necessary if solar energy is to be able to substitute fully for fossil fuels, and the means of providing such storage are under investigation in many countries. Long-term heat storage may also be required in the exploitation of solar energy for power production.

Many sources of waste heat are fluctuating or unavailable at certain times. The use of heat storage makes recovery of such sources for use possible. At present such storage is usually only economic if used for short periods, but increasing energy costs will probably lead to development of longer-term storage systems.

Methods of heat storage. Heat can be stored by raising the temperature of a body, ('sensible' heat storage); or by causing a body of material to undergo a phase change, such as melting or evaporation, ('phase-change material' storage or 'PCM storage').

Systems presently based on sensible heat storage include the hot water and steam systems already described and storage-heaters for off-peak electrical heating, which use the thermal capacity of bricks or other refractory materials to store heat at a relatively high temperature. Water is a good sensible heat storage medium, since it combines very low cost with a high heat capacity, but, without pressurization, can only be used at temperatures below 100°C. Above this point organic liquids can be used, and, for very high temperatures, metals and refractory materials. The disadvantages of sensible heat storage are the large volumes involved, (in other words the density of storage is low); and the degradation of the store as it cools. The cooling of the storage material limits the period of storage.

Phase change materials are capable of high storage densities, and since pure materials boil and melt at sharply defined temperatures PCM, or latent heat stores, can operate in a thermostatic mode, maintaining relatively constant temperatures. The large volume change associated with vaporization limits the usefulness of this phase change, and most systems are designed to operate with the solid/liquid system. The problem with these is the formation of poorly-conducting solid phase on the heat extraction surface when the store is being discharged.

Other forms of heat storage under development include such physico-chemical processes as the hydration of salts, reversible chemical reactions, and vapour absorption. In all these systems at least part of the storage is in the form of *chemical energy*. These methods offer the possibility of high storage densities, but this is gained at the expense of more expensive storage media.

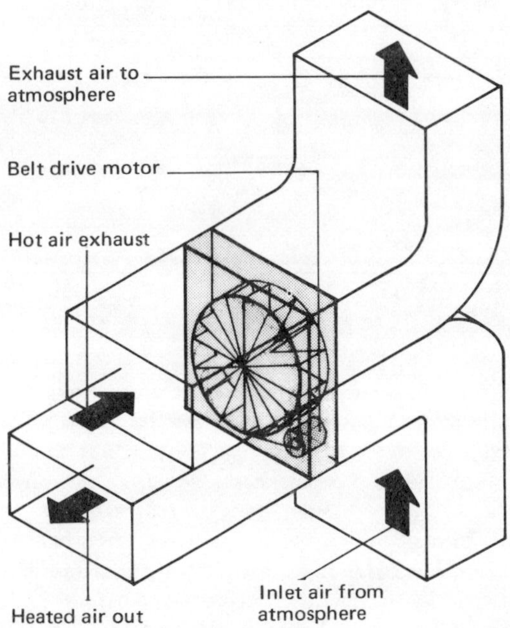

Heat wheel

heat wheel

A particular type of air to air *heat exchanger* in which the thermal capacity of a rotating wheel-like structure is used to transfer heat from a hot gas stream to a cold one. The wheel can be made of metal mesh or, for very high temperatures, ceramic material.

Heat wheels are relatively inexpensive to install, but have the disadvantage that a power source is required to turn the wheel. Because of entrainment in the structure of the wheel, a certain amount of mixing occurs between the gas streams. Despite this, these devices are gaining in popularity for industrial *waste heat recovery*.

heavy distillate

A term in petroleum refining used to describe high boiling point fractions such as *fuel oil*, or specially refined distillates, such as *lubricating oils*, obtained from petroleum *residues*.

heavy water

Familiar term for *deuterium oxide*.

helium

An inert, gaseous substance, the lightest gas next to hydrogen. It is colourless, odourless and non-inflammable. Helium gas can be liquified only at very low temperature (4.2K) and helium liquid is the only known substance that cannot, at atmospheric pressure, be frozen solid – this can happen only at temperatures below 1°K at 25 atmospheres. It makes possible the study of temperatures near to absolute zero and the phenomena of *superconductivity* and superfluidity. Liquid helium is the standard coolant for cryogenic devices.

symbol	He
atomic number	2
atomic weight	4.00
density (gas)	0.1785 kg m^{-3} (at STP)
gas liquifies at	4.2K

Helium has an extremely stable *nucleus* (2 *protons*, 2 *neutrons*) identical with the *alpha particle* emitted by radioactive substances. The radioactive decay of uranium and thorium in the earth produces helium which sometimes seeps into natural gas wells. The USA has sources of natural gas with a relatively high helium content but this amounts to no more than 7.5 per cent. Nevertheless, this is the principal source of helium. It is present in the atmosphere but only in negligible amounts. Even so, attempts have been made to extract it from air, possibly in a mixture with neon.

Some designs of nuclear reactors use helium as *coolant*, the high-temperature gas-cooled reactor (HTGR) for example. Because helium absorbs virtually no neutrons it is a very attractive coolant for a *fast breeder reactor*, a gas-cooled version of which is under development. In *liquid metal-cooled fast breeder reactors, (LMFBR)* helium is sometimes used as a 'blanket' on the open surfaces of the highly-reactive *sodium* or sodium-potassium alloy coolant.

heptane

A member of the *alkane* series of hydrocarbons. As its name suggests, it has seven carbon atoms and it exists in nine different isomeric forms. The straight-chain form, n-heptane, is used as a base reference point when calculating the *octane number* of a *motor spirit*.

hex gas

Abbreviation for *uranium hexafluoride* gas used in *isotope separation* processes.

high-Btu gas

A term generally applied to a gaseous fuel having a calorific value of greater than 700 Btu/ft^3 (26.0 MJ/m^3). The most important example of a high-Btu gas is natural gas which, according to its constituent gases, has a calorific value of between 900 and 1100 Btu/ft^3 (33.5-40.9 MJ/m^3).

high-grade heat

Thermal energy at a temperature higher than about 300°-500°C is loosely termed high-grade heat to emphasize the fact that a *premium-grade fuel* will generally be required to supply it. In contrast, *low-grade heat* can usually be supplied by non-premium fuels.

The grade of heat can be rigorously defined in thermodynamic terms by means of an *availability factor*, which is a measure of the maximum theoretical efficiency into which the heat can be converted to *work*. High-grade heat has a relatively high availability factor.

high octane fuels

Gasolines with a high *octane number*. The higher the octane number of a fuel the higher its anti-knock quality.

high temperature gas cooled reactor

A nuclear reactor in which heat is extracted from the core by helium coolant in a pressurized cooling circuit. The fuel used is highly enriched uranium carbide or oxide and its fabrication involves coating small pellets of fuel with graphite moderator and ceramic material, and firing to give small spheres capable of withstanding very high temperatures in the reactor core. Operating at temperatures higher than those attained in the Magnox, CANDU or LWR designs allows for more efficient electricity generation, and it is envisaged that the steam heat-exchanger could be eliminated by allowing the hot helium emerging from the core cooling

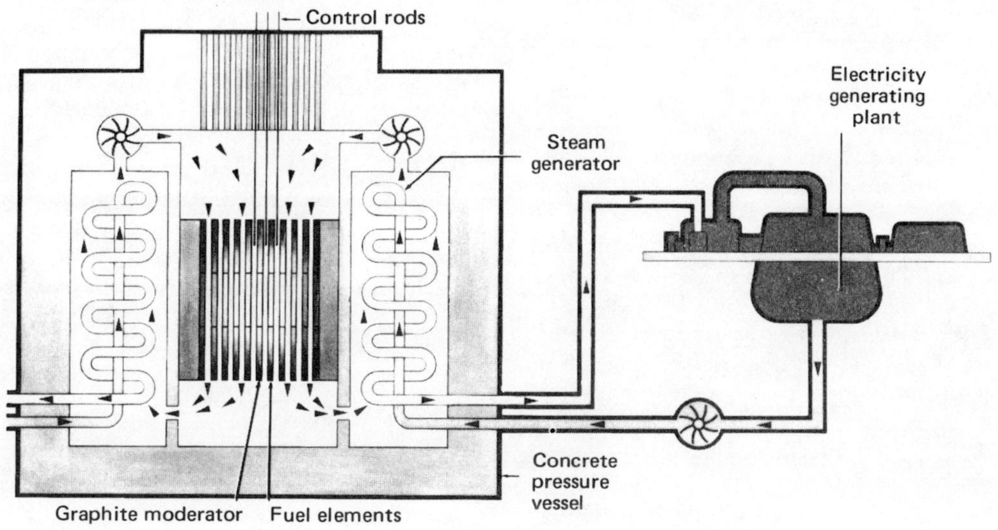

Reproduced by kind permission of the UKAEA

High temperature reactor (HTR)

circuit to drive a gas turbine directly.

With HTR technology it is hoped to take advantage of the *thorium-uranium fuel cycle* by using fertile thorium fuel in the core. Neutron bombardment transmutes thorium-232 into uranium-233, a fissile uranium isotope, which needs to be separated from thorium-232 and other fission products by reprocessing before it can be used as nuclear fuel.

High temperature gas-cooled reactor data*	
Peak power density:	6.3 kW/l
Heat output	842 MWt
Electrical output	330 MWe
Efficiency	39.2%
Fuel:	Uranium carbide particles 93% enriched, coated in graphite matrix
Weight of fuel	16.7 tonnes
Fuel burn-up	100,000 MWD/te
Moderator	graphite
Coolant	helium gas
Coolant pressure	46.7 atmospheres
Coolant outlet pressure	785°C
Refuelling	Off load

**This example: Fort St Vrain, USA*

hot rocks

A term usually restricted to high temperature rocks in the earth's crust near the surface and which are a potentially useful source of *geothermal energy*.

HRS rectifier

A device for changing the pulsating flow of sea waves into a steady flow passing through turbines, which in turn drive generators. It consists of a series of box-like compartments,

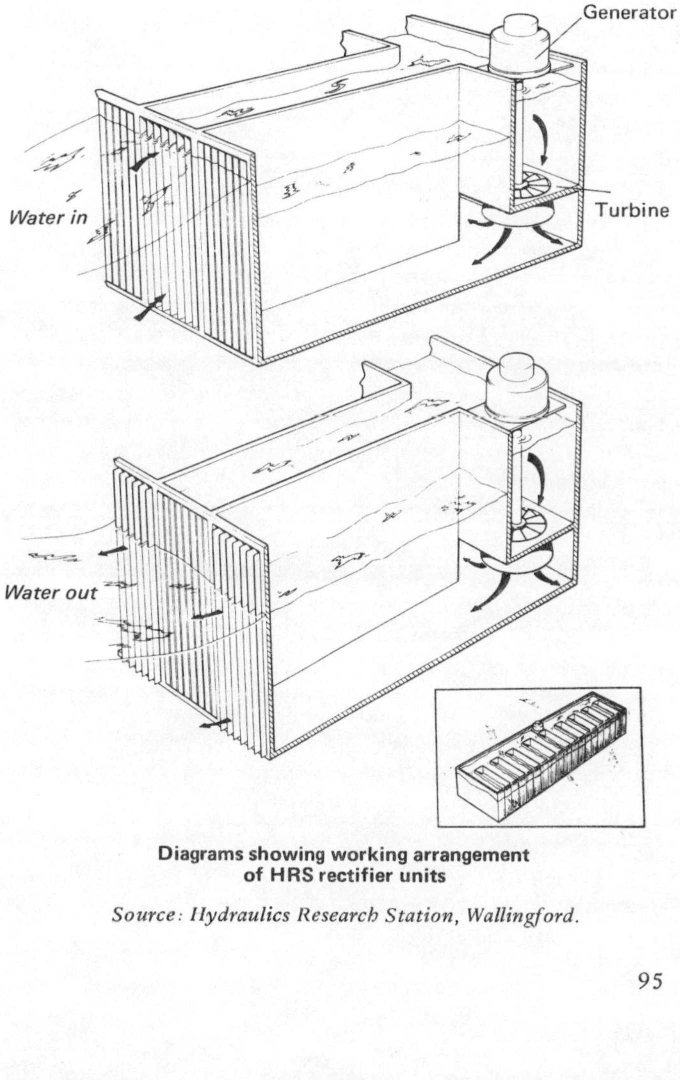

Diagrams showing working arrangement of HRS rectifier units

Source: Hydraulics Research Station, Wallingford.

each projecting through the sea surface, arranged along a line at right angles to the wave direction. Alternate compartments are high-level reservoirs and the intervening ones are low-level reservoirs. The front face of each high-level reservoir consists of vertical non-return flap valves which open inwards. There is a similar arrangement for the low-head reservoirs with the flap valves opening outwards. The valves open automatically when there is a pressure difference across them in the appropriate direction: at other times they are closed. In this way waves drive sea water into the high-level reservoirs and abstract it from the low-level ones; if the water is not used the levels attained will be those of the wave-crest and the wave-trough respectively. Sea water accumulating in the high-level reservoir system is passed through low-head turbines into the low-level reservoir system and from there back into the sea. Alternative turbines are being investigated for this particular use.

The overall design would depend on its location. If sited in very deep water out of sight of land a ship-like structure is envisaged. Closer inshore – say five kilometres offshore in 30 metres of water – concrete caissons could be used resting on the sea bed. Few novel technological problems are thought to be involved except the need to predict performance and wave loads applied to the structure.

See also *wave energy*.

HTR

Initials for *high temperature gas-cooled reactor*.

Hydrane process

A *coal gasification* process in the early stages of development designed to convert bituminous or lower rank coals into high-Btu gas (900-1000 Btu/ft^3) and based on the two-stage hydrogasifier principle. Pulverized coal is fed into a reactor so that coal particles fall freely through an atmosphere rich in hydrogen. The coal is heated to 750°C and a pressure of 70 atmospheres is used. The hydrogen is passed into the lower part of the gasifier to form a fluidized bed of char, and the gas from the fluid bed has a temperature of 975°C. It is important that only just under half of the carbon is converted so that sufficient hot char remains for feeding into a second gasifier, along with oxygen and steam, to produce the hydrogen required for hydrogasification.

The principal advantage of this method is that direct hydrogenation obviates the need for subsequent catalytic methanation of the gas stream, the gas coming from the gasifier consisting of over 90 per cent methane.

With more than 90 per cent of the methane produced directly from coal only small amounts of carbon monoxide require catalytic methanation. Other processes produce as much as half their methane through post-gasification catalysis. Total methane yield is 0.6m^3/kg of dry coal.

hydrates

Solid, snow-like compounds formed from water and methane in gas pipelines and control equipment when *natural gas* from high pressure wells has too high a moisture content. When the gas at high pressure and with a free water content is allowed to expand with a large reduction in pressure, cooling occurs with the resultant formation of hydrates. It is necessary to treat the gas before it is distributed by pipeline. One method of treatment, used by BP in their West Sole gasfield in the North Sea on the offshore production platform itself, involves passing the gas first through vertical separators so that water and *condensate* fall out freely and then through absorption towers containing triethylene glycol to further reduce the moisture content.

hydraulic fracturing

Also known as hydrofracturing. A technique for cracking and then breaking up rock underground by the use of a high pressure surface pump which develops fluid pressure in a borehole. The technique is a standard one for well-completion in oil- and gas-fields to stimulate and improve

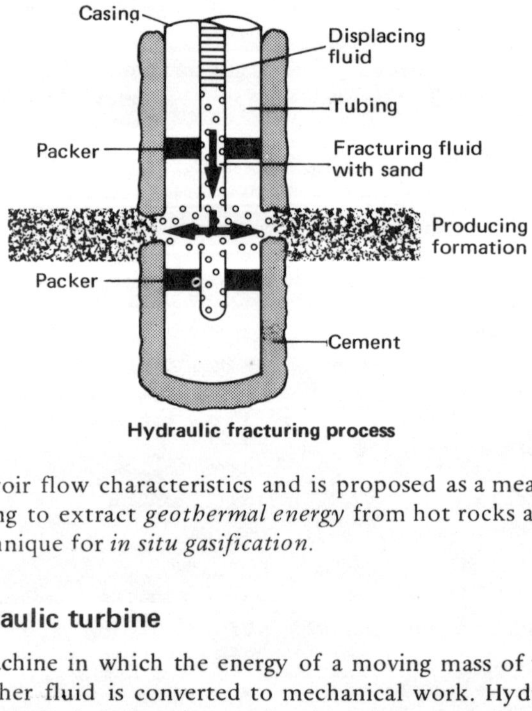

Hydraulic fracturing process

reservoir flow characteristics and is proposed as a means of helping to extract *geothermal energy* from hot rocks and as a technique for *in situ gasification*.

hydraulic turbine

A machine in which the energy of a moving mass of water or other fluid is converted to mechanical work. Hydraulic turbines find their major application in the exploitation of natural water heads, produced by damming of rivers, where they are coupled to generators to provide *hydroelectric power*. Such an arrangement is highly efficient, with up to 90 per cent of the potential energy being converted into electricity.

Water wheels used the weight and to a lesser extent the current velocity of water, but modern turbines are either of the impulse or reaction type (or a mixture of the two) utilizing jets and vanes.

'Reaction turbines' exploit the pressure head of water; 'impulse turbines' make use of the velocity of water allowed to accelerate down a long 'penstock'. In practice, of course, a reaction turbine will require the water to be moving through so that a velocity component is also exploited. By their nature, reaction turbines must run full of water, whereas turbines of the impulse type do not.

Three main kinds of hydraulic turbines are used in hydroelectric schemes – *Pelton wheels* and *Francis* and *Kaplan turbines.*

Pelton wheels. These are impulse turbines in which water passes through needle nozzles and is directed as high velocity jets at cup-shaped buckets mounted on the rim of the wheel and enclosed in a housing. The shaft may be horizontal or vertical but the trend is towards vertical shafts in large turbines because several jets can be used effectively on each wheel. Efficiency is increased, partly because of reduction of friction. Pelton wheels are used where the water head is high, from 150 metres to 1900 metres.

Francis turbines. These are reaction turbines in which water flows radially inwards into guide vanes which direct it onto the curved blades of the runner. Like the Kaplan unit described below, Francis turbines run full of water and submerged. They have a draught tube and there is a continuous column of water from head to tail race. The angle of the guide vanes can be adjusted so that the direction of water flow is changed and the turbine speed regulated. The shaft may be horizontal or vertical though the latter is more common. Francis turbines are used for medium heads of between 50 metres and 500 metres.

Kaplan turbines. In such turbines the runner is shaped rather like a ship's propeller. The propeller is driven by the water and has variable angle blades. This ability to vary blade angle with load results in high efficiency over a wide load range. A negative pressure is produced in the runner by the draught tube, the propellers acting as suction runners. Adjustable fixed nozzles in the gate ring give greater efficiency. As with Francis turbines the shaft may be horizontal or vertical but vertical shafts are more usual because the draught tube can be accommodated more readily. Kaplan units are used for low heads of up to 70 metres.

hydrocarbon

A chemical compound consisting of *carbon* and *hydrogen* only. Many thousands have been identified. The simplest hydrocarbon is methane, CH_4, the principal constituent of *natural gas.* Crude petroleum is rich in a variety of hydrocarbons, from dissolved gases with a low boiling point, through the *gasolines* and *kerosines*, to *bitumen* and *waxes* which are solids at room temperature. In *petroleum refining*, the *fractions* that are separated out each consist of a mixture of hydrocarbons boiling within a specific range.

Hydrocarbons are principally used as fuels, although a variety are employed in the chemical industry, and lubricating oils, bitumen, and waxes in particular have their specific uses.

See *Appendix 4, Table 2.*

hydrocarbon oils

Hydrocarbons which exist as oils. They may be fatty – of animal, plant or marine origin; mineral – derived from hydrocarbon deposits such as coal, petroleum and oil shales; or so-called essential oils which give certain plants their characteristic odours.

See also *petroleum refining.*

hydrocracking

The *catalytic cracking* of higher *petroleum fractions* in the presence of hydrogen to produce fractions with a lower boiling point. High temperatures (400°C) and high pressures (150 atmospheres) are required and this rather expensive process is principally used in the United States where *gasoline* demand is particularly high. The hydrogen requirement is very large, sometimes in excess of 10,000 cubic feet per tonne of oil, and consequently hydrocrackers usually have a hydrogen production plant situated next to them.

hydrodesulphurization

Also known as hydrofining or dehydrosulphurization. A process used to remove *sulphur* (mainly in the form of sulphur compounds) from *crude oil* fractions, especially in the 250°-350°C boiling range within which *middle distillates* such as *diesel fuel* and *burning oils* are obtained. The oil is passed with hydrogen, at high temperature (320°-420°C) and high pressure (25-70 atmospheres), over a *catalyst.* This consists of small pellets of cobalt and molybdenum oxides with an alumina base. The sulphur compounds are broken down and the sulphur and hydrogen combine to form hydrogen sulphide which, along with the heated oil, is removed from the reaction chamber and separated. The hydrogen sulphide itself is a valuable by-product, being readily converted to sulphur and sulphuric acid. Although this process is usually confined to distillates having a boiling point below 350°C, it can be adapted to treat heavier fractions which often contain more sulphur and sulphur compounds. Usually, however, these *fuel oils* are used in applications (eg power station boilers) that are tolerant towards contaminated fuels and in these cases the sulphur is released to the atmosphere as sulphur dioxide.

Removal of sulphur from oil fractions when it occurs as dissolved hydrogen sulphide is somewhat simpler and is usually achieved by treatment with caustic potash or caustic soda solutions. Contamination of petroleum fractions by mercaptans may be overcome in a similar way, often with the addition of an organic solvent.

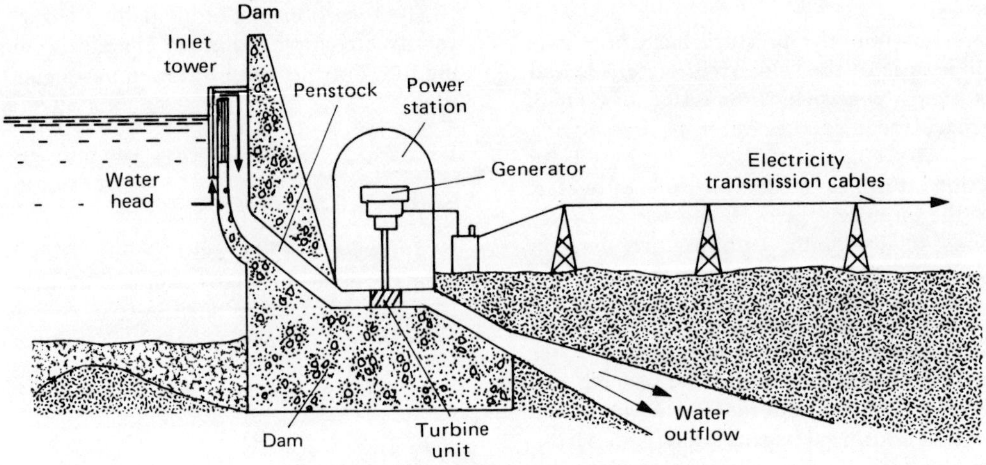

Schematic diagram of hydroelectric power station

hydroelectric power

Electricity generated by harnessing the energy in falling water. A non-depleting source of energy, hydroelectricity provides approximately two per cent of the world's primary energy needs. The majority of potential sites in the west are being exploited and there is little room for expansion to any large extent. The greatest scope for further hydro-electric installation exists in the underdeveloped countries (and Russia). However, excluding Russia, their lack of adequate markets and capital inhibits further development. If all the world's potential hydroelectric sources were utilized, approximately five to six per cent of the world's present energy needs could be met.

Nature and Generation of hydroelectric power. For many centuries man has utilized the energy inherent in running water. *Water power* provided the main source of mechanical energy until steam superseded it in the 19th century. The subsequent realization of the electrical route presented an ideal way of converting the mechanical energy associated with water.

The theoretical power of a hydro source is the product of the weight of water passing in unit time and the vertical height through which it falls, under average flow conditions. *Dams* serve to trap supplies and ensure continuity of flow while increasing the *head of water* available to drive the turbines. A small volume of water through a high head equals a large volume through a low head. The diagram shows a typical hydroelectric station in section.

Heating racks are often required in cold, mountainous regions to ensure that the water immediately upstream of the inlet does not freeze. The reservoir's water level can be regulated by allowing water to drain off through sluices in the dam, thus bypassing the power station. The water used for generating electricity passes through steel pipes or pen-stocks to the *turbines.* The water rotates the turbine blades, which in turn drive the generators to produce electricity as in a conventional power station. *Transmission lines* carry the electrical power to the centres of demand.

The development of turbines has vastly increased the potential for exploiting hydroelectricity. Turbines and generators both have efficiencies of over 90 per cent. Therefore a modern hydroelectric station can convert over 80 per cent of the energy inherent in water into electricity. In contrast fossil-fuelled power stations operate with large energy losses: they have efficiency rates of about 35 per cent.

The huge capital costs involved in building dams and power stations have generally limited the development of hydroelectric schemes to well-developed economies. Gener-ally the underdeveloped countries have neither the capital nor the demand to justify constructing and being able to operate hydroelectric schemes successfully.

Hydroelectric schemes require a continuous and regular water supply. Large artificial lakes created by damming provide that regular source of water energy. As electricity cannot be stored, damming also provides a means of storing the power.

There are three main types of hydroelectric station site. The first lies immediately downstream of a dam. Sometimes the power station is constructed within the dam wall itself. The second type, found at Niagara Falls, uses water led from a natural waterfall to a power station some distance downstream. The third type, which is common in very mountainous areas such as in Norway and Switzerland, uses water led through penstocks for several miles from lakes high up in the mountains. This arrangement makes possible the construction of power stations in more accessible places.

Advantages and Limitations. Environmental factors play an important role in the development of hydroelectric schemes. The reservoirs created by damming rivers can act as effective flood barriers and provide continuous water supplies for surrounding irrigation schemes (an obvious example is the

Aswan Dam, Egypt). Equally they can spoil the existing irrigation schemes and flood farming lands. Conservationist concern for the possible losses to wild life and areas of natural beauty by the formation of large reservoirs is a further restricting factor. (However, such reservoirs have become, in some cases, tourist and sporting attractions and the habitat for different species of wildlife.)

An unexpected disadvantage lies in the possible increased spread of disease as a result of the formation of large areas of mosquito, and similar insect, breeding grounds, (that is, large areas of stagnant water). Both the Aswan Dam and Kariba Dam (Rhodesia/Zambia) caused such a problem in their neighbouring areas.

High voltage transmission networks enable hydroelectric power to be transported over large distances to centres of demand. Thus, the demand does not have to be localized. Such networks spread the effects of the hydroelectric scheme over a wide area.

In many parts of the world aluminium smelting plants have been constructed quite close to hydroelectric power stations. In this way they can take full advantage of the cheaper electricity produced.

Silting causes problems, to differing degrees, in all reservoirs. While flushing removes some of the accumulation, the silt deposits decrease the head of water and will eventually fill the reservoir itself. The major reservoirs are estimated to have a life of between 100 and 200 years.

The demand for electricity fluctuates dramatically throughout the day and according to the season. Thus electricity producers must utilize their plants to the best possible effect in meeting fluctuating demands. One method of spreading the cost lies in *pumped storage schemes*. These provide a method of absorbing surplus electrical energy from the most efficient power stations when demand is low, and releasing the energy later at peak-demand times.

See *Map 18*.

hydrofining

Another name for *hydrodesulphurization*.

hydrogen

The lightest element known, hydrogen is a colourless gas at ambient temperatures and pressures. The hydrogen atom is the simplest of all atoms, containing only a proton and an electron. Although abundant throughout the universe, free hydrogen gas is found only in minute quantities on earth. Most of the earth's hydrogen content is found in chemically combined forms with oxygen (as water), or carbon (as various fossil fuels).

Because free hydrogen does not exist as a resource on earth it cannot be a primary fuel. It can only be produced either by the splitting of water, which requires at least as much energy as is obtained on burning hydrogen in air, or by the use of sources such as natural gas and petroleum that are themselves primary fuels. Hydrogen can only be regarded as a secondary fuel or energy carrier.

Sources of hydrogen. Hydrogen can be produced from any of the fossil fuels, but the major source at present is natural gas. *Steam reforming* of natural gas produces a 'synthesis gas' containing a high proportion of hydrogen. Light distillate feedstock ('naphtha') from petroleum refining has also been used as a hydrogen source, particularly in the UK. Partial oxidation of hydrocarbons can also be used to produce hydrogen.

At present hydrogen is deliberately produced only for chemical purposes, such as ammonia manufacture, and for welding uses. It is also a by-product of many petroleum refining operations, such as *catalytic cracking*. There is no large scale production of hydrogen for use as a fuel at present, and most by-product hydrogen is consumed on site.

Hydrogen can also be produced from coal by means of the 'water gas' reaction. Steam is passed over hot coke, alternating with air, which burns a portion of the coke to maintain temperature. The mixture of carbon monoxide and hydrogen produced is passed with steam over an iron catalyst producing a carbon dioxide and hydrogen mixture. This process, though formerly the major route to hydrogen, is now obsolete but may be revived in the future in a modified form for hydrogen production from coal. Some proposed *coal gasification* processes produce significant quantities of hydrogen.

As already pointed out the use of hydrogen as a fuel in the future relies on the development of efficient methods of production from water, using some primary energy source, directly or indirectly, to provide the necessary decomposition energy. The two methods most favoured are electrolysis and thermochemical cycles. The electrolysis of water involves the use of an electric current to split water into its constituent hydrogen and oxygen, which are released from the cell as gases. This is an established technology, with industrial units producing from 300 to 15,000 cubic metres (10,000 to 500,000 cubic feet) daily and achieving efficiencies of 60-70 per cent. There is considerable scope for improvement in the process efficiency, particularly since thermodynamics shows that an electrolyser can in theory achieve an efficiency of 120 per cent, by the absorption of ambient heat.

Electrolysis has been suggested for use in electricity systems as a means of storage, hydrogen being produced at times of low demand. Such storage would be particularly required in those networks containing substantial nuclear generating capacity and in those where supplies from unpredictable or uncontrollable sources, such as wind and tidal, form a significant proportion of the total. Hydrogen production by electrolysis has also been suggested as a means of getting to shore the energy produced by wave machines or ocean thermal energy conversion plant.

Thermochemical cycles for hydrogen require the use of a sequence of chemical reactions whose net effect is the splitting of water. (Hence this method is also known as 'thermochemical splitting'.) Although leading to decomposition at a temperature lower than required if direct heating alone is used, such processes nevertheless require

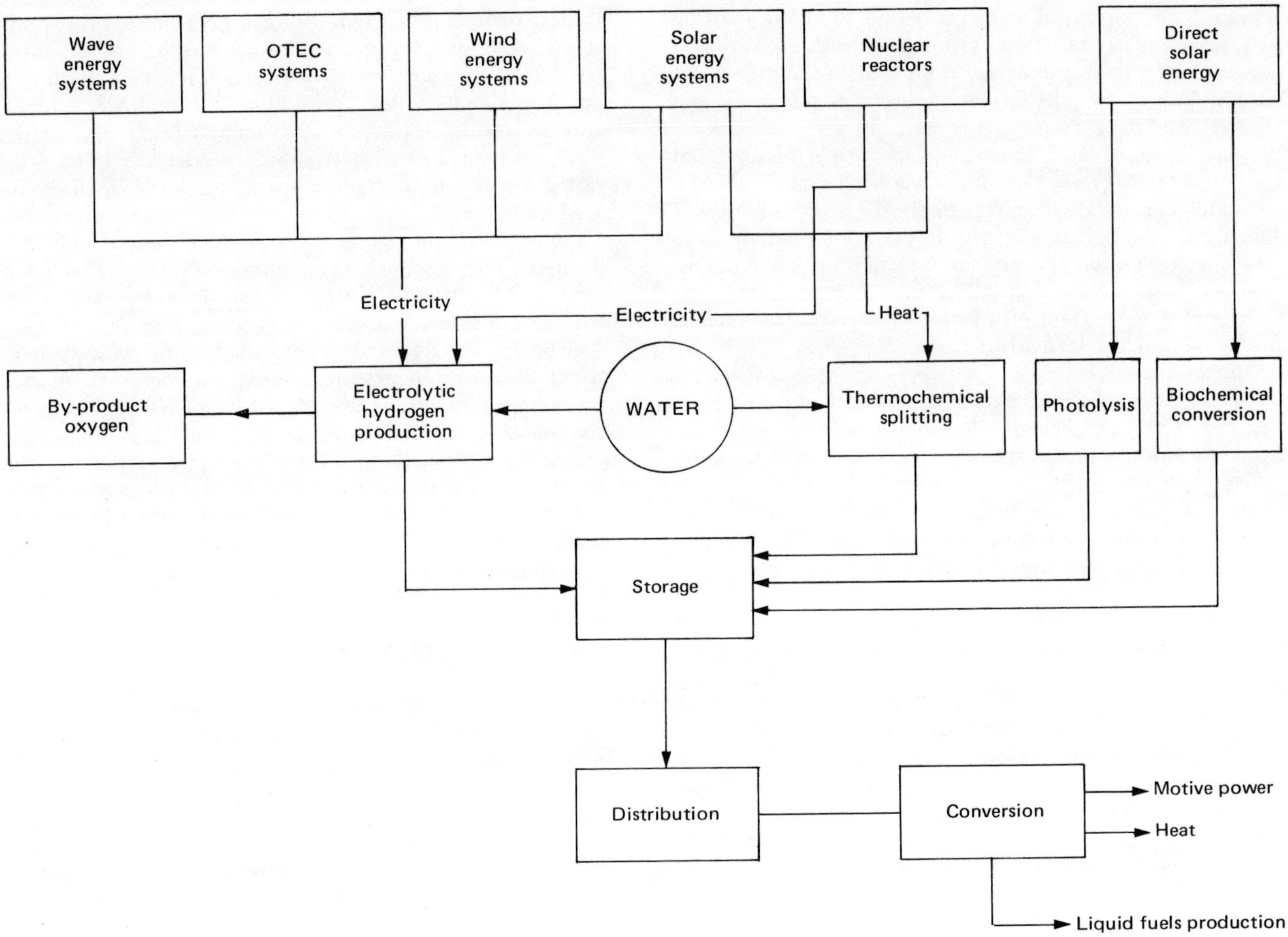

Elements of the hydrogen economy

temperatures high enough to demand the use of special constructional materials for the plant. This method has been suggested as a means of directly using nuclear heat without the need for a steam cycle. Since present nuclear reactors operate at relatively low temperatures, one of several designs of high temperature reactor would be required.

Another possible method of liberating hydrogen from water is by using the photosynthetic capacity of certain plants and *algae* (biophotolysis).

Direct fuel uses of hydrogen. Hydrogen burns in oxygen, or air, to form water:

$$2H_2 + O_2 \rightarrow 2H_2O$$

and this reaction is highly exothermic, releasing 121 MJ per kg of hydrogen. This is the basis for the use of the gas as a fuel. Unfortunately, the low density of hydrogen means that the volume required to contain a given amount of energy is greater than for any of the liquid fuels: one kilogramme of hydrogen gas occupies a volume of about 12 cubic metres (12,000 litres) at atmospheric pressure and temperature. Although this volume can be reduced by cooling and pressurization, or a combination of both, hydrogen does not liquify until cooled to a very low temperature. Hence storage of hydrogen fuel is bulky and expensive.

The main advantage of hydrogen as a fuel is the fact that the only combustion product is water, so that air pollution is confined to the nitrogen oxides that are invariably produced when any fuel is burnt at high temperature in air. In particular, carbon monoxide, produced when any hydrocarbon fuel is incompletely burnt, is absent. Hydrogen is also a potentially attractive fuel for transport purposes. High efficiencies have been obtained using modified internal combustion engines. The problem of storage is acute, however, especially for passenger transport. For unpressurized storage, the temperature of the tank must be kept below -253°C, the boiling point of hydrogen at atmospheric pressure. Losses due to evaporation must not allow an accumulation of the gas that would lead to explosion. Similar problems arise in the storage tanks required for distribution.

For direct heating purposes, hydrogen can substitute for

natural gas, although the hydrogen flame is hotter and would require changes in burner design. Because of the low activation energy required to ignite hydrogen it is possible to use catalytic burners at a low temperature (perhaps as low as 100°C), thus eliminating nitrogen oxides emission.

Indirect fuel uses. In the hydrogen-oxygen *fuel cell* hydrogen is fed to the anode and oxygen to the cathode. As a result of ionization processes in the cell there is transport of ions between the electrodes within the cell and a flow of electrons in the external circuit, producing a usable electric current. Such cells have been used in spacecraft, and the high efficiency (60 per cent) makes the fuel cell an attractive prospect as an energy converter in the hydrogen economy.

Storage and transportation of hydrogen.

GASEOUS HYDROGEN. In most currently-conceived practical systems for the *hydrogen economy*, hydrogen would be produced and used in gaseous form. However, because of its low energy storage density, it is uneconomic to store gaseous hydrogen in large quantities. The high pressure vessels required to store large quantities are at present extremely expensive. However, natural caverns, depleted oilwells, aquifers and the like might well provide adequate storage capacity in the future. It is envisaged that much of the hydrogen will be stored in the distribution system itself – in the pipelines that will deliver the fuel. This is called line packing.

It is possible that existing pipeline systems used for delivering town gas and natural gas may be adapted for the delivery of hydrogen, but the low density of hydrogen means that a large handling and pumping capacity has to be employed. Other technical problems would have to be solved, including that of 'hydrogen embrittlement', caused by reaction between the hydrogen and the metallic material in the pipes.

LIQUID HYDROGEN. Hydrogen, at low temperatures and under high pressure, becomes liquid. Liquid hydrogen could become a convenient energy carrier, capable of storage and transportation, and, indeed, has been used extensively in the space programme as a rocket fuel.

There are, however, considerable technical and economic problems to be solved before liquid hydrogen could be used on a large scale. A substantial energy input has to be invested to convert to the liquid form and the capital investment in liquifaction plant, storage and transport is very large indeed.

SOLID HYDRIDES. When hydrogen, under pressure, is passed into a container containing certain metals or alloys, the hydrogen combines chemically with the metallic material to form a hydride. Heat is evolved in the process, and has to be continuously removed to allow the process to continue satisfactorily. When the hydride formed is heated, the process is reversed and hydrogen is evolved. Typical metallic hydrides considered for storage purposes are iron titanium hydride ($FeTiH_{1.5}$) and magnesium nickel hydride ($MgNiH_4$).

Non-fuel uses of hydrogen. The high thermal conductivity of hydrogen (almost seven times that of air) combined with its low density (which leads to low frictional losses) make it a good choice for coolant in rotating electrical machinery such as generators.

Hydrogen is used in 'atomic hydrogen arc welding' in which the energetic reaction of hydrogen atoms to form the diatomic molecule is used to heat the metal.

hydrogen	
symbol	H
atomic number	1
atomic weight	1.008
melting point	14.0 K (-259.2°C)
boiling point (at 1 atm)	20.4 K (-252.8°C)
latent heat of vaporization (at 1 atm)	0.447 MJ kg^{-1}
density –	
solid (at 4.2 K)	89.0 kg m^{-3} (0.089 g/cm^3)
liquid (at 20.4 K)	71.0 kg m^{-3} (0.071 g/cm^3)
gas (at 0°C, 1 atm)	0.089 kg m^{-3} (0.089 g/l)
heat released on combustion	121 MJ kg^{-1}
flame temperature	2200°C
explosive limits (air)	4–74%
calorific value	275 Btu/ft^3 (10.8 MJ/m^3)

hydrogen bomb

A bomb whose power is derived from the process of nuclear *fusion*. If a fission bomb device or *atomic bomb* is surrounded by deuterium or tritium the vast amount of energy released by the fission blast will trigger off the fusing of the deuterium or tritium nuclei to give helium along with the release of large amounts of energy. Such an arrangement of materials, or a more sophisticated version, operating on the same principle, is known as the hydrogen bomb and the amount of power such a device could release seems limitless.

hydrogenation

The addition of hydrogen to a chemical compound. Hydrogenation plays an important part in a number of *petroleum refining* processes notably *hydrocracking* and *hydrodesulphurization*. It is also of significance in a number of *coal gasification* and *coal liquifaction* processes, as these are aimed at increasing the hydrogen/carbon ratio from that occurring naturally in coal deposits.

hydrogen economy

A hypothetical global energy supply system based on the use of *hydrogen* as an *energy carrier* and storage medium.

Proponents of the hydrogen economy suggest that the advantages of hydrogen as a clean, storable fuel will outweigh the disadvantages of low energy density and high conversion losses.

In principle, hydrogen could be generated from its major source, water, by various means: electrolysis, thermochemical splitting, and the use of algae. (These methods are discussed more fully under 'hydrogen'.)

hydrological cycle

The continuous circulation of water in liquid, solid and vapour form over the earth's surface in the atmosphere, the hydrosphere (ie water covering the earth's surface) and the lithosphere (the rocky crust of the earth). It is also called the water cycle. Water circulates between these spheres principally through gravity and solar energy. Small amounts of water are released from the ground through geothermal activity.

Solar heating causes evaporation of water from the oceans and the land surface into the atmosphere. This forms clouds, water from which is returned to the surface of the earth in the form of rain, hail, sleet and snow, often many hundreds of miles away from the area in which it entered the atmosphere. Much of the precipitation runs off the land surface directly into the oceans or into streams and lakes and eventually back into the oceans. Some flows underground and may rejoin streams and return to the sea. Some is evaporated from the land surface; some is taken up by plants and released to the atmosphere during transpiration.

The hydrological cycle links the net loss of heat from the atmosphere with the net gain of heat by the earth's surface. Water has a high specific heat. It gains and loses heat slowly so that it has an important temperature regulating rôle in soil and in the sea. Vast quantities of solar energy fall on the sea raising its temperature. Ocean currents carry this heated water from one part of the earth's surface to another having a profound effect on climate. Cold currents can have an equal significance. The Gulf Stream, for instance, carries water from the Gulf of Mexico across the Atlantic to western Europe warming the coastal waters and the air flow over regions such as the United Kingdom. This area is unusually mild for its latitude. The north-eastern coast of the United States and neighbouring areas of Canada at a similar latitude are much colder due to the cold southwards flow of the Labrador current from polar regions. Such water movements help to determine the energy consumption patterns of these regions.

On land the movement of water is important as a source of *hydroelectric power*.

See also *geothermal energy; ocean thermal gradients; tidal energy*.

hygas process

A *coal gasification* process in which the coal is directly hydrogenated in a two-stage process. Crushed coal is mixed into a slurry with anthracene oil and fed into the top of a fluidized-bed gasifier where the oil is vaporized. The dry coal, under 70 atmospheres pressure, is then subjected to treatment from a hydrogen-rich hot gas stream forced upwards from the lower section of the gasifier. The coal is heated by the gaseous mixture to over 650°C when it reacts readily with the hydrogen to form methane. Partially reacted, or unreacted, char falls downward into the second stage of the gasifier where its temperature is raised to 930°C. Further methane formation occurs, along with carbon monoxide and hydrogen produced by the reaction of the char with steam. Remaining char may be used in a steam-oxygen gasifier to provide hydrogen for the hydrogasification process.

The raw gas emerging from the gasifier passes through a scrubber to remove impurities after which *shift conversion* and *methanation* upgrade it to a gas of pipeline quality.

To date, the hygas process has been used to gasify lignite and the pilot plant in operation has a capacity to process 75 tons of feed per day.

I

IAE

Initial letters for the Institute of Automotive Engineers.

IBP

Initial boiling point. It is the temperature at which distillate first appears during a distillation test carried out in the laboratory.

ICRP

Initial letters for International Commission for Radiological Protection.

IFIAS conventions

A set of well-defined guidelines for *energy analysis* agreed at the 1974 Stockholm conference of the 'International Federation of Institutes of Advanced Study'.

impulse turbine

A prime mover in which water, steam or a hot gas under pressure is directed against the blades of a rotor causing it to rotate. The *Pelton wheel* is an example and the principle is employed in many types of turbine.

See also *hydraulic turbine; steam turbine.*

incineration of wastes

The disposal of refuse by burning in an incinerator. The incineration plant is sometimes linked to a *district heating* system and it has been argued that this should be common practice. In the United States annual solid waste production has an energy content approximately equal to 12 per cent of the total energy requirements. The quality and content of waste obviously varies with its source, be it municipal, industrial or agricultural in origin. Not all is suitable for incineration and other processes are used such as *pyrolysis, hydrogenation* and *anaerobic fermentation.*

Waterwall incinerators are used to generate steam by burning untreated solid waste materials and using the hot flue gases to raise steam in a boiler. Particulate matter is removed from the hot gases by electrostatic precipitation after they have passed through the boiler. *Fluidized-bed processes* are also used to produce hot gases for expansion through a gas turbine. The waste is first shredded and non-combustibles removed. It may then be mixed with liquid wastes and combusted. After particle removal, the hot gases are expanded through the gas turbines which are linked to a generator. Some of the hot gas is compressed and used to help the combustion process, and waste heat boilers utilize the turbine exhaust to raise steam for additional electricity generation.

See also *waste materials energy.*

induction heating

Heating an electrically conducting material with currents induced by a varying electromagnetic field. The material to be heated is effectively a secondary coil within the inductor which is the primary coil. High coil currents are used to induce maximum eddy currents in the material to be heated so high rates of heating are obtained. Induction heating finds wide application in the metallurgical industries and has several advantages over conventional heating methods. Heating is rapid because the material itself is heated directly; it can be localized by adjusting the size of the inductor coil; it can be set up quickly so reducing start up times and down time and it lends itself to automation; it is controlled easily giving consistently high quality products and is

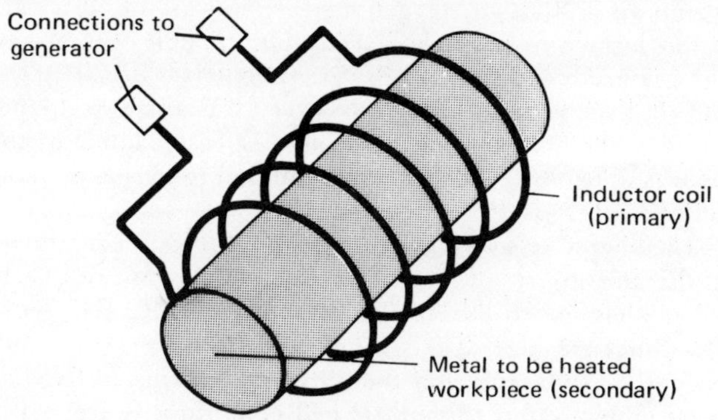

Induction heating

clean, fume free and cooler for the operatives as there is little radiated heat. It can also be energy conserving and many companies are involved in installing and developing new equipment.

industrial spirit

A *petroleum fraction* occurring in the higher part of gasoline range (100°-200°C) generally used as a solvent in industrial applications.

inertial containment

The rapid heating of reacting components in a *fusion reactor*, so that the reaction can occur before dispersal.

infra-red radiation

A form of electromagnetic radiation whose wavelength range lies just outside the visible spectrum. The wavelength range is generally reckoned to lie between 8×10^{-7} and 10^{-3} metres. Infrared radiation is a large component of radiant heat emitted by domestic and industrial heaters, and forms a considerable proportion of solar radiation.

See also *sun*.

inhibitor

A substance added in small amounts to a petroleum derivative to inhibit or prevent changes (usually oxidation) taking place which would otherwise reduce the quality of the derivative. For example, there is a tendency for gums to form in stored gasolines; inhibitors reduce this.

in-situ coal gasification

A technique for recovering gas from coal by gasifying the coal in place. Many countries have experimented with a variety of techniques but none has been found economically viable. A technique developed in the USSR has been to drill two spaced boreholes into the coal formation. The natural permeability of the coal strata, or *hydraulic fracturing*, was used to achieve communication between the two boreholes. Steam, air and an air/oxygen mixture was used to drive gas to the surface to produce several million cubic feet of gas. The gas produced was low-Btu gas (about 100 Btu/ft^3), the rate of flow varied, resource recovery was a low as 14 per cent and the heating value gradually decreased. Much of the gas was apparently lost underground and much of the coal simply burnt.

The energy crisis has led to renewed interest being shown in the possibility of in-situ hydrocarbon recovery. Technology has improved enormously since the Russian work was done; the idea of recovering coal from thin seams not currently exploitable and the prospect of men not needing to go underground to work in bad conditions is attractive. One current idea emanating from the USA suggests boring a number of holes into the coal seams. Chemical explosives placed there would be detonated to fracture the coal-bearing strata (and shale too). Oxygen would be passed down the holes and the coal ignited. The combustion area having been established at the top of the producing area, the coal would then be gasified by injection of steam and oxygen as in high-Btu gasification processes. The gases formed – largely methane, carbon monoxide, hydrogen and carbon dioxide – would be purified in a special treatment plant at the surface, so producing a gas of pipeline quality. Providing the fractured coal bed is sufficiently permeable to allow proper contact between the coal and the injected materials and sufficiently well contained by surrounding impermeable strata for complete gasification, the process should be effective in recovering hydrocarbon and other wanted materials.

The proposals overall have considerable promise, although further R and D is necessary to establish if full-scale plant would work – whether the desired fracturing can be achieved so that continuing combustion and gasification may be achieved with the right conditions of permeability and whether product recovery is adequate – and to find sites with the necessary features. Research is also needed to determine the best explosives; chemical ones are thought to be most suitable.

insolation

Radiant energy from the sun.

See also *solar energy*.

installed capacity

Also known as maximum continuous rating. The total maximum continuous rated power output of all the generating units in an electricity supply system. Required installed capacity is calculated on the basis of the anticipated maximum load, or *simultaneous maximum demand*, which the generating system is expected to meet. Installed capacity in excess of anticipated requirements is usually built in to allow for breakdown and malfunction of equipment at times of peak demand, and 20 per cent overcapacity is a common planning margin.

Installed capacity is usually measured in multiples of *watts* such as megawatts (MW) or gigawatts (GW).

insulating oil

Low volatility and viscosity petroleum distillates used in electrical equipment (eg transformers) for cooling and insulating purposes.

insulation (thermal)

Material used to reduce heat losses whether due to conduction, radiation or convection. The reduction of conductive losses can be achieved by the use of a highly cellular material in which pockets of trapped air, a poor conductor, act as the insulator. The material itself acts so as to maintain these pockets, preventing convective losses by imposing a barrier to circulation. Materials of this type include foamed plastics, such as polystyrene, and glass wool. Radiative losses can be reduced by placing a poor heat emitter, such as reflective aluminium foil, on the external surface.

The use of thermal insulation to reduce building heat losses and cooling requirements is an important part of any national or factory-scale *energy conservation* programme. Heating and cooling of buildings represents a large proportion of the energy consumption of industrialized nations.

See also *U-value*.

integrated energy supply

See *total energy*.

integrated environmental design

Design of buildings so that heat from lighting, occupants etc is sufficient to keep the temperature of the environment within the building at a satisfactory level irrespective of the outside temperature. Energy normally expended for heating purposes is thus saved.

internal combustion engine

A form of *heat engine* in which heat is transferred to a working medium, such as air, by the direct combustion of a fuel within the medium. The vast majority of internal combustion engines use air together with a hydrocarbon fuel, although in principle any power-producing gas-phase reaction could be used. This form of engine is fundamentally different from the external combustion engine in which combustion takes place outside the engine and is transferred to the working medium by conduction, or by gas flow.

Internal combustion engines are classified as rotary or reciprocating, depending on the nature of the motion. Rotary types include the open cycle *gas turbines*, used in the form of 'turbojet' and 'turboprop' engines for aircraft propulsion, and the Wankel engine for vehicles. Reciprocating types include the *diesel engine* and the *spark-ignition engine*.

The phrase 'internal combustion engine', often abbreviated to ICE, is sometimes taken to refer exclusively to *reciprocating internal combustion engines* for automotive use.

interseasonal storage

Energy storage media designed to cope with changes in demand occurring between seasons. Examples of large scale interseasonal stores are the salt cavities and LNG installations used in *natural gas storage*. Research is proceeding to incorporate interseasonal storage into *solar energy* technology.

See also *heat storage*.

inversion

The process by which *direct current* (DC) electricity is converted to *alternating current* (AC) electricity. Inversion is used in *electricity transmission* networks to provide AC power at the end of a DC link, such as might be used for underwater power lines.

ion

An electrically charged atom or group of atoms. Ions can be either positively or negatively charged, depending on whether they have been created as a result of the loss or gain of an electron from the atom or group of atoms, and are to be found in ionized gases, liquids (*electrolytes*) or in solids. Under the influence of an electric field ions can travel towards an electrode of opposite charge.

ionizing radiation

Electromagnetic radiation capable of giving atoms electrical charge, ie converting atoms into *ions*. Such radiation is a feature of *radioactive decay* and examples include *alpha*, *beta* and *gamma radiation*.

IP

Initial letters for the Institute of Petroleum.

irradiated fuel

Nuclear fuel that has been subjected to neutron bombardment and which therefore contains a mixture of *fission products*.

See also *nuclear fuel cycle*.

isobutane

A *branched chain hydrocarbon* used in *alkylation* processes.

isomerization

A process in which a *straight chain hydrocarbon* is converted into its *branched chain* analogue. This is of particular significance in petroleum refining where *alkylation* processes depend on branched chain hydrocarbons as a starting material. Straight chain alkanes such as butane may be converted to isobutane by mixing with a hydrogen chloride 'promoter' and passing the mixture over an aluminium chloride catalyst and separating out the resultant products by fractionation.

The octane number of lower boiling point gasolines (30°-70°C) may be directly increased by an isomerization process in which the gasoline feedstock is passed over a platinum catalyst at 150°C.

iso-octane

See *octane number.*

isotope

Atoms of the same element differing only in the number of *neutrons* present in the atomic nucleus. Most elements exist in nature as a mixture of a number of naturally occurring isotopes, a few of which are radioactive; in addition, a wider range of radioactive isotopes may be artificially produced.

See also *isotope separation.*

isotope separation

The separation of different isotopes of an element carried out during enrichment processes. The term is usually taken to describe the uranium enrichment process in which the proportion of fissile uranium-235 to fertile uranium-238 in a given sample of uranium is increased. Because isotopes of an element only differ from each other in terms of their masses, separative procedures must be based on physical processes that can take advantage of this small mass difference.

See also *centrifruge process; diffusion process.*

J

JET

Initials and popular name for Joint European Torus, an important *fusion reactor* project to be sited at Culham in the UK.

jet a

US term for aviation turbine kerosine, also known as Jet A-1 and JP-5.

See also *aviation turbine fuel.*

jet fuel

See *aviation turbine fuel.*

joule

An internationally recognized standard unit of energy (*SI*), defined as the *work* done by a force of one newton when it moves its point of application over a distance of one metre. A system doing work at the rate of one joule per second has a *power* of one watt; so that a watt-second equals one joule and 3600 joules is equivalent to one *watt-hour.*

Being of small magnitude, the joule is usually used as a multiple such as the megajoule (MJ,10^6 joules), gigajoule (GJ,10^9 joules) or terajoule (TJ,10^{12} joules) and, for comparison purposes, *calorific values* of fuels are often quoted in gigajoules per tonne (GJ/te).

See *Appendix 2; Tables 2 and 3.*

jp-4

US term for aviation turbine gasoline.

See also *aviation turbine fuel.*

Jurassic period

The period in the geological time scale which followed the Triassic and preceded the Cretaceous. Its strata are rich in fossils and yield large amounts of *coal* and *petroleum.*

K

Kaplan turbine

A *hydraulic turbine* of the reaction type in which a propeller-type runner is operated with adjustable blade angle to achieve maximum efficiency under varying flow rates. The blades are coupled hydraulically to the inlet 'wicket' gates. Kaplan turbines are used for heads of up to 70 metres.

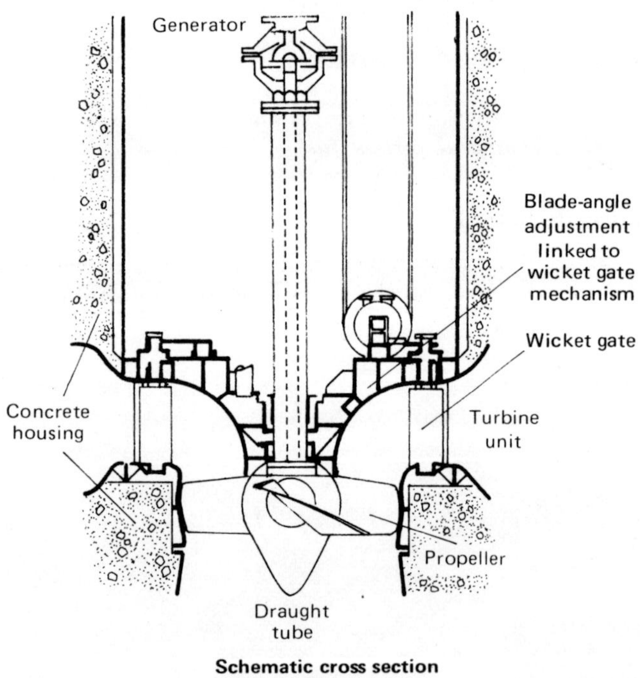

Schematic cross section of the Kaplan turbine

Kellogg process

A *coal gasification* process in which a mixture of coal, oxygen and steam are reacted together under pressure in a vessel containing molten sodium carbonate at 1000°C. The sodium carbonate acts as a catalyst in the gasification process and the *synthesis gas* produced is free of tar and oils. The process is still very much at the experimental stage.

Kelvin scale

Commonly used term for the *absolute scale of temperature*. It is arithmetically related to the centigrade scale by the relationship:

$$\text{Degrees Kelvin (K)} = \text{Degrees centigrade (°C)} + 273.15$$

kerogen

An organic hydrocarbon material with high molecular weight, found as a constituent of *oil shales*. In order to break it down, kerogen requires heating in large distillation vessels to 400°C, where it decomposes into a range of liquid and gaseous hydrocarbons which may then be treated as crude petroleum products. The kerogen content of oil shales varies greatly and shales containing less than one barrel of oil per tonne of shale material are unlikely to be exploited.

kerosine

Also known as paraffin in the UK. A *petroleum fraction* overlapping the light and middle distillate range, boiling between 150° and 300°C. It was originally known as lamp oil because of its use as an illuminant, but this is very much a minor use nowadays, when kerosine finds its principal applications in domestic heating and fuelling jet aircraft. So-called 'power kerosine', or vaporizing oil, finds a significant but diminishing use as a fuel for certain spark-ignition engines. At the lightest end of the kerosine range, and just overlapping with the heavy gasolines falls *naphtha*, an important petrochemical feedstock.

kilowatt

A multiple of the unit of power, the *watt*, and equal to 1000 watts.

See also *electric power*.

kilowatt-hour

A multiple of the unit of work, the *watt-hour*, and equal to

1000 watt-hours. The kilowatt-hour (kWh) is commonly used when quantifying amounts of electrical energy and is the amount of energy used by a one kilowatt electrical appliance operating for one hour.

See also *electric power.*

kinetic energy

The energy of a body due to its motion. The kinetic energy of a body moving in a straight line is given by $E=\frac{1}{2}mv^2$, where m is its mass and v its velocity. The kinetic energy of a rotating body is given by $E=\frac{1}{2}I\omega^2$ where I is its moment of inertia about the centre of rotation, and ω its angular velocity.

knock

The noise due to the self-ignition of part of the fuel-air mixture in the cylinder of a spark-ignition engine.

See also *anti-knock components; octane number.*

Koppers-Totzek process

A *coal gasification* process in which gasification is achieved by treating pulverized coal with a mixture of oxygen and steam at atmospheric pressure. The process was originally developed before the Second World War, and has been used commercially in the production of methanol and ammonia, as the raw gas coming from the gasifier is virtually methane-free. It can be adapted, however, to produce pipeline quality gas by subjecting the *synthesis gas* to shift conversion and *methanation.*

L

lake asphalt

Asphaltic deposits that occur as surface deposits. Examples are Trinidad's pitch lake and the tar pools in California.

See also *bitumen*.

lamp oil

Original name for *kerosine*.

laser enrichment process

An *isotope separation* process that takes advantage of the difference in vibrational frequency between two isotopes. In this process, uranium-235 may be separated from the uranium-238 occurring in natural uranium in one step, as opposed to the multi-stage *centrifuge* and *diffusion* processes. It has yet to be developed commercially, but is potentially highly significant.

latent heat

The energy, in the form of heat, required to cause a change of state of a substance. The 'latent heat of fusion' is the amount of heat energy required to melt unit mass of a pure substance. Similarly, the 'latent heat of vaporization' measures the heat required to change unit mass of pure substance from the liquid to the vapour state. The use of the word latent is to underline the fact that changes of state, such as melting and boiling, are constant temperature processes for chemically pure substances and latent heats must always be measured at constant temperature. For example, the latent heat of vaporization of water is defined as the amount of heat energy required per unit mass of water from liquid at 100°C to steam at 100°C, the measurement being conducted at standard atmospheric pressure.

Latent heats can also be defined and measured for changes of state in the solid phase, as between different crystalline forms of an element, for example.

Typically latent heats are measured in such units as $MJ\ kg^{-1}$, Btu/lb or cal/g.

latent heat storage

A method of heat storage based on the use of the latent heats of fusion or vaporization of selected compounds.

Pure compounds melt and boil at well defined temperatures, and this characteristic can be employed in latent heat storage, also termed 'phase change media' (PCM) storage, for temperature stabilization.

PCM storage media offer high storage densities but have two major drawbacks. Since the medium must solidify progressively as it gives out heat, a layer of poorly-conducting solid phase tends to form around the heat extraction coils leading to poor system performance. Secondly, thermal contraction of the solid phase on cooling tends to form insulating air pockets around the output coils.

Several approaches are under development to solve these problems, including the use of mechanical agitation and encapsulation of the material.

Substances under investigation for use in latent heat storage systems include sodium thiosulphate, sodium carbonate, ferric chloride and various waxes, all of which have fairly low melting points (between 35° and 50°C) suitable for applications in connection with solar heating. A major consideration in such systems is the cost of the storage material.

Higher temperature PCM systems are under development for industrial use.

See also *heat storage; solar energy*.

LDF

Abbreviation for *light distillate feedstock*.

See also *naphtha*.

lead

A soft, malleable and very dense bluish-white metal highly resistant to corrosion. It occurs chiefly as the carbonate, sulphate, or oxide from which the metal is isolated by reduction with carbon at high temperature. Lead has a number of energy-relevant applications and indeed the principal tonnage use for lead is in the manufacture of *lead-acid accumulators.* Large quantities of lead also go to make

symbol	Pb
atomic number	82
atomic weight	207.19
density	11350 kg m^{-3} (11.35 g/cm^3)
melting point	327.5°C
boiling point	1740°C

tetramethyl and *tetraethyl lead*, both anti-knock compounds added to motor spirit. Three of the four naturally occurring isotopes of lead are the end decay products of the natural radioactive decay chains that begin with uranium, actinium, and thorium respectively.

lead-acid accumulator

A form of secondary *electric cell* or *storage battery* in which electrical energy is stored. It is used in the vast majority of vehicles as a starter battery.

The negative plates are characteristically of cellular structure carrying spongy lead and the positive plates are a thin lead dioxide film on lead plate. The sets of electrodes are separated by plastic, rubber or other insulators and the electrolyte employed is sulphuric acid. The specific gravity of the acid is indicative of the state of charge of the battery, whose emf per cell can vary between wide limits, depending on its state of discharge, acid concentration and temperature. The emf is usually about two volts.

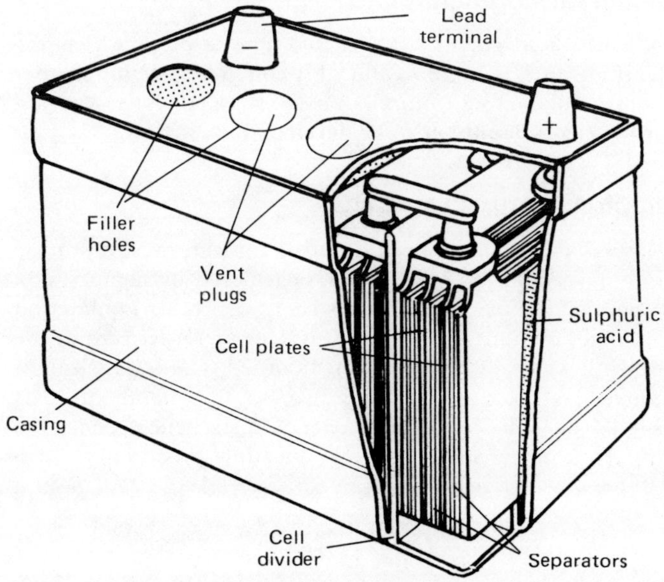

Cutaway view of a lead-acid accumulator

Despite the weight disadvantage, the lead-acid battery has remained in widespread use over many years. It is cheap to manufacture and is capable of producing a high current. If, with technical advances, a higher *energy density* version can be produced, there may be greater scope for use in battery-powered vehicles.

lean gas

1. The residual gas consisting of *methane, ethane* and some *propane* derived from *associated gas* after the *butane* and *pentane-plus fraction* have been removed. The gas may then either be transported to its point of use by pipeline, or liquified and taken by *LNG carrier.*
2. A term sometimes used to describe gas of a low calorific value, usually less than 500 Btu/ft^3

Leclanché cell

Also known as the carbon-zinc cell. A primary *electric cell* consisting of zinc and manganese dioxide electrodes immersed in an electrolyte of aqueous ammonium chloride. In order to ensure good electrical conductance, the manganese dioxide is mixed in with powdered carbon and packed round a carbon rod which acts as the electrode terminal. In its practical form, often referred to as the dry Leclanché cell, or *dry cell*, the ammonium chloride is mixed with glycerine in the form of a paste and the assembly packed in a zinc container which acts as the negative electrode. Because it is unable to provide a constant voltage for long periods, it is best suited to intermittent uses such as torches and door bells.

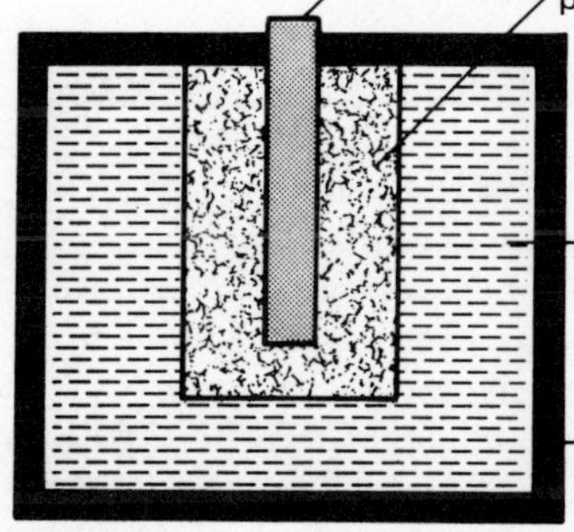

Dry Leclanché cell

light distillate

A collective term used to describe *petroleum fractions* boiling in the range 30°-250°C. It comprises the *gasolines* and most of the kerosine range and therefore the separation, blending and up-grading of light distillate is a principal feature of refinery operations. Products falling within this boiling point range include *aviation spirit, aviation turbine fuel, burning oil, motor spirit* and *naphtha.*

light distillate feedstock

See *naphtha.*

light water

Water containing the natural *abundance ratio* of hydrogen and deuterium, as opposed to *heavy water* which is pure *deuterium oxide*. In the light water reactors *PWR* and *BWR*, it is used as both *coolant* and *moderator*, but it is the least effective moderator in use because of its high rate of neutron absorption. In the *SGHWR*, light water is used as coolant but heavy water as moderator.

light water reactors

Nuclear reactors in which light water is used as the heat transfer medium. There are two main types – the *pressurized water reactor (PWR)* and the *boiling water reactor (BWR)*. The majority of nuclear plants in the United States are of these types. In BWRs, with lower pressures of 67 atmospheres, comparable steam temperatures can be obtained compared with pressures of 135-150 atmospheres in PWRs. The high pressure required in the latter has raised doubts about their safety. Light water reactors were developed originally as compact power sources for submarines; nuclear reactors consume no oxygen, unlike fossil-fuel boilers, so a nuclear submarine can remain submerged for much longer periods. In addition to possible safety considerations, light water reactors have a major deficiency in the low coolant temperatures and consequent inefficiency of electricity production from low quality heat. Because of their high power density there is also concern about the possible breakdown of emergency cooling systems should the fuel temperature in the core rise suddenly.

lignite

Also known as brown coal. A coal which has not reached an advanced state of *coalification*, falling between *peat* and *bituminous coals*. Although the degree of coalification of lignite and related brown coals varies from one deposit to another, the calorific value is usually taken to lie in the range 120-140 therms per tonne (12.7-14.7 GJ/tonne). There are large reserves of lignite, especially in Russia and North America, and present world production is running at approximately 850 million tons a year (250 mtoe). The principal use for lignite is for steam-raising in power station boilers.

liquid metal-cooled fast breeder reactor (LMFBR)

A *nuclear reactor* of the *fast breeder* type which uses liquid metal coolant. This is usually molten sodium or an alloy of sodium and potassium. Present prototypes use enriched uranium-235 or plutonium-239 as fuel and this is surrounded by a neutron-absorbing blanket of uranium-238. Because the molten coolant becomes radioactive and long-lived sodium-24 is formed in small quantities, it must be confined totally within the reactor core biological shielding. A second sodium circuit therefore has to act as the heat transfer medium carrying heat to the steam generating unit. Since sodium reacts so violently with water, the generating units have to be fabricated to the very highest standards.

The *doubling time* of LMFBRs operating in prototype form would seem to be over twenty years. It seems likely that a new fuel will have to be developed based on uranium or plutonium carbides. This should reduce the doubling time to fifteen years or so. Many other technological problems have to be solved however before LMFBRs become a commercial reality.

liquid metal-cooled fast breeder reactor data*

Peak power density	646 kW/l
Heat output	563 MWt
Electrical output	233 MWe
Efficiency	41.4%
Fuel	Mixed uranium and plutonium oxides, 20-27% effective enrichment, clad in stainless steel
Weight of fuel	4.3 tonnes
Fuel burn-up	100,000 MWD/te
Moderator	None
Coolant	Liquid sodium
Coolant pressure	1 atmosphere
Coolant outlet temperature	562°C
Vessel	Cylindrical stainless steel chamber 12m in diameter, 12m high
Refuelling	Off load

**Data based on Phenix, France*

liquid MHD generator

A form of *magnetohydrodynamic generator* in which the working fuel is in the form of a conducting liquid, rather than a conducting (ionized) gas (or plasma). Such a generator has at present yet to be developed.

liquified natural gas (LNG)

Natural gas that has been liquified for purposes of storage or transportation. Liquifaction is achieved by compression accompanied by cooling. The energy costs of liquifaction are high, amounting to the equivalent of 15 per cent of the calorific value of the gas being liquified. The re-gasification process, too, is energy intensive, so the merits of LNG should be balanced against energy costs which render the liquifaction/transportation/regasification process less than 70 per cent efficient.

liquified petroleum gas (LPG)

Propane or butane or a mixture of the two which has been liquified to facilitate storage, transport and handling. It is marketed in portable containers as *bottled gas*, or 'calor gas'. LPG is separated from crude oil during the distillation process and also during the treatment of *natural gas*.

lithium

The lightest of all metals, having a density roughly half that of water. It is found widely dispersed in small amounts as a constituent of a number of minerals and it is potentially of great importance. Irradiation of lithium with neutrons in a nuclear reactor yields a heavy isotope of hydrogen, *tritium,* which is a suitable fuel for the exploitation of *fusion energy.* Tritium already finds a use as one of the fuels used in the *hydrogen bomb*, which operates on the fusion principle.

symbol	Li
atomic number	3
atomic weight	6.94
density	534 kg m^{-3} (0.534 g/cm^3)
melting point	180.5°C
boiling point	1347°C

LMFBR

Initials for *liquid metal-cooled fast breeder reactor.*

LNG

Initials for *liquified natural gas.*

LNG carrier

A specially constructed tanker designed to transport *liquified natural gas.*

load factor

With reference to electricity generation, the term load factor refers to the ratio, expressed as a percentage, between the amount of electricity produced by a generating unit in a given period and the amount it could have produced during that period, if it had been operating continuously at its maximum installed rating. Base load plant, such as nuclear stations and the more efficient thermal power stations, operate at high load factor (up to 90 per cent) whereas peak plant, such as gas turbine equipment and pumped storage schemes, operate at low load factor, often of the order of a few per cent.

longwall mining

A commonly used system of underground mining, which may proceed on an advance or retreat basis, in which the complete seam is worked and no pillars or supports are left behind. Longwall mining has long been common practice in Europe and is now carried out in the United States to some extent. The advantages of longwall mining as compared with other methods, such as *bord and pillar*, include superior ventilation and higher rates of productivity.

low-Btu gas

A gaseous fuel with a low calorific value, ie less than 350 Btu/ft^3 (13.0 MJ/m^3). Examples include *blast furnace gas* and *producer gas* and indeed low-Btu gases are usually reused in industry for heating purposes, having been formed as the by-products of other processes.

low-grade heat

Thermal energy at a temperature below about 100°C is loosely termed low-grade. This can be supplied from non-premium sources, such as waste heat from power generation or process heating, by-product fuels such as *blast furnace gas* or domestic waste, and other inherently low-temperature sources such as *geothermal energy* and solar heat.

The grade of heat can be rigorously defined in thermodynamic terms by means of an *availability factor*, which is a measure of the maximum theoretical efficiency with which the heat can be converted to *work*. Low-grade heat has a relatively low availability factor.

low temperature carbonization

The *carbonization* of coal at temperatures of between 500°-750°C. The result of this process is the formation of a smokeless fuel, *semi-coke*, together with small amounts of gaseous and liquid products. The traditional process for making *coal gas* from coal relies on carbonization at higher temperatures, around 1000°C.

LPG

Initial letters for *liquified petroleum gas.*

lubricating oils

Oils used for lubrication purposes obtained by the *vacuum distillation* of petroleum residues. The term covers a wide range of products with a variety of applications and it is important that a lubricating oil is able to meet the required specification under a wide range of temperature conditions. Lubricating oils are usually classified according to their viscosity at a particular temperature, and one such set of classifications has been laid down by the Society of Automotive Engineers (SAE) in the United States. Seven grades are specified, known as 5W, 10W, 20W, 20, 30, 40, and 50. The first three are 'winter oils' sufficiently thin at low temperatures, and the last four are more viscous oils having viscosities specified at particular engine temperatures. Many problems associated with viscosity changes due to wide temperature fluctuations have been overcome by the introduction of 'multigrade' oils, blended in such a way as to be thin enough at low temperature but still sufficiently thick when the engine is hot. A 20W/50 multigrade oil, for instance, has the properties of a 20W winter oil at -18°C, but at 100°C has the same viscosity as a 50 oil.

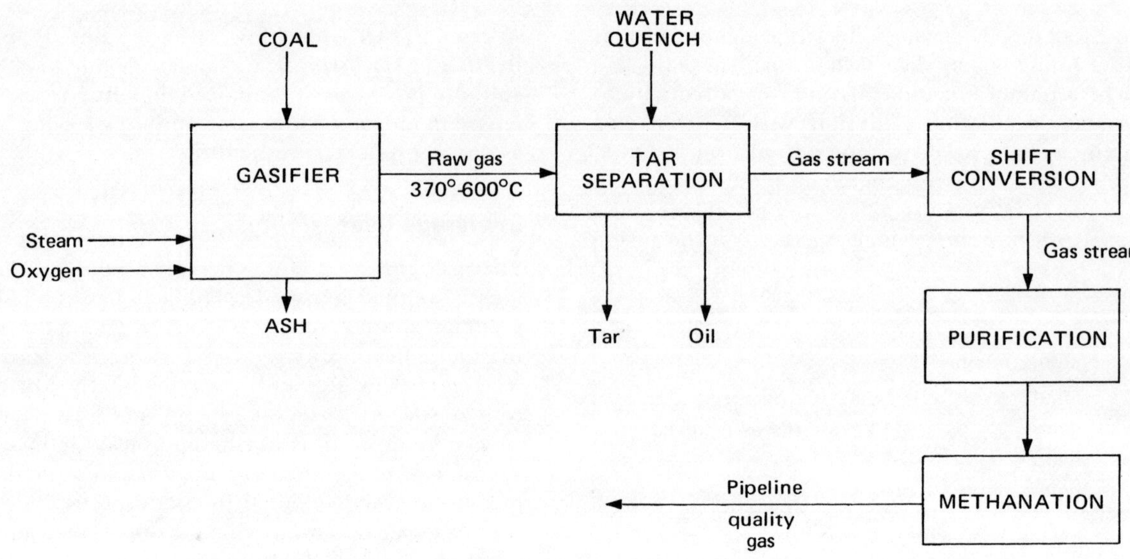

Lurgi process – schematic diagram for production of pipeline quality gas

Lurgi process

A method of gasifying *coal* in a moving bed by using oxygen (or air) and steam under pressure (up to 35 atmospheres). Since its inception in Germany in the mid-1930s when only lignite was used, Lurgi gasifiers have been adapted to take most grades of coal. Coal is fed automatically into the top of the pressurized gasifier and is distributed evenly through it. Steam and oxygen are introduced through a rotating steam-cooled grate at the bottom, where the ash is also removed. The gases pass upwards through the coal bed, the preheated coal is volatilized and the resulting char is gasified. Final reaction temperatures vary with the coal grade from 650°C to 850°C. At the high pressures used some of the coal is hydrogenated to methane, an exothermic reaction, heat from which increases the coal temperature and reduces the demand for oxygen. The raw hydrogen-rich gas is led from the gasifier at temperatures between 370°C and 600°C through a scrubber and washed before being subjected to further treatment. The synthesis gas may be used to make either gaseous products, such as pipeline gas, or liquid products as in the *Fischer-Tropsch* synthesis.

Lurgi technology is well advanced and it seems likely that future commercial coal gasification plants will take advantage of the Lurgi process.

LWR

Initials for *light water reactor.*

M

magnesium

A light, silver-white metal occurring abundantly in the earth's crust as the carbonate in the ores dolomite and magnesite. Its lightness yet comparative toughness makes it an attractive candidate for use in alloy materials, particularly with aluminium, eg in aircraft components. It is used in the manufacture of the *Magnox* alloy which fulfils an essential function and gives its name to the Magnox nuclear reactor.

symbol	Mg
atomic number	12
atomic weight	24.31
density	1738 kg m^{-3} (1.738 g/cm^3)
melting point	650°C
boiling point	1090°C

magnetic bottle containment

The use of a very powerful magnetic field to contain a very hot *plasma*, as in proposed nuclear *fusion reactors*.

magnetic surveys

Survey techniques using the distortion of the earth's magnetic field by igneous and metamorphic rocks. Oil, natural gas and minerals can be located in this way. Sedimentary rocks have no magnetic properties and the thickness of sediments can be deduced since there is a relationship between the source of the distortion and distance. Igneous rocks produce strong distortions so that magnetic surveys can reveal considerable detail about the rock systems of the prospected area. Airborne magnetometers are largely used and they have high sensitivities. Magnetic methods are used in conjunction with other methods such as *gravimetric surveys, seismic surveys* and the measurement of electrical and radioactive properties. Drilling will be carried out on the basis of survey findings.

magnetohydrodynamic (MHD) power generation

The generation of electrical power using a flow of hot gases (plasma) or conducting liquids as a conductor, the plasma passing transversely through a magnetic field. If a pair of *electrodes* are placed at right angles to the direction of the magnetic field, in a moving plasma, an electric current is induced between the two electrodes.

It has been proposed that the principle of MHD generation should be used to enhance the efficiencies of both conventional and nuclear power stations.

In a conventional power station fossil fuel is burned to produce high temperature vapours. At the high temperatures attained, the gases become conducting, due to thermal ionization. In the normal operation of the power station the heat of the gases is used solely to heat water in the boiler to produce steam for the turbines. In the MHD system extra power is extracted because the hot gases, in transit to the boiler, pass through the system of magnetic poles and electrodes, and the extra electrical power provided can, after suitable adaptation, be used to supplement the main output of the system. The efficiency of the system, overall, can be as much as 50 per cent, compared with the 35-40 per cent of ordinary power generating sets.

There are considerable difficulties involved in producing a practically viable system, but a number of large scale experimental models have been able to deliver significant quantities of power over a sustained period. (See below.)

Fuels for MHD generators. The conducting plasma can be produced by the combustion of fossil fuel like coal or natural gas, in oxygen or compressed preheated air.

The electrical conductivity of the plasma is much enhanced by the addition of a seed material, such as potassium carbonate, or caesium carbonate, that adds to the ion concentration in the plasma. The proportion of the total plasma provided by the seed material is approximately one per cent. The effect of the addition of seed material is to reduce the necessary operating temperature of the plasma; caesium reduces it from 2500K to 1500K.

Efficiency of MHD generators. The high temperature of operation of such devices provides a high theoretical operating efficiency — 50-60 per cent compared with the 35-40 per cent attainable in steam turbine generators that operate at a lower temperature.

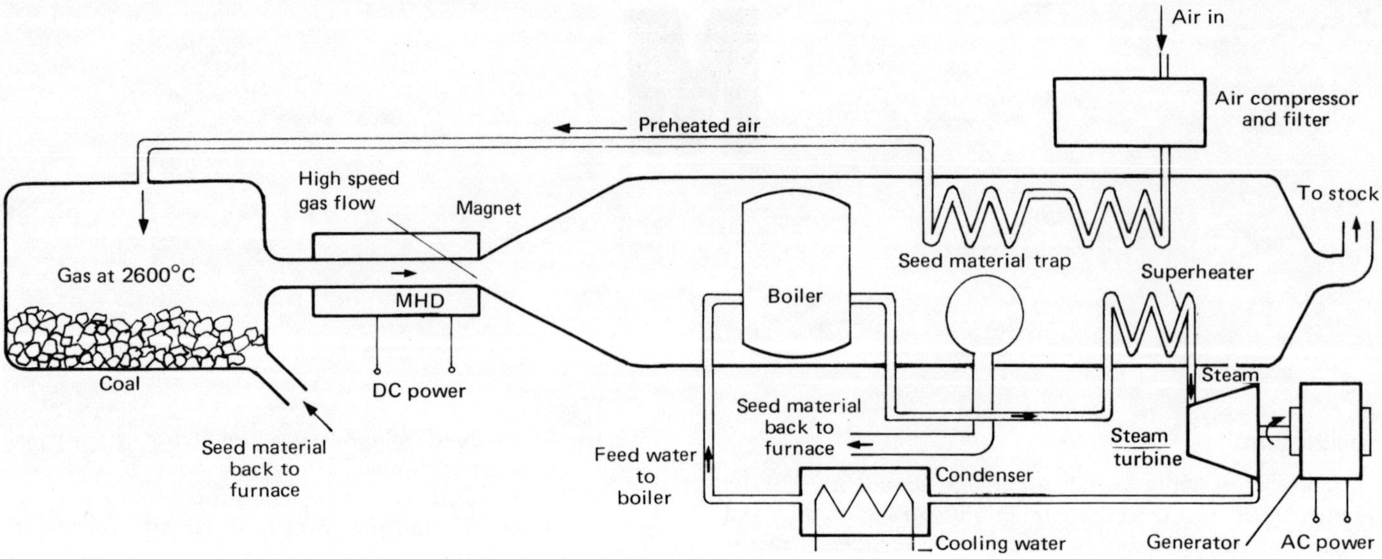

Magnetohydrodynamic power generation. MHD system used together with steam turbines to generate electric power. In a nuclear power station the nuclear reactor would replace the furnace and the system would operate in a closed cycle.

Practicability of MHD generation. The outstanding practical problem that arises is due to corrosion of the collector electrodes by prolonged exposure to very hot corrosive gases, and also the need to reclaim the expensive and potentially explosive seeding materials in the spent gases. These difficulties mean that substantial development is still required before MHD can become viable on a worth-while scale. Progress has been made, however, with the development of replenishment materials that form on the electrodes. This is done by adding refractory materials or coal dust or charcoal to the gases during the combustion process.

Large scale MHD models have been constructed and put into operation. The U-25 plant in Moscow has recorded runs of 250 hours with power outputs of up to 12MW. It is said to have operated for a total of 4000 hours and supplied electricity to the public supply system.

In the U-25, natural gas is first burned in oxygen-rich air, preheated to 1200K. The plasma temperature reaches 3000K. The plasma jet is accelerated through a nozzle to 1000m/sec and electricity is delivered at tens of thousands of volts. The power is delivered in the form of a direct current and this has to be converted to AC so that it can be used to supplement the public supply.

Types of MHD generators. Three main types of generator are envisaged – open cycle plasma, closed cycle plasma, closed cycle liquid.

Open cycle systems. In the open cycle system the fuel is burned in hot compressed air and the plasma seeded with potassium carbonate and then passed at a high velocity through the MHD generator. The plasma is then passed on to generate steam by delivering heat to the boiler of the conventional generator set. The vapour is passed out as an exhaust gas, but the seeding material is recoverable. The fuel can be coal and the use of potassium carbonate seed enables sulphur, a dangerous contaminant, to be removed.

Closed cycle plasma. In this system the plasma can be continuously recycled, with the obvious advantage that it can be cleared of contaminant. But this cannot be used when fossil fuels are being consumed because a replenished stock of fuel and oxygen is needed. It could be used, however, in nuclear fuel stations when helium could be circulated round the uranium fuel cans. However, there is the limitation that temperatures needed to make helium sufficiently conductive cannot be safely reached with presently available containers, but fuel elements may be developed to overcome this difficulty.

Closed cycle liquid metal units. These are also being considered for use with nuclear stations. The hot vapour would be replaced by suitable liquid metals. Supercooled magnets are being considered for use with practical units. To attain effective output levels, very high magnetic fields are required, and superconducting windings would enable such fields to be achieved with units of practicable size.

See also *cryogenics.*

Magnox

A magnesium alloy used as fuel cladding in the first generation nuclear reactors constructed in the UK.

See also *Magnox reactor.*

Magnox reactor

A type of nuclear reactor in which heat extraction is by carbon dioxide gas under pressure (19 atmospheres at 245°C). The heat from the gas is transferred as it is passed round tubes of water in a heat exchanger and converted

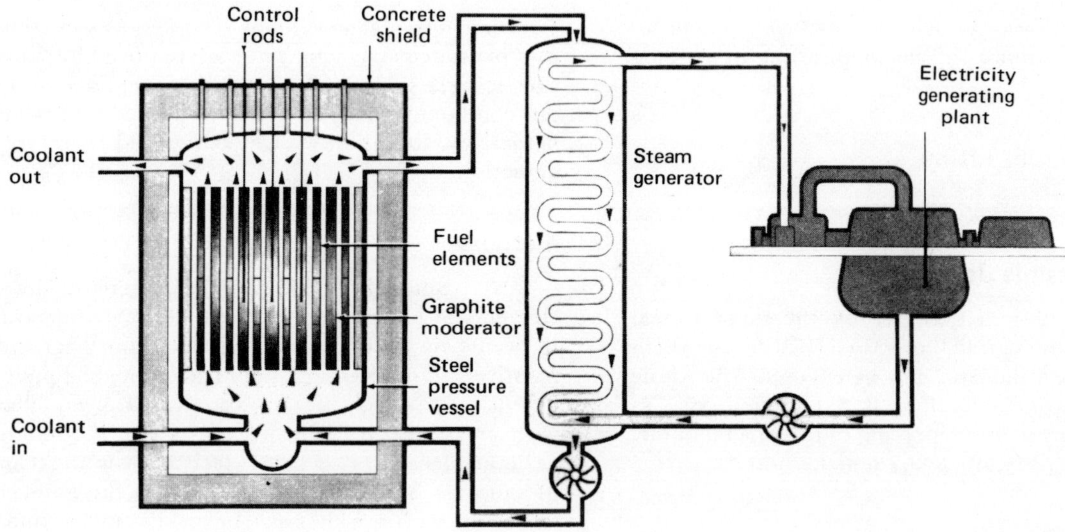

Magnox thermal reactor *Reproduced by kind permission of the UKAEA*

into steam for a turbogenerator. Heat output is 840 MWt giving about 275 MWe with an efficiency of 32.7 per cent. The fuel is natural uranium metal rods clad in 'Magnox' alloy. The moderator is purified graphite.

These are the first generation of reactors in the UK. Their reliability is high; their power density is low. Higher gas temperatures to improve the efficiency of electricity generation are not possible with this kind of fuel; natural uranium metal undergoes awkward structural changes as temperatures increase, and melts at 1130°C; Magnox melts at 645°C and may also ignite. There are also limitations on *burn-up* of Magnox fuel. Because of its behaviour at high temperatures and under heavy bombardment by neutrons, it cannot produce more than about 3-4,000 megawatt-days per tonne of uranium so refuelling has to be done frequently.

Magnox reactor data*

Peak power density	1.1 kW/l
Heat output	840 MWt
Electrical output	275 MWe
Efficiency	32.7%
Fuel	Natural uranium metal rods clad in Magnox
Weight of fuel	304 tonnes
Fuel burn-up	3850 MWD/te
Moderator	Graphite
Coolant	Carbon dioxide gas
Coolant pressure	19 atmospheres
Coolant outlet temperature	245°C
Refuelling	On load

**Data based on Dungeness A (UK)*

Natural uranium fuel has the advantage of being cheap compared with enriched fuel, but eventually the development of this kind of reactor in the UK was given up in favour of the *advanced gas-cooled reactor.*

mariculture

Farming of the sea. Fishing and related activities have long been practised, but only recently has greater thought been given to using more scientific methods and especially to integrating mariculture and energy production. Power station effluents, for instance, can have a pronounced environmental effect in raising water temperature. The growth of shellfish and other sea creatures is accelerated. In the United Kingdom, clams are cultured using the heat from power station effluent and in other parts of the world there have been numerous experiments to grow *algae*, molluscs and fishes.

Because of the large quantities of cooling water needed by nuclear power stations, effluent from these would seem particularly suited to maricultural use. Power stations sited on closed off natural inlets could warm the water so that fish farms could be established. Tanks on land could be heated similarly (geothermal heat is used for this purpose).

It has been suggested that the vertical current flows created in utilizing *ocean thermal gradients* would bring nutrients to the surface so encouraging plankton growth. Plankton is at the base of the web of food chains existing in the oceans so that production of all aquatic animal life would therefore be enhanced.

marine diesel oil

A heavy *gas oil* suitable for heavy industrial and marine diesel engines.

marsh gas

Naturally occurring gas, principally *methane*, found in marshy places as a result of decomposition of organic materials.

maximum continuous rating

See *installed capacity*.

maximum permissible dose

Maximum amount of radiation, set by the International Commission on Radiological Protection (ICRP), to which workers in the nuclear industry may be exposed. The whole body maximum permissible dose is 5 rem per year, as compared with the *dose limit* applying to the general public which is one-tenth this figure, ie 0.5 rem per year.

mcfd

Million cubic feet per day. A common unit used in the gas industry for measuring the rate of supply, or production, of gas. One million cubic feet of natural gas has approximately the same energy content as 200 barrels of oil.

mechanical energy

Energy due to the motion of bodies in a large-scale, ordered fashion. Mechanical energy thus differs from thermal energy, which is the energy of disordered, random motion of sub-microscopic particles.

Mechanical energy is exemplified by moving bodies, such as a stream of moving water; or rotating bodies, such as the flywheel of an engine.

Mechanical energy is conceptually equivalent to electrical power and, indeed, these two forms of energy can be interconverted with only small losses. The generation of mechanical energy from heat, however, is associated with large losses and this constraint also applies to the generation of electrical power from heat.

Mechanical energy is inherently difficult to store as such. Since it is not practical to have bodies moving along straight lines for this purpose, mechanical energy must be stored in the rotation of a mass. The rotation of a flywheel provides mechanical energy store, and such systems have been proposed for use in transport and for storage of off-peak electricity. Mechanical energy can also be stored by conversion to electrical energy which can then be stored in secondary electric cells.

medium-Btu gas

A gaseous fuel with a calorific value of between 350 and 700 Btu/ft^3 (13.0-26.0 MJ/m^3). Examples of medium-Btu gases include *coke oven gas* and *coal gas*, though production of the latter has now virtually ceased.

mercaptans

Foul-smelling organo-sulphur compounds sometimes occurring in crude petroleum. They are removed by washing with caustic soda solution. Very small concentrations of mercaptan compounds may be added to liquified petroleum gas or natural gas to impart a slight odour and therefore aid their detection.

mercury

A silver-coloured metallic element, unique amongst the elements in being liquid in the 0-100°C temperature range. It occurs in nature as the sulphide (cinnabar) and is comparatively rare. Mercury liquid has a high vapour pressure and it is extremely toxic so it should always be handled with great care. Its principal uses are in thermometers and as liquid electrodes (eg brine electrolysis in the manufacture of chlorine and caustic soda). It has the ability to form amalgams with a wide range of metals, and so finds a use in dental preparations, and mercury is also employed in a variety of batteries. It would doubtless be used more extensively in electric cells if its price were not such a prohibitive factor.

symbol	Hg
atomic number	80
atomic weight	200.59
melting point	–38.9°C
boiling point	356.6°C
density	13,550 kg m^{-3} (13.6 g/cm^3)

mercury cell

A primary *electric cell* which, apart from more conventional uses, has a variety of interesting applications such as power sources for hearing aids and heart pacemakers. It has the advantage over the dry Leclanché cell that it does not deteriorate, provides a constant emf over a long period of time and has a higher energy density. It consists of zinc and mercuric oxide electrodes in contact with a potassium

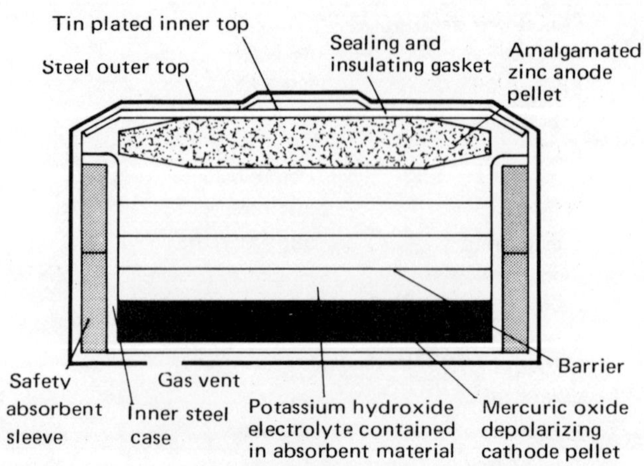

Mercury cell

hydroxide/zinc oxide electrolyte. A small amount of graphite is mixed with the mercuric oxide electrode to improve conductance. Its principal disadvantage is that the mercury content of the cell is responsible for high production costs.

merit order

A hierarchical sequence used in the operation of electricity supply systems to determine when particular generating units should be brought into use according to changes in demand. The position of a generating station in the merit order depends on a number of factors including its *thermal efficiency* and the cost of the fuel used to run the plant.

metal-air batteries

Electric cells employing a metal for the negative electrode and oxygen from the air as the positive electrode. Many different metals have been tried including aluminium, iron, cadmium and lead but the most promising results to date have been gained with zinc though much research and development remains to be done. These metal-air systems can function as primary or secondary cells and are proposed for use in battery-powered vehicles.

metallurgical coke production

The production of *coke* of a quality suitable for metal processing traditionally carried out by the *carbonization* of suitable *coking coals.* Because reserves of low-sulphur coals for the metallurgical industries are in potential short supply, new methods of coke production are being investigated which enable higher sulphur content coals to be used. Preparation of coking coals by the 'blending' of other coals of insufficiently high coking quality is also being pursued.

methanation

The process of up-grading the mixture of carbon monoxide and hydrogen produced by coal and oil gasification processes to a *high-Btu gas* consisting principally of methane. A nickel catalyst is employed. The basic reaction is:

$$CO + 3H_2 \rightarrow CH_4 + H_2O$$

The process is highly exothermic, ie heat is produced, and it is important to control the temperature sufficiently so as to encourage the formation of methane and to avoid undue heat damage to the surface of the catalyst.

methane

A colourless, odourless gas and the simplest member of the alkane series, with the chemical formula CH_4. Also called marsh gas (because it forms in swamps through the anaerobic decomposition of vegetable remains) and firedamp (because it is a fire hazard in mines where it forms an explosive mixture with air). Methane occurs naturally as a constituent of *natural gas* (averaging around 85 per cent by weight). Methane is also derived from the activated sludge treatment of sewage and there are many schemes for producing it from animal and plant crop wastes by *anaerobic fermentation.* An even more recent proposal is to produce methane from marine sediments that have a high organic content. Suitable regions are enclosed marine areas such as the Black Sea, and the Great Lakes of North America. Sediment would be raised from the sea bed, the organic matter separated and methane obtained by heating.

Methane is an excellent fuel, having a high calorific value of 1000 Btu/ft^3 (37.1 MJ/m^3) and burning with a clean flame. It is used as a starting point in the synthesis of many organic chemicals and *hydrogen.* It reacts with steam at high temperatures in the presence of catalysts to yield carbon monoxide and hydrogen (*synthesis gas*) which can then be converted to higher *alkanes, methanol* and other alcohols.

When methane is burnt incompletely in air carbon black is formed. This is an important constituent of rubber and a pigment in printing ink.

See also *biogas.*

methanol

Also called methyl alcohol. A colourless organic liquid, and the simplest alcohol, with chemical formula CH_3OH. Its calorific value is about half that of gasoline.

Formerly produced largely by the destructive distillation of wood, methanol is now mainly obtained from carbon monoxide and hydrogen (*synthesis gas*) derived from *natural gas* feedstock.

Methanol is also produced by the alcoholic fermentation of organic waste materials of plant and animal origin. In the Second World War it was manufactured from such material and used as a fuel additive, increasing the octane rating of the fuel. Methanol is used primarily as a chemical feedstock, but it has been suggested that it could be used to a greater extent as a fuel, especially if blended with gasoline. It has also been considered for use in *fuel cells.*

methyl alcohol

Alternative name for *methanol.*

MHD

Initial letters for magnetohydrodynamic.

See also *magnetohydrodynamic power generation.*

middle distillate

A term used to describe petroleum fractions falling in the boiling range 200°-350°C. It comprises the upper end of the *kerosine* range, *gas oil* and light *fuel oil.* Whilst much middle distillate goes to providing products such as *diesel fuel*, some is cracked to give lighter fractions for which there is a higher demand, for example *gasolines*.

migration

The movement of *petroleum* within the source rock and through the permeable carrier beds to a trap, which acts as a reservoir, or to the surface. Primary migration is a slow process that takes place due to the compacting forces within the source rock under gravity. Molecular diffusion may play a part as well. It is most likely that petroleum migrates as a solution in ground water. Water tends to move laterally down a fluid gradient from areas of high fluid potential to those of lower potential as a sedimentary basin is developing. Because petroleum is less dense than water it will tend to rise and the upward migration of the petroleum continues unless it is restricted by an impermeable layer of rock. It will follow this layer until it is confined by a trap or it reaches the surface. *Anticlines*, *faults* and *stratigraphic traps* contain the petroleum accumulations discovered.

See also *survey methods*.

moderator

Material placed in the nuclear reactor core to slow down fast neutrons and thus make them more likely to occasion a nuclear *fission*. Typical moderators are *graphite* and *heavy water*, materials consisting of light atoms which are poor neutron absorbers.

Moderator materials may also be used as reflectors: extended beyond the region of the fuel elements to surround the reactor core they will cut down on leakage of neutrons by bouncing escaping neutrons back into the fissioning material in the core thus increasing the neutron density there. Adding or taking away reflector therefore affects the rate of the chain reaction and some reactor designs make use of this in their control systems.

monazite

The most important naturally occurring source of *thorium*. Monazite ore may contain up to 18 per cent thorium.

motor spirit

(Also known as motor fuel, petrol, and in the United States, gasoline.) Fuel used in spark-ignition internal combustion engines. It consists of *light distillates* from the 30°-200°C boiling point range specially blended to give the appropriate anti-knock characteristics. In addition to the *branched-chain hydrocarbons*, *alkenes* and *aromatics* so introduced, non-hydrocarbon additives are also used such as TML (tetramethyl lead) and TEL (tetraethyl lead) to improve fuel ignition qualities.

See also *octane number*.

moving-bed catalyst

A catalyst which is free to move with the reaction chamber, thus ensuring a greater degree of mixing with the reactants. Moving-bed catalysts are used in variations of the *Fischer-Tropsch process*.

mtce

Initials for million tonnes (or tons) coal equivalent, a unit commonly used for recording, measuring, or projecting primary energy use. One million tonnes coal equivalent approximates to 0.60 million tonnes coal equivalent.

See *Appendix 2; Table 3*.

mtoe

Initials for million tonnes (or tons) oil equivalent, a unit in common use in measurements of primary energy consumption. One million tonnes oil equivalent is approximately the same as 1.67 million tonnes oil equivalent.

See *Appendix 2; Table 3*.

mud

The common name for drilling fluid which takes its name from the fact that muddy water was once used for the purpose. Muds are usually mixtures of finely particulate materials such as special clays and barytes.

MW, MWe, MWt

Shorthand terms for megawatt (10^6 watts), megawatt-electrical and megawatt-thermal respectively. The distinction between MWe and MWt is an important one as the former refers to electrical energy and the latter to thermal energy. For instance, the heat output of a nuclear reactor or power station boiler may be given in MWt units, but the electrical output of the turbogenerator which utilizes this energy output is quoted in MWe units.

N

naphtha

Sometimes known as 'heavy benzine' or 'heavy gasoline' or simply light distillate feedstock, naphtha is a *petroleum fraction* boiling in the 70°-200°C range. It is mainly used in the preparation of blending agents in *motor spirit* and as a petrochemical feedstock. Synthetic natural gas is made from naphtha feedstock by *steam reforming*.

naphthenes

Also known as cycloalkanes or cycloparaffins, naphthenes are *alicyclic hydrocarbons* with the general formula C_nH_{2n}. Examples include cyclopentane (C_5H_{10}) and cyclohexane (C_6H_{12}) both of which occur naturally in crude petroleum. Cyclohexane is readily dehydrogenated in the presence of a platinum catalyst to yield *benzene* and this is employed in petroleum refining to improve the octane number of *motor spirit*.

naphthenic

A classification of crude petroleum used by the US Bureau of Mines in their *correlation index* to denote a crude oil having a high density.

natural asphalts

See *bitumen*.

natural gas

The general term given to a mixture of predominantly *hydrocarbon* gases found in subsurface rock reservoirs. The main constituent gas is *methane* (often consisting of 85-95 per cent of the total) together with ethane, propane and butane. Highly valued as a clean and convenient fuel, natural gas is widely employed in the domestic sector, particularly in western Europe and the USA (the main producing country).

Origins. The origins of natural gas have long been the subject of speculation. *Associated gas* is usually the volatile portion of *crude oil* found in varying proportions wherever crude oil is discovered and probably has its origins in marine material. *Non-associated gas* is apparently unrelated to liquid oil accumulations and may be derived from vegetable matter. (See the entry for *petroleum*, specifically Origins and Nature.)

The liquids and gases even in associated fields are not necessarily derived from a common origin. But their different migration paths and processes may terminate in the same rock reservoirs. These reservoirs are not hollow caverns but formations of rock in the pore space of which gas and, or, oil and water enter, and usually separate out, accumulating at respective levels. The degree of separation depends on the relative proportions of liquids and gases, the viscosity of the petroleum and the porosity of the rock. Associated gas may often be found in the form of a *gas cap* overlying the oil-bearing strata.

Nature. All the hydrocarbon natural gas mixtures are *alkanes* (general formula, C_nH_{2n+2}). *Methane* (CH_4) is the main constituent, usually 85 to 95 per cent of the total. The table shows the range of constituents and calorific values of natural gas.

Constituents	% of total volume 'wet'	(range)	'dry'
Hydrocarbons			
methane	84.6		96.0
ethane	6.4		2.0
propane	5.3		0.6
iso-butane	1.2		0.18
n-butane	1.4		0.12
iso-pentane	0.4		0.14
n-pentane	0.2		0.06
hexanes	0.4		0.10
heptanes	0.1		0.80
Non hydrocarbons			
carbon dioxide		0–5	
helium		0–5	
hydrogen sulphide		0–5	
nitrogen		0–10	
argon		less than 0.1% of helium content	
Calorific values: 900–1100 Btu/ft^3 (33.4–40.9 MJ/m^3)			

Source: E N Tiratsoo, *Oilfields of the World*. Scientific Press 1973.

The proportions of methane, etc, will vary from region to region.

The group of products which occur at the lighter end of the liquified hydrocarbons is known as *natural gas liquids.*

A *dry gas* has a very low proportion of extractable liquid hydrocarbons, less than 0.1g/1000 ft^3 of gas (non-associated gases are usually dry). A *wet gas* contains 0.3g/1000 ft^3 or more of liquid hydrocarbons. A *lean gas* falls between the two.

The hydrocarbon reservoir is said to be saturated when there is more than enough gas present to saturate the liquid under the prevailing conditions of heat and pressure, and gas separates as a free substance. (An undersaturated reservoir has no free gas present.) During production the penetration of the reservoir reduces pressure and the solution will rise to the surface with gas emerging first. Gas within the solution, 'dissolved' gas, imparts pressure to raise the oil while reducing its viscosity.

The *gas/oil ratio (GOR)* describes the amount of gas held in a solution at normal temperature and pressure conditions. Usually expressed in cubic feet per barrel (ft^3/bbl) of oil, it can vary from hundreds to several thousands of ft^3/bbl. Beyond 10,000 ft^3/bbl a well is usually defined as a gas well.

Gas statistics are normally expressed in cubic feet (ft^3) or cubic metres (m^3). ($1m^3 = 35.3\ ft^3$).

The *calorific values* of natural gas usually lie in the range of 900-1100 Btu/ft^3 (33.4-40.9 MJ/m^3). Natural gas has a higher calorific value than *coal gas* and burns with a slower but more powerful flame.

Production and Consumption. At the well-head, impurities from the reservoir rock and drilling operations, along with any liquid water, are extracted by mechanical means. Water vapour is removed to prevent the formation of *hydrates*, (which choke transport apparatus). The higher hydrocarbons are removed to leave a dry gas (suitable for fuels and feedstock), either by compression, absorption or adsorption methods.

Various components are separated out for use in different industries; eg *ethane* for the petrochemical industry; *propane,* and *butane* are liquified and sold as *liquified petroleum gas (LPG).*

Sour gas is gas contaminated with hydrogen sulphide (H_2S) at more than 5 g/m^3. Hydrogen sulphide is actively corrosive and poisonous but after *desulphurization* the recovered *sulphur* is sold as a valuable by-product. A *sweet gas* has less than one part per million of hydrogen sulphide. Nitrogen decreases the calorific value of natural gas and must be removed where it occurs in quantity. However this also has a high value in the petrochemical industry, specifically in the manufacture of fertilizers. The Groningen field exploits gas with a nitrogen content of 14 per cent, but after conversion only 2.5 per cent remains.

An odour is imparted to the gas at the final stage, for safety, as the gas is otherwise virtually odourless.

LIQUIFIED NATURAL GAS (LNG), LIQUIFIED PETROLEUM GAS (LPG). For storage, and long-distance transport, gas is often liquified (1 ft^3 of liquid methane weighs 26.5 lb, which equals approximately 625 ft^3 of gaseous methane). Both LNG and LPG can be transported by tanker, and LPG transport by road and rail is well developed. Refrigeration is needed to keep temperatures low and the gas in liquid form. Concentrated liquid-gas products have a low flash point and their vapours form an explosive mixture when in contact with air. Great care is necessary during tanker refuelling operations.

An important factor in the world gas supply has been the large amount of wastage through 'flaring off'. Gas in fields with low GORs is often considered unworthy of commercial exploitation. Thus gas not required for reinjection, to maintain the oil-extraction pressure, is often ignited and flared off at the well-head. The wastage was and is enormous. (For example, Saudi Arabia's reported flare off wastage amounts to 14 thousand million cubic metres a year, which is equivalent to approximately 12 million tonnes of crude oil.)

The general usage of natural gas has increased steadily from the beginning of this century. Thus the development of viable commercial projects for gas, including its use for conversion to feedstock products, growing concern over the depletion of fossil fuels and limited governmental intervention has encouraged gas usage and reduced the proportion wasted through flaring off.

Reserves. Data on natural gas reserves is less reliable than information on crude oil reserves, but subject to similar conditions of general assessment (see *petroleum*, specifically Reserves). As the majority of consumption is local, data have diminished international significance. Wastage, through flaring off and reinjected gas, does not show in the figures (see *Appendix 1; Table 3*). These figures give a regional account of estimated reserves and a comparison in million tonnes of oil equivalent. The international significance of gas reserves will be greater in the future, however, as shipments of gas by LNG carrier and transmission via intercontinental pipelines become more important. Examples of such trade include the export of gas from Indonesia to Japan, from Russia to Western Europe, North Africa to Italy, and possibly Mexico to the United States. Indeed, exploitation of the large reserves of Siberian natural gas is only likely to take place given favourable export opportunities, which in turn may depend on a cooperative political climate.

Uses. Natural gas is known as a *premium fuel*, as it is a clean, safe and convenient source of heat. It has a special value in domestic use as it requires no local storage facilities and causes virtually no atmospheric pollution. (Sulphur is one of the major pollutants; however, natural gases from most regions, including the USA and UK, are sulphur-free.) Natural gas is also used for steam-raising and direct heating in industry.

Dual-fired boilers can be run off gas enabling suppliers to regulate the load during the peak demand periods of other energy consumers. As a premium fuel it has advantages in operations such as glass, ceramics and steel-making where a clean fuel is needed.

In the USA and USSR, natural gas is used to power electricity generating stations and in Europe some power stations are equipped to burn natural gas. However, increasingly powerful lobbies are emerging to prevent the use of natural gas for generating secondary fuels.

Natural gas is widely used as a chemical feedstock providing a wide range of petrochemicals. The main derivative is *ammonia* for nitrogenous fertilizers. (Some of the previously wasted flared natural gas in the Middle Eastern countries is now being used to produce fertilizers on a greater scale.)

Liquid petroleum gas, previously used mainly in the agricultural and domestic sectors, is being utilized increasingly as feedstock for the petrochemicals industry. It is also employed for domestic and commercial uses where piped gas is unavailable.

See *Appendix 1; Tables 3 and 4 and Maps 11 and 12.*

natural gasoline

See *condensate.*

natural gas liquids

Hydrocarbons occurring in natural gas accumulations which are liquid at less extreme conditions of temperature and pressure. Propane, butane, and pentane occur as natural gas liquids and those with a relatively low vapour pressure, such as pentane and hexane, are often termed *condensate* or *natural gasoline* in this context.

natural gas storage

Storage of natural gas to meet peaks in demand and to act as a safeguard in case of interruptions in supply. The two principal methods of gas storage on a large scale are liquified natural gas (LNG installations) and leached out salt cavities, and such stores are used to smooth out interseasonal demand swings. Diurnal and weekly variations are dealt with by smaller, local storage units such as water-sealed holders, high pressure storage vessels ('bullets') and pipe nests. Line packing, ie using the transmission mains for storage purposes, is also employed.

In the future it is planned to use depleted gas fields as large-scale gas stores and a principal advantage here is that the 'store' will be strategically positioned with regard to the gas transmission system.

neutron

A *nucleon* with no charge, composed of a *proton* and an *electron* and with a mass very slightly greater than that of the proton. Neutrons occur in all atomic nuclei apart from hydrogen-1, and atoms with the same numbers of protons, but different numbers of neutrons are referred to as *isotopes* of the same element. When *nuclear fission* occurs it is the continuing ejection of neutrons from the nuclei undergoing fission (themselves having undergone neutron bombardment) that is responsible for the *fission chain reaction* which in its uncontrolled form may result in a catastrophic release of energy in the *atomic bomb*. Control of the chain reaction is made possible by the natural phenomenon of a small percentage of *delayed neutrons* which, unlike *prompt neutrons*, are not emitted until a very short time after fission has occurred. The probability of a neutron causing fission is increased if it is slowed down from its speed of ejection from the fissile material, and for this reason *moderators* and *reflectors* are introduced into reactor cores. On collision with these materials, the neutrons are not absorbed but bounce off, simply losing some kinetic energy. In contrast, *control rods*, made of strongly neutron-absorbing material, are used to regulate the neutron density in the core and hence the fission chain reaction.

NGL

Initial letters for *natural gas liquids.*

nickel

A relatively abundant metallic element, occurring in nature in a number of ores often accompanied by iron, magnesium, cobalt, and copper. Nickel is a hard, silvery-white metal, is a reasonably good conductor of electricity and heat, and has ferromagnetic properties. It is widely employed in alloys and finds energy-related use as a catalyst in a number of petroleum refining processes, and as electrode material in the nickel-cadmium and nickel-iron electric cells.

symbol	Ni
atomic number	28
atomic weight	58.71
melting point	1453°C
boiling point	2732°C
density	8902 kg m^{-3} (8.90 g/cm^3)

nickel-cadmium accumulator

A *secondary cell* similar in principle and design to the *nickel-iron accumulator.* The major difference is that cadmium oxide, rather than iron oxide, is the active material used in the negative electrode.

nickel-iron accumulator

Also known as the Edison Battery. A secondary *electric cell* in which the electrodes consist of interlocking sets of metal plates immersed in aqueous potassium hydroxide electrolyte and assembled in a steel container. The positive electrode comprises nickel metal and nickel hydroxide held in a steel frame, and the negative electrode iron oxide similarly contained. The cell voltage is 1.2 volts and, unlike the *lead-acid accumulator*, the electrolyte is not used up as part of the cell reactions. Only a small amount of electrolyte is needed, therefore, and this enables a more compact electrode arrangement to be employed.

non-associated gas

Natural gas which does not occur in association with oil. Such gas is usually fairly dry and consists mostly of *methane*, with small amounts of *ethane, propane* and *butane*. *Acid gases* such as hydrogen sulphide and carbon dioxide may also be present. Some accumulations, such as the large Groningen field in Holland contain a significant proportion of nitrogen. The fields in the southern sector of the North Sea, which up to recently have been supplying all the United Kingdom's natural gas, are non-associated gas deposits, remarkably free of nitrogen or appreciable acid gas impurities.

No 2 fuel

Name used in the USA for *gas oil* used as a fuel for domestic central heating installations.

nuclear binding energy

Term used to denote the immensely powerful, short range force which binds the particles of the *nucleus* of an *atom* together against the disintegrative force of the positively charged particles of the nucleus (the *protons*).

It is the result of part of the mass of the individual *nucleons* being converted into energy on coming together to form a nucleus (according to the formula $e = mc^2$) and the potential energy of the nucleus – *nuclear energy* – lies in this binding force.

nuclear cell

Also known as atomic battery. A small power source in which the energy of the particles emitted from the nuclei of radioactive elements is converted to electricity. Such batteries may have high or low voltages but most can supply only very small currents and because of their cost are restricted to certain specialized applications, such as supplying power to low-consumption electronic devices. They are very reliable, however.

One type of high voltage atomic battery (see diagram below) has a radioactive source, *tritium* gas, absorbed on a coil of *zirconium* metal wound round a tube insulated by a glass terminal seal. The terminal is sealed into an evacuated nickel casing which has a thin inner coating of carbon as a collector. The number of emissions per second from the source that are captured by the carbon collector determines the current. The intensity of emission for tritium reduces to half over 12 years. Strontium-90 was used in early types of the battery but this is much more highly toxic than tritium.

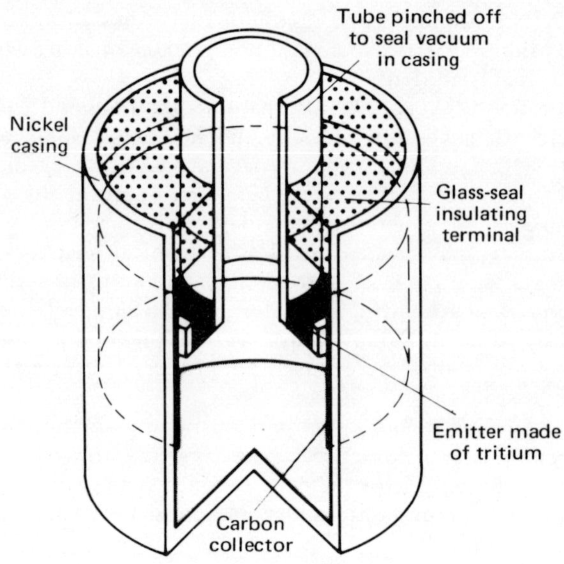

Schematic representation of a high voltage nuclear cell

There are several types of low voltage battery, each employing a different principle. The thermoelectric type uses the heat produced by radioactivity and employs a *thermopile*. The temperature difference created between the source junctions and the casing junctions produces current and while energy output per unit weight is about the same as for conventional electric cells using chemical energy, usable energy is being released all the time with a nuclear battery, which limits the life of the cell.

Usable fuel sources are restricted by considerations of *half-life*, heat output and safety. *Plutonium-238* and strontium-90 are frequently used as emitters and lead-telluride alloys for thermocouples operating over a temperature range of from 200°-480°C. When shielding is not crucial, as in unmanned space vehicles, thermoelectric nuclear batteries are most convenient. Power output is from 5-60 watts.

Gas ionization nuclear batteries use a *beta particle* emitter (eg tritium) to ionize a gas (eg argon) sited between a pair of electrodes (eg lead dioxide and magnesium) across which there is created a potential difference. Movement of the ions within the electric field produces a current flow. Voltage values of about 1.5 are achieved; cells may be arranged in a series to boost this.

Thermionic fuel cells employ differentially heated electrodes (temperatures of 1200°-1800°C for the emitting electrode made of tungsten, and 500°-900°C for the collecting electrode of molybdenum or nickel, for example). Electrons emitted from the hot electrode flow back to the emitter from the collector via an external circuit. The space between the electrodes contains caesium vapour. The hot electrode may be heated using *solar energy* or energy from *nuclear reactor* or *radioisotope* sources. Special arrangements are made using vaporization/condensation principles in a pipe to cool the collector. Thermionic fuel cells may be used in the large scale generation of electricity from nuclear sources.

Scintillator photocell batteries operate through the conversion of nuclear energy to light energy which is then converted to electricity. The battery consists basically of silicon photocells and a promethium-phosphor light source. Cadmium sulphide is a suitable phosphor. A tungsten alloy shield is needed to keep radiation levels tolerably low.

nuclear energy

Nuclear energy is energy that has its origins in the *nucleus* of an *atom*. It is released as a result of the regrouping or rearrangement of the nuclear particles, as in the processes of *radioactive decay, fission* and *fusion*. In the nuclear reaction, part of the nuclear mass itself is changed into energy. Thus nuclear energy is to be distinguished from chemical energy which has its origins in the rearrangement of the extra-nuclear particles of the atom, as in the burning of coal or wood.

The potential energy of the nucleus lies in the immensely powerful, short range force binding its particles together and giving it stability. This force arises from the conversion of part of the mass of the nucleus into energy in accordance with Einstein's formula $e = mc^2$, where e is energy in ergs (10^7 ergs = 1 joule), m is mass in grams and c is the velocity of light in centimetres per second (cm/s). Since light travels at a velocity of 2.998×10^{10} cm/s, the value of the constant c^2 is 8.988×10^{20}. Clearly, vast amounts of energy are released by the conversion of very small amounts of mass. For example, 1000 g mass can theoretically be converted to 8.988×10^{23} ergs or 8.988×10^{16} joules, the equivalent of 2.497×10^{10} kilowatt hours, or enough energy to power a 100W light bulb for over 28½ million years.

The amount of energy in a nucleus can be calculated by measuring the amount of its mass that is 'missing', that is, the amount of its mass that has been converted into nuclear binding energy. The nucleus of the lightest element, hydrogen, has no mass missing. Then, as the elements become heavier, increasingly more mass is missing for each particle of their nuclei up to the elements in the neighbourhood of iron (*atomic number* 26). After that, less and less mass is missing as the elements get heavier and heavier. Therefore if by any means nuclei of high or low mass number are converted to those nearer a medium mass number, energy is released.

This transformation of one nucleus into another with the accompanying release of energy does take place. It happens under extremely high temperature conditions such as exist in the interior of the sun or other stars where light nuclei fuse together to form heavier nuclei, which is nuclear fusion; it also happens when the unstable nuclei of the heaviest elements break down and are transmuted into lighter nuclei, as in *radioactivity* and nuclear fission.

The stability of a nucleus depends on the grouping and compatibility of its main particles, principally *neutrons* (with no electric charge) and *protons* (with a positive electric charge). These particles are packed tightly together and the protons, being positively charged, repel one another violently. Without the conversion of part of the mass of the nucleus into energy which overwhelms these electrostatic forces, the nucleus would have no stability. The degree of stability of each nucleus depends principally on the relative strength of its binding energy over the repulsive force of its protons. It is thus related to its atomic number. For example, nuclei of even mass number tend to be most stable; helium 4 (2 protons, 2 neutrons), oxygen 16 (8 protons, 8 neutrons) or calcium 40 (20 protons, 20 neutrons). And nuclei of uneven mass number are least stable.

In the very heaviest naturally occurring nuclei the repulsive forces among the protons are beginning to match up to the force of the nuclear binding energy. These nuclei are under great internal stress and sooner or later will break down. They are the nuclei of the radioactive elements, radium, actinium, *uranium* and *thorium*. Their disintegration happens spontaneously and can also be induced. It happens gradually in the process of radioactive decay, or rapidly, in fission.

The process of radioactive decay. Radioactive nuclei disintegrate by the loss of a heavy particle containing two protons and two neutrons, and also by loss of a far lighter particle with negligible mass, the negatively charged electron.

With each disintegration energy is released. It is released as energy of motion (*kinetic energy*): the *alpha particle* (the particle with two neutrons and two protons and therefore two positive charges) emerges from the nucleus at high speed and great force tearing electrons off surrounding atoms. This activity uses up its energy and it does not travel far. The *beta particle* (the very light electron with one negative charge) also emerges at high speed but the ionizing effect of its radiation is less than in the case of the α-particle. It is capable of travelling a greater distance and can penetrate several millimetres thickness of metal foil. After the emission of alpha and beta particles, the nucleus emits shortwave, electromagnetic waves or photons called *gamma rays*. With no electric charge, gamma rays cause a comparatively small amount of disturbance as they pass, lose energy gradually and travel a long way. They are extremely penetrating.

The energy released during radioactive decay can be used to power small devices and instruments. It can be converted directly or indirectly into electricity in *nuclear cells* whose output is low but reliable and which have long operating lives. They can also be lightweight and therefore portable. Shielding is necessary to contain the ionizing radiation which is damaging to all living organisms.

The Nature of Fission Energy. A great deal more energy is released when an unstable heavy nucleus bursts apart into two lighter nuclei. This is the most violent kind of disintegration a nucleus undergoes. It is called nuclear fission. It can happen spontaneously or through the unstable nucleus absorbing a stray neutron. The result of the splitting of the nucleus is usually about 2/5 of it flying off one way and 3/5 another with two or three odd neutrons also shooting out together with a release of energy. For example, the energy added by a neutron to a nucleus of *uranium-235* (92 protons and 143 neutrons) is enough to cause fission. Uranium-235 becomes uranium-236 which immediately splits into nuclei of lesser mass, possibly strontium-90 (38 protons, 52 neutrons) and xenon-143 (54 protons, 89 neutrons). This accounts for 233 of the nucleons of uranium 236 so in this case there are three stray neutrons.

Any one or all of the stray neutrons released in this fission reaction can cause fission of further nuclei and so a *fission chain reaction* may be set up. The existence and behaviour of the neutrons emitted in the fission process is of great importance in solving the problems of converting the energy from fission into nuclear power. A *nuclear reactor* is a device for setting up, sustaining and controlling a fission chain reaction. Other fission products only create problems. Some of the 200 or so which have been identified are strong absorbers of neutrons and will therefore damp down the chain reaction. All of them are highly radioactive and therefore dangerous to life. Some have a very short *half-life* therefore producing a high level of *radiation*; some a very long half-life therefore presenting great waste disposal problems.

The Nature of Fusion Energy. Nuclear fusion occurs at very high temperatures (millions of degrees Kelvin) and is therefore sometimes termed thermonuclear fusion. The source of the sun's energy and so, directly or indirectly, of all forms of energy on earth is in the fusion reactions taking place in the sun's core where the hydrogen nuclei fuse together to form helium and the resulting loss of mass is transformed into energy.

At such very high temperatures all substances have become gases whose atoms are stripped of their electrons, creating *plasma* consisting of nuclei and free electrons in a high energy state. In this plasma, the light hydrogen nuclei, which consist of one positively charged proton, can be thrown together with such violence that the strong repulsion between two positively charged particles is overcome and they fuse together to form eventually the heavier nucleus of helium, releasing energy and throwing off high energy particles in the process.

Energy released from fusion is the energy converted from the mass lost in the transformation of light nuclei into heavier nuclei and the energy of the particles emitted in the process. The amount of energy released per fusion reaction is much smaller than in fission reactions, but the masses of the nuclei involved is also much smaller; hydrogen nuclei of mass number 1 compared with uranium of mass number 235. Pound for pound of fuel, a fusion power reactor, if it were possible to construct one, would deliver about five to ten times as much power as a fission power reactor. Theoretically, a pound of hydrogen could produce 3.5×10^7 kilowatt hours of energy. However, although the *hydrogen bomb* has been exploded and it is possible to produce on earth, even if only briefly, the very high temperatures needed for some fusion reactions, it has not yet proved possible to start a fusion chain reaction and keep it going at a constant, controlled rate. Thermonuclear reactors are still only theoretical laboratory experiments.

nuclear fission

See *fission.*

nuclear fission chain reaction

See *fission chain reaction.*

nuclear fuel cycle

Term used to indicate the sequence, to some extent a cycle, of processes which nuclear fuel undergoes before it reaches the *nuclear reactor*, in the reactor and in reprocessing which retrieves some of the material to be made into new fuel.

Nuclear fuel originates in the earth's crust where there are ores containing the radioactive metals *thorium* and *uranium.* Thorium has only recently been used as a reactor fuel. The principle source of *fission energy* has been, and is, uranium.

Deposits of uranium-bearing minerals occur widely but the highest-grade ores known have already been exhausted and ore now being worked is of much lower grade, 0.4 per cent and less. There is far more uranium, 4,000 million tonnes, available in theory from sea water. It has proved possible to extract it and the energy cost of doing so is not high but the basic process needs far more development and will be feasible only in a few areas of the world. At present therefore the first stage in the production of nuclear fuel is mining, either open-cast or underground.

All the ore extracted, with its very small proportion of uranium, is sent to mills near the mine where it is crushed and ground into a fine sand which then goes through a series of chemical processes to leach out the uranium, separate it and finally concentrate it as a mixture of oxides, commonly called 'yellow cake'. At least 250,000 tonnes of ore is needed to produce 1,000 tonnes of uranium.

In the form of yellow cake, uranium is minimally radioactive since it has a very long *half-life*, although chemically toxic like lead. It can be transported in steel drums to fuel fabrication or enrichment plants all over the world.

The first process undertaken at both fabrication and enrichment plants is to purify the yellow cake which is still only about 85 per cent uranium. This is generally done by dissolving it in HNO_3, filtering, and treating the uranyl nitrate solution by solvent extraction methods. Every care must be taken to establish a high standard of purity at this and all later stages of fuel fabrication since any impurities in reactor fuel absorb *neutrons* needed to sustain the *fission chain reaction* and, especially in *fast breeder reactors*, to breed *fissile* from *fertile nuclei.*

The form in which the fuel is fabricated and whether it is first enriched or not is determined by the requirements of different designs of nuclear reactors. The first generation of gas-cooled reactors use natural uranium fuel, as do the Canadian Deuterium Uranium (CANDU) reactors which use the highly efficient but expensive *moderator*, heavy water. Most reactor designs, however, prefer enriched fuel, that is, fuel that has a greater proportion of *fissile nuclei* than the

nuclear fuel cycle

Reactor	Enrichment	Fuel	Cladding
Magnox	none	natural uranium metal	magnox
CANDU	none	natural uranium oxide	zircaloy
AGR	2%	enriched uranium oxide	stainless steel
PWR	3%	enriched uranium oxide	zirconium
BWR	2.2%	enriched uranium oxide	zircaloy
SGHWR	2.3%	enriched uranium oxide	zirconium
LMFBR	20-27%	mixed uranium and plutonium oxides	stainless steel
HTGR	93%	enriched uranium carbide particles or enriched uranium and thorium carbide particles	ceramic coatings

0.7 per cent found in natural uranium. The extra fissile nuclei required are produced by isotope separation processes in uranium enrichment plants and in breeder reactors by the transmutation of fertile to fissile nuclei usually through neutron absorption.

Uranium enrichment plants. Isotopes of the same element are identical in every respect except that there is a minute difference in mass between them. This difference is the basis of all methods of separating one kind of isotope from another. Some methods, like mass spectroscopy, are suitable only for small-scale production in laboratories of radioisotopes for example, for research and medical purposes; some, like the *gaseous diffusion process*, are developed for large-scale production of highly-enriched (at least 90 per cent fissile nuclei) uranium needed for weapons materials. Gaseous diffusion plants built for military purposes have also been the main source of enriched fuel for nuclear reactors, although the reactors require fuel enriched only to two or three per cent. These plants are very large and their energy cost is very high (covering half a square kilometre, for example, and requiring in full operation, 2,000 megawatts of electricity). Also they are not easily converted from producing highly-enriched weapons fuel to producing two or three per cent enriched reactor fuel. New designs for diffusion plant are therefore being developed and also a new technology, the *gas centrifuge process*, for commercial production of nuclear fuel. Gas centrifuge plants require one-tenth the electricity needed in diffusion plants and are more compact. On the other hand they are more expensive to build. Another method being developed commercially is the nozzle process, simple and cheap to build but using a lot of electricity. The most attractive technology for the future however is the *laser enrichment process*. Using this, separation of uranium-235 from uranium-238 could probably be completed in one stage instead of the several thousand stages of the diffusion or centrifuge cascades.

All the isotope separation processes require the uranium to be in the form of a gas, so the solid uranium oxides of the yellow cake are converted into uranium hexafluoride, UF_6, commonly known as hex, a highly poisonous and corrosive gas but the simplest compound of uranium that can be vaporized and with the advantage that fluorine occurs only as a single nuclide so the diffusion rates of the molecules in this gas depend only on the mass difference between the isotopes of uranium. In present practice, the hex is then pumped through a great number of units arranged in cascade. Large-scale cooling is necessary to remove the heat generated in the gas by pumping and the hex must also be kept below atmospheric pressure to avoid leaks and to keep it from condensing. It emerges finally as a large amount of depleted hex, with less than 0.25 per cent uranium-235 and a very much smaller amount of enriched hex. Uranium enrichment at present is a very complicated and costly process.

The importance of the possible development of a commercial breeder reactor is that it by-passes such complications and serves as its own enrichment plant. In it, a surplus of fissile nuclei can be bred. When a nucleus of uranium-238 absorbs a neutron it is transmuted into a nucleus of the fissile plutonium-239; similarly thorium-232 produces uranium-233. This is a process that happens to some extent in all reactors: the only purpose of the first nuclear reactors was to produce plutonium-239 for weapons purposes (the heat generated was a waste product). *Thermal reactors*, however, cannot produce more fissile nuclei than they consume because fission by thermal neutrons produces too few new neutrons to provide a significant surplus for breeding. The breeder reactor is designed to run on fast neutrons, which produce a higher number of new neutrons per fission. To ensure fission from fast neutrons the neutron flux density (the number crossing a given area of core in unit time) must be high and the fuel must have a high proportion of fissile nuclei. The breeder *reactor core* has highly-enriched fissile fuel surrounded by a blanket of fertile fuel. There is no moderator. The number of neutrons produced per fission is then sufficient to keep the chain reaction going and to produce more fissile nuclei than the reactor consumes. This fissile fuel can be retrieved when the reactor fuel is reprocessed.

Fuel fabrication. Uranium metal exists in three phases and in the transition from the alpha to the beta phase, at about 660°C, it expands and distorts and becomes brittle. At 1130°C it melts. It is awkward to handle but can be fabricated into reactor fuel by ordinary metal-working processes. Because of its limitations however, and those of the cladding used for it (magnox alloy melts at 645°C when it may also catch fire), uranium metal is not used in the later-designed reactors. These use uranium dioxide (UO_2) fuel which starts as a powder made from uranyl nitrate solution from uranium mills (when the proportion of uranium-235 to uranium-238 will be the same as in natural uranium); from enrichment plants (when the proportion of uranium-235 will be greatly increased); or from reprocessing plants.

This powder is compacted and sintered into dense, cylindrical pellets and ground to size. Solid uranium dioxide has a melting point of 2800°C, much higher than uranium metal, but very low thermal conductivity, much lower than uranium metal. If the pellets are very small and of high

nuclear fuel cycle

density thermal conductivity is improved. (Dense pellets also retain gaseous fission products more effectively.) To guard further against heat build-up inside the fuel, UO_2 fuel elements have a much smaller diameter than uranium metal rods. UO_2 pellets are stacked in fuel 'pins' – very narrow, thin-walled metal tubes which can be as little as 6mm in diameter. The tubes are then filled with inert gas, sealed and assembled into fuel elements, in some reactors as many as 100 to an assembly.

Plutonium metal is even more awkward than uranium metal. For reactors, plutonium fuel is made from the plutonium dioxide. Again it is much more difficult to handle than the uranium oxide and must be treated with great care to avoid leakage of toxic material and the possibility of inadvertently achieving *criticality*. It is processed by the same methods as uranium oxide and the two oxides are sometimes mixed, for fuel for the liquid-metal fast breeder reactor, for example.

All fuel elements must be made with extreme precision to withstand damage when undergoing intense irradiation in the reactor and to remain stable and intact. The same standards of extreme precision and purity have to be applied to cladding as to fuel. It is as a consequence of neutron-absorbing impurities that cladding gradually corrodes and leaks radiobiologically dangerous fission products into the coolant, moderator and surrounding structures, also causing them to become radioactive.

Because of the long half-lives of uranium-235 and 238, fresh fuel assemblies, like yellow cake and enriched uranium, have a low level of radioactivity. This is slightly increased by the presence of the shorter-lived uranium-234, 0.8 per cent in enriched uranium, but is low enough for fuel assemblies to be handled, with proper protection, and shipped by ordinary transport. Care is taken to prevent shipping cases coming into contact with water (which would act as a moderator) or being stacked in such a way that the pile becomes critical.

Fresh fuel elements are fed into the Magnox, CANDU and AGR while the reactor is running. The PWR, BWR, HTGR, SGHWR, and LMFBR are shut down for refuelling. At the same time the spent fuel elements, heavily shielded, are extracted by remote control and dropped into cooling ponds.

Fuel reprocessing. The irradiated fuel leaving the reactor contains, besides unused uranium, many new ingredients created by *fission* reactions and *radioactive decay:* plutonium-239, 240, 241 and 242 and other elements of the *actinide series*, americum, curium and neptunium; also hundreds of fission products (and their decay daughters), including krypton-85, strontium-89 and 90, iodine-129 and 131 and caesium-137. It is also continuing to produce a great deal of heat from radioactive decay.

If fuel elements are kept for as long as 100 days in cooling ponds where they are cooled and shielded by a great mass of water, many of the shorter-lived isotopes will have decayed into insignificance and much of the heat been dissipated. Fast breeder reactor fuel though, will still be producing as much heat as Magnox fuel immediately on discharge. It is then possible to move the fuel elements on to the next stage, reprocessing the fuel to retrieve the valuable uranium and plutonium. The fuel elements are transferred (by remote control) from the cooling ponds to massive water-filled drums fitted with cooling systems to prevent the build-up of decay heat and transported with many precautions to reprocessing plants.

In the reprocessing plant, the fuel elements are moved on a remotely controlled process line through a series of cells with walls of two-metre thick concrete. Magnox fuel elements are stripped of their cladding which is sent for storage in heavy concrete bins. The fuel rods are dropped into nitric acid which dissolves them. The cladding cannot be stripped off oxide fuel pellets because under irradiation they swell and become wedged into their pins. Instead the whole element is chopped into pieces which are dropped into nitric acid which dissolves out the remains of the fuel pellets. The cladding, like the Magnox cladding, then goes into shielded storage.

The dissolved fuel is mixed with a succession of solvents which separate the uranium and plutonium from the fission products and then from each other. They emerge as uranyl nitrate solution and plutonium nitrate solution and thus can be made into new fuel elements for reactors or used for weapons fuel. Plutonium, and therefore also mixed oxide, fuel is highly radio-toxic and likely to achieve accidental criticality. It must be handled with great care. But it is not impossible that at least small amounts of highly fissile material could be stolen and its possession used as a terrorist threat or leading generally to a spread of nuclear weapons increasing the danger of nuclear warfare. Strict security precautions against such possibilities are an internationally recognized necessity.

Radioactive waste. Reprocessing is not the final stage of the nuclear fuel cycle. Left over after the recovery of as much uranium and plutonium as possible, is an acid solution containing fission products and actinides formed from uranium by successive neutron captures. The solution is hot and intensely radio-toxic. It cannot ever be disposed of by any conventional method: the level of radioactivity of the fission products may drop below that of natural uranium in less than a thousand years but the actinides will continue to be radioactive for hundreds of thousands of years.

Present practice is to concentrate the solution by allowing it to boil by its own decay heat and then to store it in stainless steel tanks. The concentration process must be closely controlled to prevent dangerous chemical reactions, for example, or runaway heat output. The storage tanks must be elaborately constructed, with agitation systems to prevent solids settling and water jackets, which include leak detectors, to remove decay heat. There are filters for the air around the liquid but if a tank boiled dry, they would not be sufficient to prevent the escape of radioactivity to the atmosphere. One tank costs about £4 million and takes four years to build. It must be replaced after 20-25 years.

An array of storage tanks, which can be interconnected

to spread the heat-load increased by incoming waste, is sealed into steel-lined, concrete buildings. And there the radioactive waste stays until a more permanent method of disposing of it can be found.

Proposed final solutions to the problems of *radioactive waste* management include: sending the waste by rocket into space; burying it in the core of the earth through the edges of the tectonic plates at the bottom of the ocean, or under Antarctic ice, or in chemically and seismically stable geological formations, rock salt, clay or hard rocks, either on land or under the ocean.

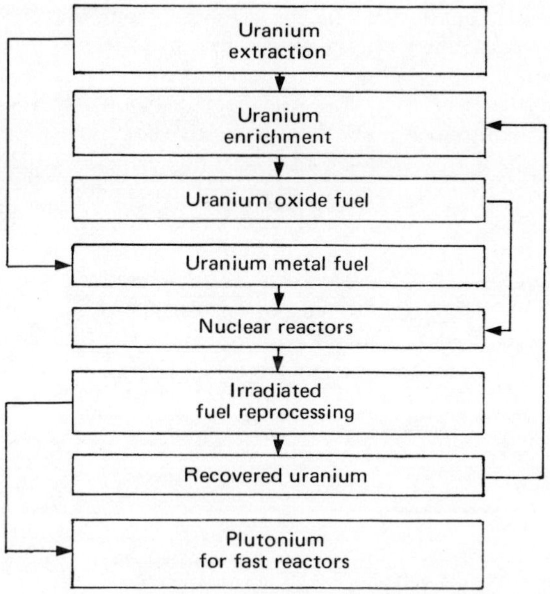

Nuclear fuel cycle

nuclear fusion

See *fusion*.

nuclear reactors

Devices in which a nuclear *fission chain reaction* can be sustained and controlled. Power reactors were developed originally, and some still are used, for the production of nuclear materials having military potential, and for fundamental research and radio-isotope production. Their main use, however, is for producing large quantities of heat which can be converted, usually through the medium of steam, into the mechanical energy required to power a turbo-generating unit and thus to generate electrical energy. The nuclear reactor is equivalent to the boiler in a fossil-fuelled thermal power station. The development of reactors employing nuclear *fusion* reactions is at a very early development stage.

In a reactor, slugs or pellets of nuclear fuel – usually *uranium* or *plutonium* – are bombarded by *neutrons*. The reactions created depend on the energy of the neutrons and the nature of the nuclei of the material under bombardment. For example, uranium-235 nuclei have a high probability of undergoing fission if hit by a slow (thermal) neutron. Considerable heat and two or three free neutrons are released. These neutrons, if slowed down by a *moderator*, for example, *heavy water, graphite*, or *light water*, produce further fissions of uranium-235 nuclei and so the fission chain reaction is sustained. Uranium-238 alone cannot sustain a fission chain reaction but its nuclei are transmuted into plutonium-239 when hit by fast neutrons. Plutonium-239 is fissile and so will contribute to the chain reaction. So-called fast reactors use thorium-232 or uranium-238 as fuel. Only a small proportion of these fuels are fissioned directly. These are mostly transmuted, respectively to uranium-233 and plutonium-239. Reactors employing them are called *fast breeder reactors* and such fuels are called *fertile fuels* in contrast to the *fissile fuels*.

The fission chain reaction is regulated using *control rods* inserted into the reactor core. These are made of neutron-absorbing materials such as *boron* or *cadmium*. As they are withdrawn the neutron density increases, criticality is achieved and the reactor starts to operate. The rate of fission reactions thereafter, and consequently the heat output, is controlled by withdrawing or inserting the control rods; when all are fully inserted the reactor is shut down.

Nuclear reactors release a lot of heat which must be removed if the reactor core is not to melt. *Coolants* employed are gases (eg *helium* and *carbon dioxide*) and liquids (eg *heavy water, light water* and molten metals such as *sodium*).

Many different reactor designs have evolved since Fermi's *atomic pile* went critical in 1942. They vary according to use and to fuels, moderators, coolants and many other factors not least 'political' and economic considerations. Most nuclear reactors in use or being constructed are of the thermal or burner type. The United States has favoured light water as moderator and coolant in its designs; Britain has developed gas-cooled, graphite-moderated types and Canada has specialized in heavy water-moderated and cooled reactors. *Light water reactors* are of two main kinds – *boiling water reactors (BWRs)* and *pressurized water reactors (PWRs)*. In the former, steam is produced directly from the light water coolant which boils in the reactor core at a pressure of about 67 atmospheres. PWRs are pressurized at up to 150 atmospheres and heat is transferred from the primary core circuit to a secondary circuit where steam is produced. *Gas-cooled reactors* employ gases which have lower heat transfer capacities than liquids. Adequate circulation of gas for efficient heat removal consumes considerable energy and GCRs operate at low steam pressures and temperatures. *Advanced gas-cooled reactors (AGRs)* and *high temperature gas-cooled reactors (HTGRs)* use higher temperatures and pressures. The latter type are more efficient at raising steam for electricity generation and the possibility of using them for generating process heat for industrial use is receiving attention. Heavy water reactors enable higher fuel burn-up than graphite-moderated types.

Factors affecting nuclear reactor design criteria include type of fuel, control rods, moderator and flow of coolant which all have to be accommodated within the reactor core. This has to be contained within a sufficiently strong and corrosion-resistant reactor vessel and the whole must be housed in a thick biological shield to prevent contamination of personnel or the environment. Fuel elements and other parts that have to be removed periodically must be accessible but no leakage of radioactivity must occur. Equipment accordingly has to be operated by remote control. The materials used must be able to withstand prolonged neutron bombardment and produce as little interference as possible in the nuclear reactions. Additionally they must be resistant to high temperatures, pressures and continual contact with corrosive substances. High quality stainless steel and other special alloys (eg *zircaloy* and *Magnox*) have been developed for cladding and other parts. The particular materials used depend considerably on operating criteria.

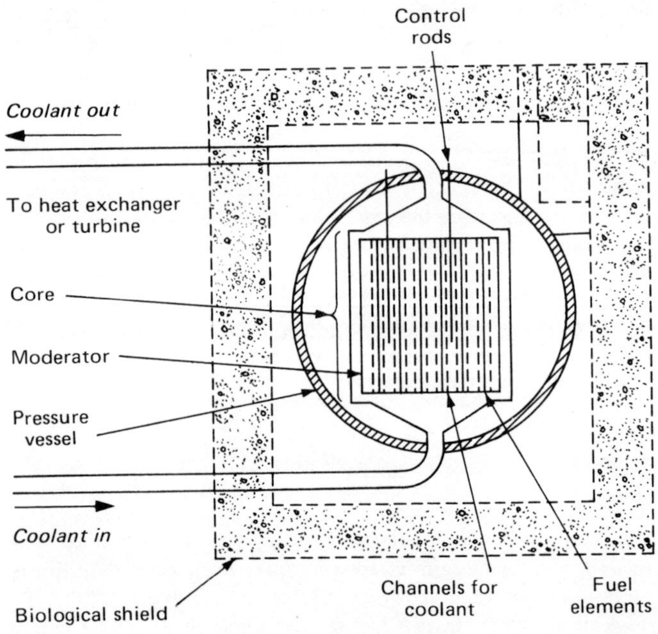

nuclear reactor core

A given power output may be obtained either from a large core with a low power density or a small core with a correspondingly high power density. This will depend on the type of fuel. Natural uranium, for instance, has fewer fissile nuclei than enriched uranium or plutonium fuel. Enrichment is very expensive but a smaller building costs less. The higher the core temperature the more useful the heat produced, but heat resistant fuels and reactor parts will cost more than those operating at lower temperatures. Similar considerations will apply to pressurization: safety, too, is of great concern with pressurized systems.

Generally nuclear reactors have an efficiency in terms of heat energy conversion to electricity of around 30 per cent, equal to that achieved with conventional thermal power stations. But they cannot be switched on and off so quickly or their power output adjusted so rapidly as with conventionally fuelled power stations. They are most suited to supplying base-load requirements therefore. Some types, for example PWRs, supply relatively low quality heat and are comparatively inefficient for electricity production purposes. If ways can be found of increasing heat utilization then obviously efficiency will increase.

Much of the planning of the nuclear industry is concentrated on the development of *fast breeder reactors*. There are fundamental differences in design from the *thermal reactors* described briefly above. Because the fission chain reaction in a breeder core is dependent on fast neutrons, no moderator is used to slow the neutrons down to thermal speeds. And since more fast neutrons than slow neutrons are required to produce a fission, a higher neutron density is needed and the reactor core must be smaller than in a thermal reactor. Structural and other materials such as coolant must be kept to a minimum so that neutrons are not slowed down unnecessarily before they have a chance to hit fissile nuclei. In current prototype designs, liquid sodium is used as coolant to carry away the intense heat. But helium, which has certain advantages, has been suggested for a *gas-cooled fast breeder reactor* which is being developed.

See also *calandria; ceramic reactor; fuel element; liquid metal-cooled fast breeder reactor (LMFBR); pebble-bed reactor; reactor core.*

nuclear reactor core

See *reactor core.*

nucleon

A term used to describe a constituent part of an atomic nucleus, ie a *neutron* or *proton.*

nucleus

The positively charged core of the atom, consisting of *protons* and *neutrons* packed tightly together. It constitutes almost all the mass of the atom but occupies a very small part of its volume. Nuclear *fission* involves splitting the nucleus apart, to yield new smaller nuclei, free neutrons, and a release of energy.

O

ocean thermal energy conversion (OTEC)

The extraction of useful energy from the oceans, using temperature differences between warm surface water (heated by absorption of solar energy) and colder, deep water to generate power.

Minimum temperature differences of 15°C are needed in practical systems, and such temperature gradients are found in tropical ocean regions between latitudes of 10°N and 10°S. Low wind velocities must prevail for OTEC systems to be effective.

Meaningful quantities of power have been generated in ocean areas where the temperatures range from between 25° and 30°C at the surface to between 4°-7°C at a depth of 750 metres. In the basic solar sea thermal power plant warm water is drawn into a broad diameter tube at the top of the system, where it produces steam under low pressure in the boiler module, and is cooled in the condenser by the intake of cold water at the base of the system. The efficiency of such a system (operating typically between 30°-5°C) is calculated at 2-3 per cent. Such efficiencies are low compared with heat engines using conventional fuels, but since fuel costs are zero, and the system needs only to be designed for low working temperatures and pressures, the cost of power is reckoned to be economically feasible.

Proposed practical operations include open cycle and closed cycle systems. Open cycle ocean thermal energy conversion units, such as the *controlled flash evaporation process (CFE)*, rely on seawater as the working fluid to produce low pressure steam vapour which is harnessed to generate power. Closed cycle ocean thermal energy conversion units employ low boiling-point fluids, other than sea-water itself, as working fluids in generating power. Examples of proposed closed cycle systems include the UMASS system (using *ammonia* and/or *propane*) and the CMU system (ammonia).

Economic feasibility of OTEC. The installation and plant costs are estimated to be comparable with those of nuclear power stations, but higher than fossil-fuelled power stations, though with the advantage of zero fuel costs. The savings that can be made in future designs will depend upon the development of more sophisticated heat exchangers possibly using titanium instead of the presently envisaged *aluminium.*

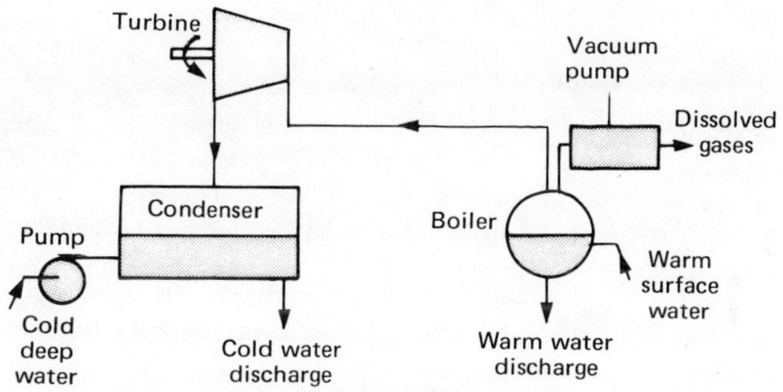

An open cycle sea thermal plant

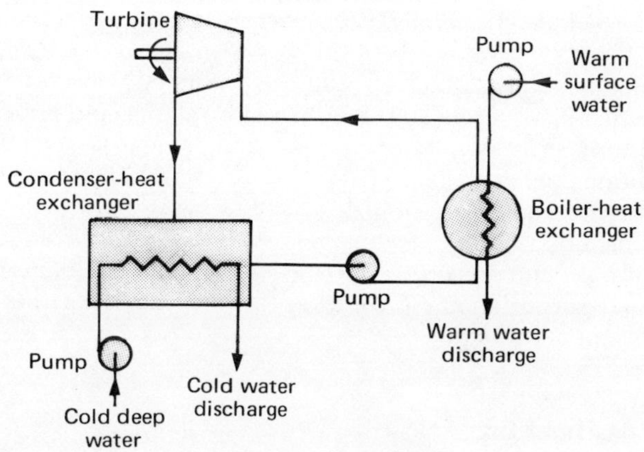

A closed cycle sea thermal power plant

Utilization of OTEC power. Because of the likely siting of OTEC plants (with power ships envisaged in some locations), it is most likely that they would be used to supply power for energy-intensive industrial processes such as production of ammonia for fertilizers of aluminium production. It is also envisaged that hydrogen might be released from the sea-water, using OTEC power to feed a future *hydrogen economy.*

A side benefit derived from OTEC is the bringing to the surface of nutrient-rich deep sea-water. This can be used to promote greatly increased plankton production which results in rapid growth of mollusc crustacea and fishes and a number of schemes to take advantage of such *mariculture* benefits have been proposed.

The map 'Possible ocean thermal gradient plant locations' illustrates ocean regions where ocean gradients of sufficient differential exist.

See *Map 22.*

ocean thermal gradient

Difference in temperature between warm surface ocean water and cold deep water, used to run a *heat engine* to produce useful power.

See also *ocean thermal energy conversion.*

octane number

Also called anti-knock value or octane rating, this is the volume percentage of *iso-octane* in a matching blend of two reference fuels, iso-octane and n-heptane. The former has a good anti-knock rating of 100 and n-heptane a poor rating of 0 on the arbitrary scale used.

Motor spirit is specially blended to avoid the phenomenon of 'knocking', ie the premature ignition of the fuel/air mixture in a spark-ignition engine. *Straight chain hydrocarbons* have poor anti-knock qualities whereas the *branched chain hydrocarbons* contribute to a high octane rating; hence the use of n-heptane (straight chain) and iso-octane (branched chain) as reference points. Motor spirit is usually sold on the basis of its *research octane number* (RON). Premium grades of 97 to 100 are commonly marketed. Numbers in excess of 100 are achieved by adding anti-knock additives such as *tetramethyl lead* and *tetraethyl lead.* In addition to branched chain hydrocarbons, alkenes and aromatics contribute to high octane rating in motor spirit. Petroleum refining techniques such as *isomerization, alkylation* and *catalytic reforming* are employed to increase the proportion of these constituents present in the gasoline fraction.

oil gasification

The conversion of liquid petroleum products into gaseous fuels or feedstock. Oil gasification was practised increasingly in the 1960s as the gas industry began to move away from coal as its primary fuel. Petroleum products, and *naphtha* in particular, were extremely cheap and conversion processes were developed which made oil gasification an economic proposition. Later on, techniques to convert light distillate into a *synthetic natural gas*, in contrast to a *town gas* substitute, were developed and most SNG manufactured today is made from petroleum derivatives. In the future, coal-based SNG may predominate as oil feedstock becomes increasingly scarce and expensive. Oil gasification processes producing hydrogen for ammonia manufacture, established for some years, will doubtless continue as long as oil is required as a source of hydrogen.

oil in place

The quantity of oil estimated to be in a *reservoir.* Not all of this oil can be recovered, amounts varying from about 10 to 80 per cent – the so-called *recovery factor.* Total oil in place multiplied by the recovery factor constitutes the total *recoverable oil reserves.*

oil refining

See *petroleum refining.*

oil shale

A fine-grained argillaceous rock containing *hydrocarbons* in the form of a waxy material known as *kerogen.* When distilled out the hydrocarbons produce crude *petroleum* which on further refining can be used as conventional crude oil.

Enormous deposits (primarily from the lower *carboniferous period*) exist in western USA, and lesser deposits in Brazil, USSR and western Europe. The capital costs arising from extraction and conversion are extremely high and as yet shale oil cannot compete successfully with conventional fuel sources. The proportion of shale to inorganic rock is so low that enormous quantities have to be mined for a reasonable oil yield.

Origins and Nature. Oil shales were formed in numerous differing geological eras, from at least the Cambrian era, and possibly earlier, to the present day. In contrast to *tar sands*, oil shales are hard rocks. The inorganic material is usually marlstone mixed with varying amounts of clay and sand, but the shales vary considerably in composition and kerogen content. Kerogen has a high *molecular weight* of over 3000 and as such is not soluble in the usual petroleum solvents.

Processing. Oil shale is mined, like coal, and subjected to temperatures of around 400°C in a retort or large distillation vessels. The kerogen breaks down into liquid and gaseous *hydrocarbons* which are condensed, if necessary, and extracted in oil form. Further upgrading is needed before the shale oil can be used in conventional refineries. The Scottish shale industry, which operated from the early 1860s until 1962, employed this basic method for extracting oil from shale. Roughly eight per cent by weight of the shale was processed as crude oil (over 100,000 tonnes per year).

The large volumes needed to be mined, and the massive accumulations of spent shale to be disposed of, are both economically and environmentally prohibitive. Mining costs alone reflect a large portion of the cost of oil from shale.

Recent developments, particularly in the USA, involving in situ mining, may overcome the problems of waste disposal and some of the despoliation associated with large scale mining, while reducing mining costs. A large underground chamber within the shale is fractured by explosives or hydraulic pressure and natural gas is pumped in, ignited and fed with air. The combustion process spreads and separates the oil as in above ground retorting. Oil can be drained off into prepared sumps or extracted through wells as in conventional oil production. This process is technically difficult and dangerous. Although pilot plants have been in operation for some time there is little available information as to the success of such methods. Certainly there must be uniform and sufficient supplies of oil to justify the site expenses of large scale production. Again, environmental factors should be of major consideration. For example, underground watercourses could become contaminated by oil overspill or explosives used in the fracturing. This is especially problematical should nuclear power be employed, as is being suggested, to fracture the chamber. Subsidence may also be a problem.

Reserves. Estimates concerning the amount of recoverable shale oil are far more confused and varied than other fossil fuel figures. The amount of *oil in place* is certainly massive, possibly a thousand times that of proven conventional oil reserves. At present the majority of this oil in place is economically irrecoverable, as the capital costs and energy losses (through extraction and conversion) are extremely prohibitive.

The two major areas of research cover firstly the development of in situ mining techniques, and secondly, programmes working towards the discovery of further rich sources of shale oil.

The world's richest deposits are found in the USA, the Green River shales of Wyoming, Colorado and Utah, particularly in the Piceance Creek basin (some 450 km^2) containing 80 per cent of the region's potentially recoverable oil. Brazil, Estonia, China, Scotland, England and Germany have oil shale resources some of which have been worked with varying success.

The Chattanooga shales in the USA and Billingen shales in Sweden are important sources of low-grade *uranium*, and are not exploited therefore as sources of oil.

The World Energy Conference puts recoverable shale oil reserves at 230 thousand million tonnes (as compared with an estimated 129 thousand million tonnes of petroleum and natural gas reserves). These figures were qualified by explaining that information concerning world oil shale reserves is confused and in some countries non-existent, and that extraction costs and difficulties greatly reduce the amount which is economically recoverable. The quantity of oil per unit volume of rock varies dramatically.

Uses. When upgraded and refined, shale oil can be used exactly as conventional petroleum products.

Although declining fossil fuel reserves and increased prices have furthered interest in oil from shale, its use as a fuel source has been long known. For example, in the Italian district of Medina, during the 17th century, shale oil was utilized for street lighting.

There is a reportedly large shale oil industry in China (but little information is available). In Leningrad, USSR, shale oil is used in the manufacture of gas, and in Estonia, pulverized oil shale is employed directly for electricity generation. Using oil shale directly has the advantage of reducing the amount of processing needed, but the quantities of shale employed must be huge. (Corrosion in the boilers due to large amounts of residual ash would also be problematical.) There may be a future for using oil shale as a raw material for the chemical industry. Such usage of oil shale would contribute to an energy saving as other conventional materials would not then be needed.

The overall view indicates that oil shales will add very little to the world energy supply in the near future.

The production of oil from *coal* and from *tar sands* seems a more probable means of energy supply. In both cases the conversion technology is better developed and the raw material richer.

oil-water contact or level

In petroleum exploration this term refers to the depth at which the interface between oil and water accumulations occurs in a reservoir.

olefins

Alternative name for *alkenes*, aliphatic hydrocarbons whose molecules contain a double bond and having the general formula C_nH_{2n}. Examples include ethylene, propylene, and butene. Olefins are more reactive than *paraffins (alkanes)*, have high *octane numbers* and are therefore blended into motor gasolines.

OPEC

Initial letters for Organization of Petroleum Exporting Countries.

open-cast mining

A method of mining, also called strip mining, in which material is stripped from the surface without recourse to mine shafts or tunnels. It is widely used in the coal mining industry, particularly in the extraction of lignite, and has been responsible for great improvements in productivity as the required labour input is small. The main part of the capital cost of strip mining lies in the huge excavators and draglines used in the mining operation. Strip mining can have extreme environmental consequences, and not only as an aesthetic blot on the landscape. It has been responsible for serious land subsidence and, if the coal mined has a significant sulphur content, open-cast mining can lead to serious pollution by the sulphur being leached out of the mine debris by rain.

(open cycle) ocean thermal energy conversion units

Systems using seawater through an open cycle that utilize existing temperature gradients in the world's oceans to produce power through *ocean thermal energy conversion (OTEC).*

Seawater is drawn into the apparatus and the difference in temperature between the solar heated surface water and the colder water from the ocean depths effectively runs a *heat engine* to produce power.

The earliest experimental ocean thermal energy conversion units were of the open cycle type. For example, in 1930 Georges Claude installed such a plant off Matanzas Bay (Cuba) which operated for two years using a 14°C differential between warm and cold water intake to produce a 22KW output.

Steam was produced in a low pressure flash evaporator chamber, and passed through a *turbine* to power a generator, before being condensed in a direct contact condenser, using cold seawater.

The design failed as a practical proposition because of the corrosive properties of the seawater and because of the need for a prohibitively large turbine. These difficulties are largely overcome in later developments of this original design.

Recent developments in ocean thermal energy conversion include the use of working fluids, such as *ammonia* and *propane*, which are not corrosive, are more efficient in terms of heat transfer and have high vapour pressures at ambient temperatures. These fluids have to be contained in closed cycle ocean thermal conversion units.

See also *controlled flash evaporation process.*

OTEC

Initials for *ocean thermal energy conversion.*

outstep well

Another name for *appraisal well.*

oxidation

Any chemical reaction that involves the loss of electrons from an atom but also defined as the addition of oxygen or other strongly electronegative element to a compound or to another element. Combustion is an oxidation reaction the control of which is of great importance in the efficient utilization of fossil fuels.

oxygen

A colourless, odourless and tasteless gas comprising 20 per cent of the atmosphere by volume. Combined with other elements, oxygen makes up nearly 50 per cent by weight of the earth's crust, and is thus the most abundant element by far. *Combustion* of fuels is almost exclusively based on reactions of carbon, or hydrocarbons, with oxygen resulting in formation of carbon dioxide, water and *heat*. It is an essential ingredient of all life-enhancing processes on earth, such as respiration in animals and photosynthesis in plants.

Pure oxygen may be isolated from air by liquifaction and subsequent distillation.

symbol	O
atomic number	8
atomic weight	16.00
melting point	54.8 K (−218.4°C)
boiling point	90.2 K (−183.0°C)
density —	
liquid (at 54.8K)	1140 kg m^{-3} (1.14 g/cm^3)
gas (at 0°C, 1 atm)	1.43 kg m^{-3} (1.43 g/l)

PQ

paraffin

1. Colloquial UK term for kerosine or burning oil.
2. Common alternative name for the group of aliphatic hydrocarbons more correctly called *alkanes*.

paraffinic

1. A classification of crude petroleum used by the US Bureau of Mines in their *correlation index* to describe a crude oil of low specific gravity.
2. A term loosely used to describe mixtures of hydrocarbons containing a high proportion of *alkanes* (paraffins).

paraffin wax

A wax made up principally of *alkane* hydrocarbons of high molecular weight, separated out from lubricating oils by crystallization. It is important, when making lubricating oils from petroleum, to remove waxes, otherwise at low temperatures in particular the performance of the lubricant will be poor. This process is known as *dewaxing*.

Waxes have a variety of applications including candle making, electrical insulators, as a waterproofing material in papers and textiles, and in the manufacture of polishes.

passive solar heating

A method of building design such that maximum use is made of incident solar radiation for the maintenance of acceptable internal temperatures in winter without the use of 'active' devices such as solar collectors, and without causing problems of summer overheating. The aim of such design is the attainment of a self-regulating, reliable system with no mechanical moving parts which is suitable for general domestic installation.

Elements of such design include a large south-facing glazed area with some form of thermal insulation shutter for use at night; the use of simple *sensible heat storage* in the building fabric; the exploitation of thermal convection currents for heat distribution in winter as well as summer cooling; the design of the building structure to give maximum summer shade; and the use of simple reflective blinds to reduce summer overheating.

pass-out steam turbine

A type of *steam turbine* in which steam is extracted from an intermediate stage, generally for heating purposes. The steam is condensed at a relatively high temperature (80°-120°C) and the heat given up is transferred to a distribution medium such as hot water or low-pressure steam. Condensate is returned to the boiler. Pass-out turbines are one of several types of prime mover that can be used in *combined heat and power* (CHP) schemes.

Compared with the *back pressure steam turbine*, the pass-out turbine has the advantage that the power/heat output ratio can be varied within wide limits. However, this operational advantage is achieved at some expense since the cooling system must be retained.

See also *district heating; total energy systems.*

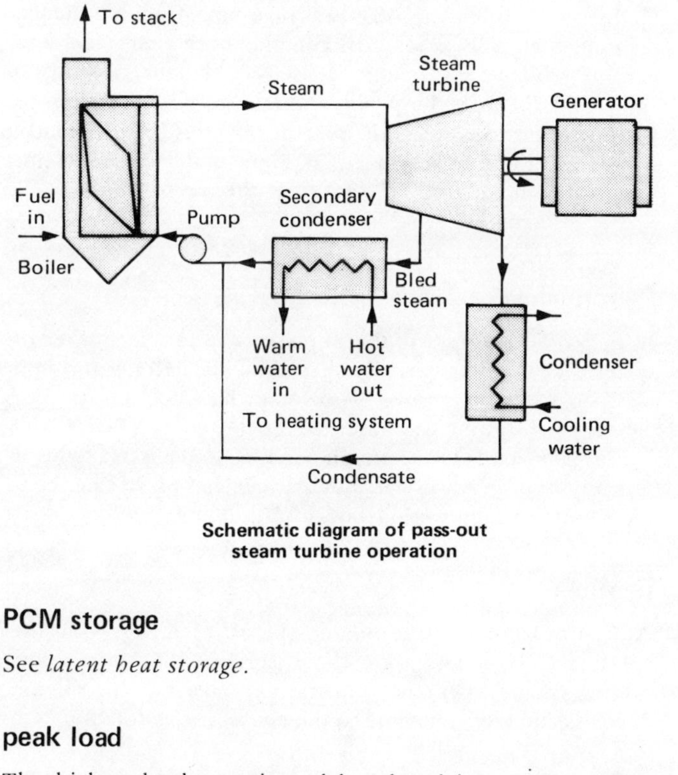

Schematic diagram of pass-out steam turbine operation

PCM storage

See *latent heat storage*.

peak load

The highest load experienced by electricity or gas supply

systems, usually occurring in extreme weather conditions. Storage media are installed to help meet peak loads but even so *peak load shaving plant* may be required to satisfy demand, particularly in electricity generation and transmission where storage is inherently more difficult and expensive.

See also *simultaneous maximum demand*.

peak load shaving plant

Equipment specially installed in electricity and gas supply systems to meet peak loads at times of high demand. Shaving plant usually has a low capital installation cost but is comparatively expensive to run. Examples include *gas turbines* used to generate electricity and small *oil gasification* plants in gas supply systems used to manufacture gas when the rate of supply from other sources is insufficient.

peat

A low-grade *fossil fuel* laid down over the last million years as the decomposed remains of vegetable matter. It has a low calorific value of around 75 therms per tonne (8.0 GJ/tonne) and is chiefly used as a heating fuel in parts of Russia and in Ireland. It is also used in the Irish Republic as a fuel for power station boilers.

pebble-bed reactor

A type of *high temperature gas-cooled reactor* developed in Germany in which the core is a huge bin of pebble-shaped ceramic fuel balls. Each ball contains both fissile fuel and carbon moderator. Helium is used as coolant and this is blown up through the pebbles which are slowly shaken to the bottom of the bin as the reactor operates. If sufficiently irradiated, they are removed; if some usable fuel remains they are re-introduced via the top of the reactor.

See also *ceramic reactor*.

Pelton wheel

A *hydraulic turbine* of the impulse type in which a jet of water is directed at specially shaped buckets attached round the periphery of a wheel so turning the wheel on its axis. This type of turbine was patented in 1880 by L A Pelton and is normally used where a high head of water is available as in many hydroelectric schemes. Efficiencies of 90+ per cent are possible and power outputs of 60,000 kW.

pentane

A hydrocarbon of the *alkane* (paraffin) series with the formula C_5H_{12}. There are three isomers each with a low boiling point. Pentane occurs naturally in petroleum deposits and gives its name to the *pentane-plus fraction*.

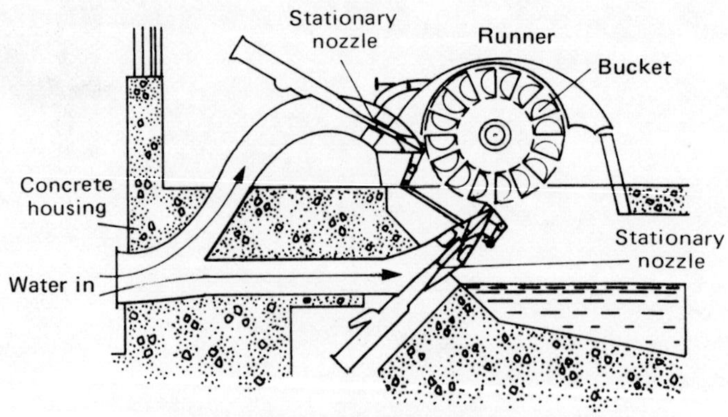

Section through an impulse (Pelton wheel) type of hydraulic turbine unit (much simplified)

pentane-plus fraction

A range of low-boiling point liquid hydrocarbons occurring naturally in petroleum and natural gas liquids. The lightest members are *pentane* and its isomers, followed by hexane, isomers of hexane, and higher boiling point hydrocarbons.

PER

Abbreviation for *process energy requirement*.

petrochemical feedstocks

Feedstocks derived during *petroleum refining* and which are used for the production of a variety of chemicals. For instance, *naphtha*, or light distillate feedstock, is widely used in the petrochemical industry.

petrol

Popular UK term for *motor spirit*.

petroleum

Petroleum (a general term meaning 'rock-oil', indicating that it is found in the earth's crust and is superficially different from animal or vegetable oils) is a complex mixture of hydrocarbons, consisting predominantly of *alkanes* but also containing benzenoid hydrocarbons and naphthenes. The raw material is known as *crude oil*, and crude oils from different parts of the world vary considerably in viscosity, specific gravity and levels of impurities. The constituents of petroleum vary from region to region, but in general it has a carbon content of 83-87 per cent, a hydrogen content of 11-14 per cent, minor amounts of oxygen, nitrogen and sulphur and traces of metals. The ash content is uniformly low at about 0.05 per cent.

Origins. Petroleum occurs in sedimentary rocks deposited under marine conditions. The process by which crude oil is formed is believed to be by the gradual decay and compression of various marine deposits. Fine-grained muds rich in planktonic remains are deposited offshore in an oxygen-poor environment. Slow decomposition by anaerobic bacteria turns the marine remains into an amorphous material known as sapropel. As the muds are buried and compressed by subsequent sediment accumulations, the sapropel is converted to petroleum compounds by physical and chemical processes which are not fully understood but which must occur at temperatures of less than 200°C to preserve the organometallic compounds (porphyrins) found in most crude oils.

Large reservoirs of petroleum will exist only in those areas where thick successions of marine strata were laid down. In addition, migration from such a source bed to a suitable reservoir bed must be possible and escape at the surface must be prevented by a suitable seal or *caprock*. Finally lateral migration of the hydrocarbons must be prevented by traps within the reservoir beds. The three most common structures in which crude oil and associated gases are found are: *anticlinal* traps formed by folds in the earth's strata (80-90 per cent of the world's proven oil reserves are located in anticlinal traps); *fault* traps produced when the earth's crust fractures under stress, and *salt dome* traps produced when deep beds rich in salt are forced upwards under high pressure.

Nature. The predominant hydrocarbons found in crude oil are of the *alkane* series. (General formula, C_nH_{2n+2}, where n can range from one to several hundreds thus determining whether the hydrocarbon is in a gas, liquid or solid state.) Different products derived from crude oil contain a mixture of a limited range of hydrocarbons. *Gasolines*, for instance, are made up of hydrocarbons with between five and ten carbon atoms per molecule. (See *Appendix 4; Table 2.*)

Basically, the longer the molecular chain the less combustible the hydrocarbon is. *Branched chain hydrocarbons* are more controllable, in internal combustion engines, and are used in the manufacture of *high octane fuels.* The many products can be divided conveniently into four groups:

Petroleum products

Group	Examples	Uses	Boiling point °C
Gaseous products	Propane Butane	Portable cooking and heating appliances	–45 to 0
Light distillates	Gasolines Naphtha	Motor spirit Petrochemicals	30–250
Middle distillates	Kerosines Diesel oil Gas oil Heavy gas oil	Diesel road vehicles, jet aircraft, domestic heating, gas turbines	200–350
Residue	Heavy fuel oil Lubricating oil Bitumen	Industrial fuel, ship's bunker fuel, lubricants, road making	greater than 350

Units of measurement. Quantities of oil are expressed in terms of either volume (commonly barrels) or mass (tonnes). A barrel of oil contains 35 Imperial gallons, 42 American gallons or 159 litres. The number of barrels of crude oil per tonne varies from about 6.9 to about 7.6 depending on the density of the particular oil.

Since many petroleum products are used as fuels, calorific value (a measure of the heat released on combustion) is an important property. Some typical calorific values are given below:

	Therms/tonne	GJ/tonne
Gasoline	445	47
Gas oil	430	45
Heavy fuel oil	405	43
Crude oil	425	45
Coal (bituminous)	275	29

See *Appendix 3; Table 4.*

Exploration and production. Prospecting for oil is becoming an increasingly specialized field as more sophisticated techniques are developed in an effort to reduce the number of dry wells drilled. Since the more obvious oil fields have been found (on land and in shallow water) exploration for new oil fields has moved to more difficult areas (Alaska, North Sea) where drilling costs are very expensive and thus there is a great incentive to improve prospecting techniques. The two basic geophysical techniques in current use are gravimetric surveys and seismic surveys. Gravimetric methods involve identifying anomalies in the gravitational pull of the earth caused by variations in the density of rocks. Seismic methods estimate the depth and position of potential oil-bearing structures by measuring differences in seismic wave reflections generated above an area by small explosive charges, air guns etc. More advanced seismic methods known as bright spot techniques are being developed whereby it may be possible to distinguish between echoes from liquid, gas and rock and thus reduce the number of dry wells drilled. Advanced aerial photography from both aircraft and spacecraft using infrared techniques has also aided prospectors in identifying potential oil-bearing structures (eg NASA's Earth Resource Technology Satellite, ERTS). Computers are used to sort and interpret the collected data. Data interpretatation is often as expensive as data collection, but, although the overall prospecting costs are high, they are small compared with the cost of drilling an offshore well in hostile waters.

Following the identification of a potential oil-bearing structure by one of the methods above, an exploration well will be drilled and, if oil is found, further appraisal wells will be drilled to assess the extent of the oil reservoir. If oil is present in commercially viable quantities, production equipment will be moved to the site. Production wells will then be drilled into the oil-bearing strata. (Since offshore oil platforms are very expensive, production wells are deviated considerably from the vertical so that as much of the field as possible is exploited from a minimum number of platforms.) Initially, oil and gas can be raised using the intrinsic pressure within the reservoir caused by water below the oil and gas in solution or above the oil. However, at some stage in the production phase, secondary recovery methods will be used to maintain oil flow and maximize oil production. These methods involve the reinjection of water or natural gas into the reservoir and can increase oil recovery of a field from 25-30 per cent to 40-50 per cent of the total oil in place. From the field crude oil is then transported by pipeline or tanker to a refinery.

Uses. The main uses of petroleum products are for transport, for domestic heating and by industrial, agricultural and bulk consumers.

Transport. For motor vehicle, air and sea transport petroleum products are the major fuels (only in the railway sector is there significant competition with petroleum products as a transport fuel) and bitumen is widely used for road surfaces. Although nuclear powered submarines and electrically-charged battery vehicles have been developed, these are likely to have very little impact in the near future while oil from conventional sources is available.

Domestic heating. Hot-water boilers and hot-air furnaces burning gas oil are widely used to centrally heat houses and to a much smaller extent kerosine is used directly in 'paraffin heaters'. In this sector petroleum products compete with gas and electricity for convenience as fuels.

Industrial sector. The major uses of petroleum products in the industrial sector are for steam-raising to produce process heat; for firing industrial furnaces to melt and heat treat metals, and to heat kilns drying pottery, bricks, cement etc; as a lubricant for bearings, gears and hydraulics; and as a feedstock for the petrochemical industry from which a wide range of products is obtained. These include plastics and fibres, solvents, detergents, waxes, bulk chemicals for other industries, and pharmaceuticals.

Agricultural. Petroleum products are used for machinery and tractor fuels, heating and crop drying, and, indirectly as fertilizers and associated chemicals.

Bulk consumers. Electric utilities and chemical companies often buy in bulk direct from refineries at a reduced unit price, and, to reduce distribution costs, chemical complexes and oil-fired power stations are often sited close to refineries.

World oil consumption figures	
World oil consumption 1976	= 2878.8 million tonnes
Consisting of	
Total Western Hemisphere	= 1094.7 million tonnes (38.0%)
Total Eastern Hemisphere	= 1784.1 million tonnes (62.0%)
World total excluding USSR, Eastern Europe and China	= 2342.8 million tonnes (81.4%)

Source: BP Statistical Review of the World Oil Industry 1976

Reserves.

World 'Published Proved' Oil Reserves at end 1975

Country/area	Thousand million tonnes	Share of total	Thousand million barrels
USA	5.1	5.6%	38.9
Canada	1.1	1.2%	8.2
Total North America	6.2	6.8%	47.1
Latin America	5.0	5.5%	35.4
Western Europe	3.4	3.8%	25.6
Middle East	50.1	55.5%	368.3
Africa	8.7	9.6%	65.1
USSR	11.0	12.1%	80.4
Eastern Europe	0.4	0.5%	3.0
China	2.7	3.0%	20.0
Other Eastern Hemisphere	2.9	3.2%	21.2
World	90.4	100.0%	666.1

Source of data: USA – American Petroleum Institute; Canada – Canadian Petroleum Association; All other areas – Estimates published by the 'Oil and Gas Journal' (Worldwide Oil issue 29th December 1975)

Notes: 1. For the USA and Canada the data include oil which it is estimated can be recovered from proved natural gas reserves. 2. The data exclude the oil content of shales and tar sands.

See *Appendix 1; Tables 5, 6, 7* and *Appendix 4; Tables 1 and 2.*

petroleum coke

A solid carbon residue which may be formed as the result of particular *petroleum refining* processes. Previously used as a solid fuel, petroleum coke is now a useful source of *graphite*.

petroleum fractions

The constituents of crude petroleum that distil at different boiling points or over given boiling ranges during the fractional *distillation* of crude oil. This is carried out in a *fractionating tower* or column and fractions within the required boiling range are taken from predetermined points along the column. The lightest constituents, gases and light gasolines, are the most volatile and come off at or near the top of the tower, fractions having higher boiling points coming off progressively at points further down the tower.

Further separation of distillates obtained from the primary fractionation is achieved by sending the distillate to side-strippers where a more sophisticated fractionation is carried out to obtain separate fractions covering a narrower boiling point range.

Petroleum fractions obtained from the fractionating process may be classified in one of several rather loosely defined ways eg light distillate, middle distillate, heavy distillate and residue or as petroleum gases, gasoline, kerosine, gas oil and residual oil. All these categories constitute boiling point ranges which themselves overlap one another to a large extent.

petroleum gases

Hydrocarbons that are gaseous under normal conditions occurring naturally in crude oil deposits or as a constituent of natural gas liquids in natural gas accumulations. The term petroleum gas usually refers to propane or butane, or a mixture of the two.

See also *liquified petroleum gases.*

petroleum refining

The treatment of crude petroleum so as to separate out a number of *petroleum fractions* suitable for specific applications such as feedstock, transport fuels, heating oils, or lubricants. The first refining stage may constitute sufficient processing for some oil products, but others, such as gasolines and lubricants have to undergo further blending and refining treatment before they reach the necessary standard. There has over recent years been a shift away from *kerosines* to *gasolines* as the petroleum products most in demand and this has necessitated new refinery techniques as the market for gasolines in many areas exceeds its natural occurrence in crude petroleum. Removal of sulphur and heavy metals is also essential for most oil products. Cracking techniques, including *catalytic cracking* and *thermal cracking* have therefore been developed to breakdown the larger hydrocarbon molecules making up the *middle distillate* range into smaller ones suitable for blending into *motor spirit*, and for use as *petrochemical feedstock*. The blending of refinery products to meet the specific requirements of the modern spark-ignition engine also requires the introduction of additives such as organolead compounds not produced as part of the normal refinery process.

It is now common practice for oil refineries to be situated close to centres of consumption, rather than at the supply source, but in the formative days of the petroleum industry and certainly up to the time of the Second World War it was usual to refine petroleum at the point of production, eg Middle East, Caribbean area. The economics of oil production and refining dictated the need to transport large quantities of crude oil in tankers of ever-increasing size long distances to huge, modern refinery complexes such as those situated at Rotterdam in Holland. It seems likely that the Middle East producing countries, aware of their newfound power and anxious to build up their home-based industries, will in future be far keener to sell refined petroleum products rather than straightforward crude.

photobiological energy conversion

The use of plants to convert the energy of sunlight into chemical energy through *photosynthesis.*

See also *photosynthetic energy.*

photosynthesis

The process by which solar energy is trapped in the *chloroplasts* of green plants by *chlorophyll* and becomes locked as chemical energy in the carbon compounds formed from carbon dioxide and water. Water is split into its components, hydrogen and oxygen, and the carbon dioxide is reduced; carbohydrates and eventually other complex organic molecules are formed, the oxygen being released to the atmosphere. The overall reaction may be summarized as:

$$6CO_2 + 6H_2O \xrightarrow[\text{energy}]{\text{solar}} C_6H_{12}O_6 + 6O_2$$

In reality it is a complex series of enzyme-controlled reactions with intermediate compounds being formed initially. Photosynthesis is of particular importance because its end products cannot be synthesized by other means from simple chemicals. All living things are ultimately dependent on it.

See also *cellulose; energy crops; photosynthetic energy.*

photosynthetic energy

Energy locked in plant cells as chemical energy due to *photosynthesis* and which has been derived from solar energy. Photosynthesis, in effect, involves the chemical reduction of carbon dioxide taken in from the atmosphere by hydrogen derived from the splitting of water, utilizing the energy of sunlight trapped in the membranes of the *chloroplasts* by the green pigment *chlorophyll. Carbohydrates* are formed after a series of complex enzyme-controlled chemical reactions.

All life, with very few exceptions, is dependent on photosynthesis for animals cannot synthesize complex organic chemicals from simpler inorganic raw materials. They must eat plants to survive. In turn, carnivorous creatures eat the herbivores. The chemical energy in their bodies is therefore derived in the first instance from solar energy trapped in the sugar molecules formed in photosynthesis.

The energy in fossil fuels is derived from past photosynthesis; photosynthesis contributes currently to crop production for human and animal feedstuffs, to the supply of industrial raw materials, such as timber, and the growth of a few other plant products for fuel.

Undoubtedly there is scope for increasing the production of energy through photosynthesis. Photosynthetic energy is renewable and the products containing it can be stored with few problems.

Since it is a solar energy dependent process, the production of photosynthetic energy is obviously greatest where the annual input of solar energy is highest, namely in sub-tropical regions. (See Map 17.) The quantity of solar energy falling on land suitable for cultivation is many times current energy use. But there are a number of factors limiting the photosynthetic output from a particular region. Additionally, the possible secondary use of energy which is derived photosynthetically depends on how much solar energy is available on an all-the-year round basis, how efficiently the particular plant species convert solar energy and how much of the energy incorporated in photosynthesis can then be extracted.

Temperature (high and low) is an important limiting factor affecting the rate of photosynthesis, as is water shortage and the lack of soil nutrients. Photosynthesis is reduced below 10°C and almost ceases below 5°C. Water shortage often occurs at times when incident solar energy is high. The degree of uptake of the available solar energy is dependent on the level of photosynthetic activity, the arrangement of leaves (for example, how much they overlap) and the distribution of photosynthetic products through the plant tissue.

Nearly half of the incoming solar radiation (essentially that in the visible part of the spectrum) is suitable for photosynthesis, but at low light intensities only about a fifth of this can be fixed. Other limiting factors pertain as the light intensity increases, such as atmospheric CO_2 levels.

The majority of temperate crop species attain maximum photosynthetic activity at comparatively low light levels (approximately 0.2-0.3 kW m^{-2} of radiation total) so that at full light levels perhaps less than one per cent of the total light energy available is fixed. Some tropical species (eg sugar cane) can photosynthesize even where the light intensity is high. They are poor photosynthesizers at temperatures of 20°C or below, however, achieving maximum photosynthetic levels at 30°C or higher.

Obviously the length of the growing season and the type of crop grown will affect the conversion rates referred to. Perennial crops – eg grasses or evergreen conifers – will tend to have the highest annual outputs in temperate regions. In the tropics high rates can be achieved throughout the year with many plantation crops (sugar cane is an example).

With cereal and root crops, which grow for only part of the year, the total energy intake is dependent on the efficiency of uptake during a restricted period. In such circumstances the storage organs of root crops are able to take in the plant's photosynthetic products over a longer period than are cereals, though rice crops grown in areas where light intensity is high achieve high rates of uptake (two to three per cent of incoming light energy during the growing period).

In temperate regions photosynthesis from crops is well below the potential levels. A NASA/NSF study quotes a figure of three per cent of the total land surface of the United States being able to provide photosynthetic energy equivalent to the total electrical energy requirements for 1985 if crop plantations were operated at a three per cent conversion efficiency of the total solar energy available. Estimates for the UK suggest that the total cultivable land surface with a closed cover of perennial grass and assuming one per cent fixation of total incipient solar energy could provide half the country's energy requirements. In practice, even if it were possible to use the whole of the cultivable land surface in this way, photosynthetic production is very much lower. Soil and climatic factors, temperature, light intensity, soil nutrients, water availability, and variation in canopy density all contribute to reducing the photosynthetic efficiency to less than a quarter of the potential levels. Only one tenth of the energy requirements could be met. Collecting, storing and processing the photosynthetic crops would also be energy consuming.

At the moment cultivable land is used to grow crops for human and animal consumption. Timber is used for industrial materials and energy supply, and crop residues, such as sugar cane (bagasse), to supply energy for process heat. But, in the case of food crops, it is usually only the crop residues that find use as energy sources. Crop production is highly consumptive of support energy – in fertilizers, fuel for running farm equipments etc – and this is largely fossil fuel derived. Any assessment of possible energy production from crop sources must take this into account (see *energy ratio*). Where crops are grown for animal consumption the low output of photosynthetic energy is compounded by the poor secondary use of this through the animal. It may consume more energy in support energy than is obtained in edible energy. The use of photosynthetic energy for meat production is a wasteful way of using such resources. It would be better to eat the plant crop directly or even to convert the crop into protein synthetically for consumption.

It is doubtful whether the growing of crops for photosynthetic energy production purposes will ever have a large-scale application as a source of industrial energy. It could have importance on a local basis, however. Similarly there may be local agricultural and domestic applications. However, the use of crop wastes or of crops such as timber which can be grown on otherwise unsuitable land could prove a valuable energy source on a local scale.

Additionally, more thought could be given to growing crops for both food and fuel. In the United States and Australia in particular, studies are taking place into the utilization of natural crop species such as eucalyptus and various grasses. Such energy farming activities would not conflict with food farming in that it need not take place on land cultivable for regular crop species. The products from these energy farms could be pyrolyzed (see *pyrolysis*) or fermented anaerobically.

The culture of *algae* (eg Chlorella sp) as possible sources of protein has been carried out experimentally for some time. It is now suggested that a suitable culture medium could be obtained using organic waste materials and the algae cultivated for their food and energy content. The wastes would also be purified.

Perhaps by selective breeding more productive crops could be produced, with increased food and fuel yields. Some agricultural trends might even be reversed. There has been a tendency, for example, to grow shorter stalked varieties of cereals, partly to avoid weather damage, but also to reduce the output of straw which many farmers have regarded as a nuisance.

Better use could be made of the straw produced currently from growing cereals. Half of the trapped photosynthetic energy fixed by cereals is contained in the straw. Straw can be used as animal feed or bedding, for energy purposes, to enrich the soil by ploughing in, and as a raw material for industrial use (for example, in the manufacture of insulating boards). In the UK more than a third of the straw is burnt, the energy so wasted representing a quarter of that country's energy requirements for agriculture. Even its use as a heat provider for grain drying would be cost and energy saving. Throughout the world there is wastage of residues from forestry and timber processing activities. Obviously the collection and use of such wastes would be energy consuming and careful energy budgeting would be necessary before any attempts were made to establish plant on any scale. It would seem, however, that the local use of waste materials of this nature would make sense. Perhaps materials could be burned to provide heat for cooking and heating purposes or pyrolyzed to provide liquid fuels with more flexible usage.

Cellulose rich residues can be fermented anaerobically to produce *methane*, (see *anaerobic fermentation*). Bacterial action can also be used to break down starch and cellulose to glucose and thence to *alcohol*. In the Second World War starch-rich crops were fermented for alcohol production and this was added to *aviation spirit*. There is scope, too, for digesting animal wastes on farms for methane production. Pig, chicken and cattle manure is suitable for such treatment and easily collected where rearing is intensive (see *waste materials energy*).

One other area of research which the energy crisis has precipitated is that of trying to mimic photosynthesis by creating artificial systems. The success of such attempts is closely linked to our understanding better the complex changes that take place during photosynthesis. For instance, using chloroplast membranes and bacterial hydrogenases, hydrogen gas can be produced by splitting water. Electricity may also be produced photoelectrochemically by such reactions.

Investigations of nitrogen fixation in plant roots has shown that the availability of photosynthetic products translocated to the roots is a limiting factor rather than the activity of the nitrogen-fixing bacteria. This and other related investigations will make it possible to increase nitrogen fixation and to choose alternative crops which have been discovered recently to have symbiotic nitrogen fixing activities in their roots. In this way it may be possible to reduce the application of nitrogen-rich fertilizers and thus save energy.

photothermionic generator

See *solar cell*.

photothermoelectric generator

See *solar cell*.

photo-voltaic cell (photo-cell)

A device which converts electromagnetic radiation into electrical energy. The principal component of a photovoltaic cell is a crystal of semi-conductor material. Radiation striking the semi-conductor creates a difference in electrical potential (voltage), the magnitude of which depends upon the intensity of the radiation.

Semi-conductors are special materials in which a small quantity of energy will excite electrons and give them sufficient energy to jump from 'filled bands' to 'empty bands'. Electrical conductivity in a material depends upon electrons moving in partly filled bands, much as traffic can flow only in partially filled roads. Bands constitute ranges of energy levels and depend upon the atomic structure of the material.

A crystalline semi-conductor such as silicon comprises a regular array or lattice of atoms held together by an equally regular array of electrons. These electrons fill bands having the lowest energy. Deliberately adding (or 'doping') impurity atoms, possessing an excess or deficit of electrons per atom as compared with the rest of the lattice, will disrupt the electronic structure of the lattice. Such defects facilitate electrons being promoted to unfilled bands.

A very slight excess of the electrons in an n-type semiconductor, or deficit of electrons in a p-type semi-conductor will markedly increase its electrical conductivity. Surplus electrons and positive holes are both mobile within the lattice structure. However, the holes move in opposite directions to the electrons in a given electric field.

A junction formed between p- and n- type semi-conductors will give rise to a potential difference on contact, since some electrons diffuse from the n-type semi-conductor to the p-type semi-conductor. The junction itself with its potential difference separates the electrons and holes, as they have to move in opposite directions in its electrical field.

Photons landing sufficiently close to the junction break off electrons from the lattice leaving behind an equivalent number of holes. Normally equal numbers of electrons and holes in the same material will recombine rapidly but the presence of the junction's field diverts the electrons and holes in opposite directions. This movement of electrons comprises the current and voltage which may be connected to an external circuit.

Photo-cells are not very efficient in converting solar energy into electricity. For example, *silicon cells*, comprising p-type semi-conductors with a thin (10*u*) surface layer of n-type semi-conductor, have an 11 per cent conversion efficiency. Possible ways of improving efficiency include using semi-conductors which have a wider absorption band, or using optical systems such as a set of dichroic mirrors to separate light of different wavelengths. The separated beams of light would then activate photo-cells at their optimum wave-length bands.

The disadvantage of any of these systems to improve efficiency is that with today's technology they further increase the cost of collection. An alternative approach involves concentrating the light, but the resultant higher operating temperatures interfere with the action of the cells. High melting point semi-conductors could provide one solution.

With today's cost of silicon cells about 1000 times the capital cost of a fossil fuel power station per kW installed, photo-cells are not at present viable except in special circumstances, for example where cost is irrelevant or power requirements are modest. However, technological advances in the future could cut the cost of silicon cells by a factor of 100. Already the less efficient cadmium sulphide cell costs considerably less than a silicon cell. Research continues into alternative combinations of semi-conductor materials and less expensive manufacturing methods.

pinch effect

The self-compression of an electric arc by its own magnetic field.

See also *fusion reactor.*

pipeline capacity

The carrying capacity of a pipeline designed to carry gaseous or liquid fuels usually measured in million cubic feet per day (mcfd) of gas or barrels per day (b/d) of oil.

pipeline quality gas

A phrase used to describe *high-Btu gas* produced by *coal gasification* processes suitable for distribution by pipeline to consumers already using *natural gas.* The first stages of coal gasification usually produce carbon monoxide and hydrogen, a *medium-Btu gas* mixture, and this mixture requires conversion by catalytic *methanation* to a pipeline quality gas essentially consisting of *methane.*

pitch

A collective term covering the dark-coloured, highly viscous residues obtained from the distillation of *coal tar*. The term should not be confused with *bitumen* (UK) or *asphalt* (USA) which are petroleum residues.

pitchblende

See *uraninite.*

plasma

A gas or vapour in which the atom or molecules are ionized, consisting of positive (or negative) ions and electrons. Such plasmas are electrically conducting.

See also *fusion; fusion reactor; magnetohydrodynamic (MHD) power generation.*

plasma MHD generator

A form of magnetohydrodynamic generator in which the working fuel is in the form of a conducting (ionized) gas or plasma. The plasma is normally 'seeded' to increase its conductivity.

plate power (or plated power)

The US term synonymous with *power rating.*

platforming

A shorthand term for 'platinium reforming', a *catalytic reforming* process used to convert straight-run *gasoline* into a product richer in branched-chain hydrocarbons and aromatics, thereby increasing its octane number. The catalyst used in this process is platinum (about 0.5 per cent by weight) dispersed on a highly purified alumina carrier. The platforming process produces large quantities of hydrogen which may be used elsewhere on the refinery complex site for *hydrocracking* or *hydrodesulphurization.*

The *catalytic rich gas process*, in which *naphtha* undergoes *steam reforming* to give gaseous products is essentially a platforming process.

plutonium

A highly toxic radioactive metal which is not definitely known to occur in nature but is possibly present in the earth in minute quantities. It has many isotopes, the most important of which is the *fissile* plutonium-239 which is made in breeder reactors from uranium-238 by absorption of neutrons. The isotopes of mass numbers 240, 241 and 242 are produced by side reactions in *nuclear reactors*; plutonium 241 is fissile. The presence of these isotopes makes the waste products from breeder reactors highly dangerous and complicates their disposal.

The fertile isotope plutonium-238 is used as a fuel source in the *nuclear cell.*

See also *nuclear fuel cycle; nuclear reactor; radioactive waste; uranium.*

symbol	Pu
atomic number	94
atomic masses	238, 239, 240, 241, 242
melting point	640°C
boiling point	3200°C
density	19,814 kg m^{-3} (19.8 g/cm^3)

plutonium-238

The first isotope of *plutonium* to be identified, it is fertile and has a half-life of 86.4 years.

See also *nuclear cell.*

plutonium-239

A fissile isotope produced in nuclear reactors from uranium-238. It has a half-life of 2.44 x 10^4 years.

See also *plutonium.*

poisoning

In nuclear reactor terminology, the formation of a fission product which is a strong absorber of neutrons. If not carefully taken into account, this phenomenon can reduce *reactivity* to such an extent that the *fission chain reaction* ceases altogether. An example is poisoning by xenon-135, a product of uranium fission, which is a particularly strong neutron absorber.

pollution

See *air pollution; environmental effects; sulphur emissions; thermal pollution; water pollution.*

potential energy

The energy possessed by a body or system that is due to its position or state. Potential energy can arise due to gravitational forces, as in the case of a *head of water*; electrostatic forces, as in the case of the potential energy stored in an electrical capacitor; mechanical forces, as in a compressed spring; and chemical forces, as in the case of energy stored in chemical compounds and storage batteries.

See also *kinetic energy.*

pour point

Of a petroleum oil, this is the lowest temperature at which the oil will flow or pour after it has been cooled and left without disturbance for a predetermined length of time under specified conditions. The pour point for a low density African crude might be $-51°C$ and that for a high density South American crude 15°C. Pour point is one of the indicators of the probable flow properties of the crude oil in its transmission through pipelines and loading and unloading to and from tankers. It is an important element in a crude oil analysis, and is also used as a measurement of the flow properties of refined oil products.

power

The rate at which *work* is done, or the rate at which energy is produced or consumed. It is most commonly used in relation to electricity output and consumption, for instance when referring to the *installed capacity* of a generating system or the *power rating* of an electrical appliance. The unit of power is the *watt* or, more commonly, a higher multiple such as the kilowatt (kW) or megawatt (MW).

power density

The rate of production of heat in the nuclear reactor core, per unit volume. A common unit of measurement is kilowatts per litre (kW/l). The *Magnox reactor*, and the *advanced gas-cooled reactor* developed in the UK, both have relatively low power densities and so the demands made on the *coolant* and the cooling circuit are less than in reactor designs, such as the *BWR* and *PWR*, where the power density is much higher. The power densities to be taken into account in any commercial *fast breeder reactor* designs will be markedly greater than those encountered in the PWR and this has given rise to doubts about their likely safety.

power kerosine

Also known as vaporizing oil. A kerosine fraction used as a fuel in spark ignition engines such as agricultural tractors and stationary engines. As diesel engines are now more commonly installed for such purposes, the use of power kerosine is diminishing and it is being replaced by diesel fuel.

power rating

The rate at which a machine or appliance uses electrical energy. Power ratings may range from the order of one hundred watts (light bulb, refrigerator, television) through two or three kilowatts (immersion heater, electric fire) up to several megawatts (large compressors).

power stations

Colloquial term for plants producing electricity on a large scale.

See also *thermal power stations.*

premium fuel

A fuel which is able to command a higher price in the market place than a competitive fuel capable of performing the same task. In this sense a fuel may warrant the description 'premium' because of its relative cleanliness, convenient handling, or ease of storage. However, the term 'premium' as commonly used does not necessarily tie in with this strict economic definition. The phrases 'premium markets' and 'premium uses' are often loosely used in conjunction with premium fuel when describing a convenient fuel (irrespective of price), a particular application for which there is no readily available replacement (eg motor spirit), and specific uses where efficiency of conversion is maximized by burning one fuel in preference to another (again irrespective of price). For instance, natural gas is often described as a premium fuel; it is convenient and clean and capable of being used highly efficiently, but it is often the case that buyers do not have to pay a higher unit price for gas as compared with competitive fuels.

pressure vessel

Reactor vessel of a nuclear reactor in which coolant is circulated under pressure and which houses the reactor core and other reactor internals. It is a large container of welded steel or pre-stressed concrete. The term is also used for the pressurizer in a PWR which maintains the pressure of the reactor coolant system and copes with changes in the coolant volume.

pressurized water reactor (PWR)

A type of nuclear reactor in which heat generated in the reactor core is extracted by light water circulating at high pressure (up to 150 atmospheres) in the primary circuit. The high pressure prevents the water from boiling in contact with the fuel elements even though the temperature is 320°C. The water acts as a coolant and moderator. Heat is transferred from the primary circuit to the secondary circuit in a heat exchanger and the steam produced (at a lower temperature and pressure than the water in the primary circuit) drives a turbogenerator. This contrasts with a *boiling water reactor (BWR)* in which steam is produced in the reactor core and ducted straight to the steam turbine. Heat output is 3250 MWt giving about 1050 MWe with an efficiency of 32+ per cent. The fuel is uranium dioxide pellets, three per cent enriched, and clad in zirconium alloy (zircaloy), each fuel pin being only one centimetre in diameter. There are slight differences in operating temperature and pressure and reactor details between different manufacturers; the figures given in the table are typical. Since the reactor vessel operates at high pressure, its construction is critical from a safety viewpoint. It is made up of high quality welded steel and has a heavy lid of the same material held on by huge bolts. The tubes through which coolant enters and leaves the pressure vessel are thick-walled and welded to the pressure vessel.

More PWRs are in operation than any other type of nuclear reactor, but there is considerable controversy about their safety, particularly concerning the emergency core cooling systems designed to prevent the reactor core from overheating should an accident occur. The reactor vessel itself is enclosed in thick concrete shielding, the walls of which form the sides of the reactor well which is flooded with water when the core is refuelled. The whole of the primary circuit is also shielded.

Despite their popularity, PWRs are relatively inefficient producers of heat for electricity generation.

pressurized water reactor (PWR) data*

Peak power density	102 kW/l
Heat output	3250 MWt
Electrical output	1050 MWe
Efficiency	32.3%
Fuel	uranium dioxide, 3% enriched, clad in zirconium alloy (zircaloy)
Weight of fuel	99 tonnes
Fuel burn-up	21800 MWD/te
Moderator	light water
Coolant	light water
Coolant pressure	150 atmospheres
Coolant outlet temperature	318°C
Refuelling	Off load

* *This example: Zion 1, USA*

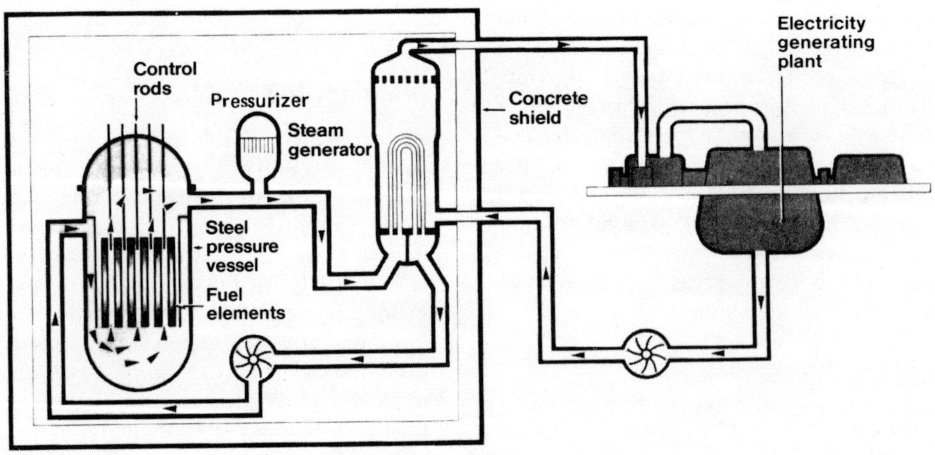

Pressurized water reactor *Reproduced by kind permission of the UKAEA*

primary cell

An *electric cell* which is not rechargeable. Examples of primary cells commonly in use are the *Leclanché cell, mercury cell,* and *alkaline primary cell.*

primary energy

Energy contained in *fossil fuels* such as coal and petroleum and energy derived from renewable sources such as the sun, wind and waves. All the energy which we use comes from these primary sources, though very often the energy may be supplied in the form of *secondary fuels* such as electricity, manufactured gas, or coke. National energy consumption, and future projections of this, are usually measured in terms of primary energy usage using a common unit such as million tonnes coal equivalent (mtce) or million tonnes oil equivalent (mtoe), though it is often more instructive to examine demand for *delivered energy*, or even better, *useful energy.* In this way it is possible to pinpoint feasible *energy conservation* measures which may lead to a drop in primary energy needs.

primary fuels

Naturally occurring fuels which may be used to provide energy directly without undergoing any manufacturing or conversion processes. Examples include coal, petroleum, natural gas and wood; by convention, refined petroleum products are usually included under this definition, as is naturally occurring uranium. *Ambient energy sources* are defined as primary sources of energy but are not primary fuels as such as they are not capable of undergoing direct combustion. They may be used to provide heat or motive power directly, or undergo conversion to *secondary fuels.*

primary recovery

The recovery of oil and gas from a reservoir by utilizing natural forces. The pressure on the hydrocarbons in the reservoir rock must be greater than that at the lower end of the well for oil or gas to flow out. The type of oil, permeability of the reservoir rock, and reservoir pressure will all affect rate of oil flow. As oil is withdrawn the reservoir pressure will decrease and with it the outflow of oil unless other natural mechanisms can sustain the drive. *Water drive, solution gas drive* (also called depletion drive), *gas cap drive* and *gravity drainage* are natural production mechanisms.

process energy requirement

The amount of energy taken from stocks or resources in the direct production of a commodity or service, excluding such indirect energy inputs as the energy used to process or transport raw materials and to construct factories and equipment. The process energy requirement, often abbreviated to PER, is useful to compare the energy implications of alternative processes. In an *energy audit*, the use of PERs shows where *energy conservation* methods can best be applied.

See also *energy analysis; gross energy requirement.*

producer gas

Also known as air gas, a relatively clean, low-grade fuel gas employed mainly in industry for heating materials and in furnaces. The gas is produced by passing a blast of air through red hot fuel. Although coke, anthracite or coals with poor coking properties, are best, other solid fuels such as sawdust, leather waste and even bagasse (the fibrous material left after sugar has been extracted from sugar cane) can be used as fuel.

The carbon in the fuel undergoes two separate reactions in forming producer gas. First, the oxygen in the air oxidizes the carbon to yield carbon dioxide:

$$C + O_2 \rightarrow CO_2$$

Then, as the carbon dioxide rises through the bed of hot fuel, the carbon reduces it to carbon monoxide:

$$CO_2 + C \rightarrow 2CO$$

Carbon monoxide is the main combustible component of producer gas. However, depending upon the composition of the fuel used, the thickness and other conditions of the fuel bed and whether the air blast is damp, producer gas may also contain hydrogen and traces of methane. In addition, because nitrogen accounts for approximately 80 per cent of the air, it is the main but inert constituent of producer gas (about 65 per cent). Indeed the presence of nitrogen as a diluent accounts for the low *calorific value* of producer gas, around 125 Btu/ft^3 (4.7 MJ/m^3).

See also *semi-water gas; water gas.*

prompt neutrons

Neutrons leaving the fissioned nucleus at the precise moment of fission, as opposed to *delayed neutrons* which are emitted just after fission occurs by fission products.

propane

A gaseous hydrocarbon, and member of the alkane series, occurring naturally in crude petroleum and natural gas liquids. Along with *butane*, it is readily liquified and such a mixture is often referred to as *liquified petroleum gas* (LPG), a convenient and transportable heating and cooking fuel.

proton

A positively charged *nucleon*, the number occurring inside an atomic nucleus determining the *atomic number* of the element in question. In all but one case, protons occur along with *neutrons* in the nuclei of atoms. The exception is the hydrogen-1 atom which contains one proton and no neutrons; hence a positively charged hydrogen atom, ie a hydrogen *ion*, is often correctly referred to as a proton.

pulverized fuel

Solid fuel crushed and ground before being fed into large boilers, eg at coal-fired power stations.

pumped storage schemes

Electricity generation schemes installed to help meet

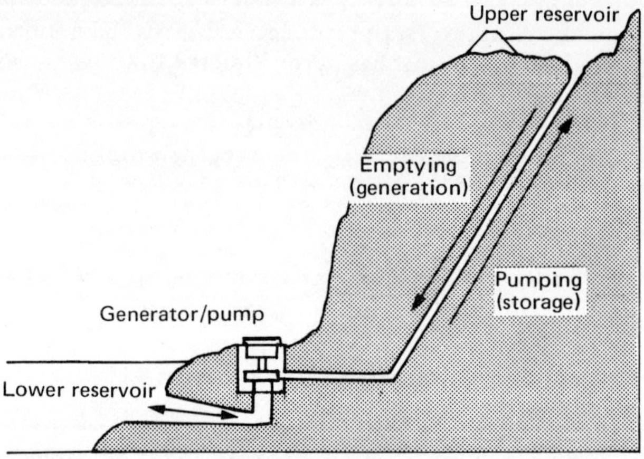

sudden demand fluctuations and peak loads. Electricity generated during periods of low demand is used to raise water from a low to high reservoir, and during peak periods the water is allowed to fall back to the lower reservoir, via a hydroelectric turbogenerator. Despite the apparent energy inefficiencies involved in such schemes, they can prove economically attractive when compared with constructing conventional peak load thermal plant operating at a low *load factor.*

PWR

Initial letters for *pressurized water reactor.*

pyrolysis

The thermal decomposition of a chemical compound, or mixture of compounds. Examples of pyrolysis with energy-specific applications include the *thermal cracking* processes formerly carried out in the petroleum refining industry and some coal gasification and liquifaction processes in which the coal feedstock is subjected to high temperatures leaving behind a pyrolyzed *char.*.

quad

An abbreviation for 'quadrillion Btu's', or 10^{15} Btu, a unit used in the United States as a measure of primary energy consumption. Present US primary energy consumption is approximately 80 quads per year.

Approximate Conversion Factors
One quad is equivalent to:
40 million tonnes of coal 24 million tonnes of petroleum 1 trillion (10^{12}) cubic feet of natural gas 0.5 million b/d of oil for a year

R

rad

Unit of Radiation Absorbed Dose. One rad is the amount of radiation that will cause 1 kg of material to absorb 0.01 joules of energy.

radiant energy

A form of energy that is capable of being transmitted through empty space (ie a vacuum). Although *alpha* and *beta radiation* fit this description, the term is usually restricted to *electromagnetic radiation*.

Radiant energy is given off by any hot body, and the light and heat we receive from the sun is a result of the sun's high temperature. Lower temperature bodies give off less energy and produce longer wavelength radiation. Thus as a body cools its colour changes from white (as in the case of an incandescent lamp filament) to red (eg an electric fire). On further cooling only invisible *infra-red radiation* and microwave radiation are emitted. Heat transfer by means of radiant energy is an important part of many energy technologies.

Radiant energy can also be produced by internal energy transfers within atoms and molecules. This principle is employed in the various forms of discharge lamp and in the laser.

radioactive decay

Transmutation of one isotope of a radioactive element into another isotope of that element by the emission of alpha or beta particles, or by fission with the emission of neutrons into isotopes of different chemical elements. The new isotopes, called decay daughters, are often also radioactive and will decay in their turn. This sequence is called the decay chain, which ends with the formation of a stable isotope.

radioactive isotopes

Isotopes whose nuclei exhibit *radioactivity*. Some radioactive isotopes occur naturally: the radioactive elements *uranium* and *thorium*, for example, are found in minerals in the earth's crust. Most elements can be made to produce radioactive isotopes by irradiation with neutrons. Radioisotopes are used in medicine and industry.

radioactive waste

Material, in gaseous, liquid or solid form, which is a by-product of all the processes in the *nuclear fuel cycle* and which emits ionizing radiation, damaging to all living organisms. For management purposes it is classified as low level, intermediate level and high level waste.

The base line for the setting of standards for the management of man-made radiation is the natural radiation level resulting from cosmic rays, *uranium* and *thorium* in the ground and certain radioactive isotopes in human bodies, particularly potassium-40. This background radiation varies but is estimated to be about 0.1 *rem* a year. Guidance on acceptable dose levels above this base line is given by the International Commission on Radiological Protection (ICRP), founded in 1928, which co-ordinates and assesses radiobiological research. It is a complicated and difficult task because of the nature of the damage done by ionizing radiation, the effects of which are not immediately apparent, except in a nuclear disaster, but appear gradually within weeks, years or generations. Absorption of *alpha particles* or *neutrons* or irradiation by *gamma rays* interferes with the way a cell divides and reproduces itself. The immediate damage is usually minute and invisible but can set up a chain of consequences which multiply the original disruption and lead, in ordinary body cells to cancers and, in reproductive cells, to genetic abnormalities.

The radioactivity of nuclear fuel is not particularly intense before it reaches the *nuclear reactor*. However, one of the products of the natural alpha decay of uranium in the earth is radium-226 which will remain in the ore after the uranium has been extracted, so uranium mine tailings will remain dangerously radioactive for thousands of years. Radium-226 produces the inert radioactive gas radon-222 which, when it decays, produces polonium-218, a solid which being momentarily electrically charged, clings to any neighbouring dust particle. Without highly efficient ventilation in the mines uranium miners have proved to be highly susceptible to lung cancer through breathing in this dust. Precautions must also be taken to prevent the waste ore being blown across the countryside or polluting rivers. The same care must be taken with the large volume of liquid waste, chemically toxic as well as radioactive, left over from the production of uranium yellow cake.

Uranium enrichment plants produce waste which is more hazardous because of its chemical toxicity than because of its radioactivity and can therefore be buried below ground. Land burial is also considered acceptable for low-level solid waste – miscellaneous rubbish like used protective clothing – from other processes in the fuel cycle.

Once the nuclear fuel is inside the nuclear reactor the level of radioactivity increases spectacularly. Neutrons released by *fission* make everything around them radioactive. The *reactor core* is surrounded by heavy biological shielding but even so some radioactivity escapes into the environment. Leakages from operating reactors are known as routine or running releases and are strictly monitored. One of the principal discharges is of argon-41: air blown round the concrete biological shield to protect it from the direct heat of the reactor core is discharged into the atmosphere and some of the nuclei in this air will have absorbed neutrons and become radioactive. Large amounts of argon-41 are released in this way but since it has a half-life of only 110 minutes it decays before sinking to ground level. It has not been considered necessary so far to contain the argon-41, nor the carbon-14 discharged in the form of carbon dioxide by gas-cooled, graphite-moderated reactors, or the tritium from heavy water-moderated reactors. The sodium *coolant* of the liquid-metal fast breeder reactor, however, must be strictly confined within the biological shield. Under neutron activation it becomes the intensely gamma-active sodium-24.

Fuel cladding may develop flaws and leak gaseous or volatile fission-products into the coolant including iodine-131 which, inhaled or ingested in milk, concentrates in the thryroid gland. Iodine-131 has a half-life of only eight days but iodine-129 has a half-life of 16 million years and it has been proposed that there should be controls of the emission of this isotope.

All waste material from nuclear reactors and reprocessing plants is radioactive and must be dealt with accordingly. Low level liquid waste – washing water or water from cooling ponds – may be run off into the sea; although there are limits on its discharge. Low level solid waste may be buried.

Intermediate level waste does not require special cooling to remove decay heat but is highly gamma-active and so must be shielded. It includes metallic components from the reactor core and the cladding stripped from fuel elements before reprocessing. Usual practice at present is to store it in heavy concrete vaults. Some permanent method of disposal for any of the waste contaminated with plutonium must eventually be found.

High level waste, the by-product of nuclear fuel reprocessing, must also never be allowed to return to man's environment. The acid solution left behind after the uranium and plutonium have as far as possible been recovered contains, as well as fission products which will remain dangerously radioactive for many hundreds of years, actinides like plutonium which will be intensely radiotoxic for hundreds of thousands of years. It also generates a dangerous level of decay heat. Current practice is to place it in elaborately cooled, double construction stainless steel tanks fitted with air filters, leak detectors and agitators to prevent any solids from settling, sealed into heavy concrete vaults.

One proposal for the increased security and better interim management of high level waste is to fabricate it into glass blocks (after a sufficient cooling period to prevent the glass being melted by decay heat). The advantages of vitrification would be greatly increased if the actinides could first be removed from the fission products. The vitrified blocks of fission products could theoretically be safely buried in uranium mines after less than a thousand years. If the actinides could be made into fuel elements and irradiated in fast reactors they would be converted into fission products and therefore also have the possibility of final disposal. However, the problems of separating the actinides, fabricating the fuel elements and reprocessing them seem at present to be insurmountable.

If the vitrification of high level waste including the actinides is successful, there still remains the problem of its ultimate disposal (any storage system requires maintenance and is always vulnerable to accidents or acts of war). Shooting it by rocket into the sun, melting it into ice caps, and inserting it into the core of the earth are some of the solutions proposed. The most likely proposal however is to bury it in stable geological formations on land or under the ocean.

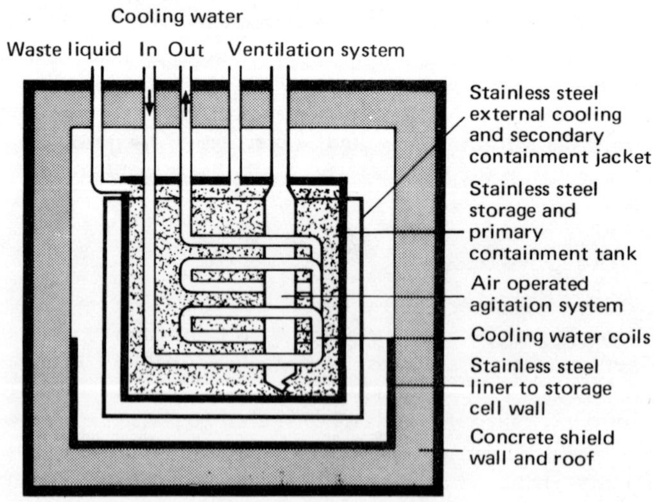

Fission product storage tank

radioactivity

The decay of a radioactive isotope yielding another isotope of the same, or different element, which may in turn disintegrate further until a stable isotopic state is attained.

See also *nuclear energy*.

radium

A radioactive constituent of *pitchblende* and *carnotite*, discovered by Madame Curie whilst investigating the nature of pitchblende ore. The rate of *decay* of the isotope radium-226 is a standard by which rates of radioactive disintegration may be measured. One gramme of radium-226 undergoes 3.7×10^{10} disintegrations per second, and such a rate of decay is defined as the *curie*. Radium-226 is itself a decay daughter occurring in the uranium-238 decay chain. It further decays by loss of an alpha-particle to give radon-222.

radon

A highly radioactive inert gas occurring in nature as radon-222, a *decay daughter* in the radioactive disintegration of uranium-238. It is an *alpha particle* emitter, having a *half-life* of 3.8 days and is a considerable hazard to those engaged in uranium mining, as its decay daughters are, too, radioactive.

reaction turbine

See *hydraulic turbine; steam turbine*.

reactivity

In nuclear terminology, a measure of the rate of a *fission chain reaction* in a *nuclear reactor*. 'Adding reactivity' by withdrawing the neutron-absorbing *control rods* from the reactor core and increasing the neutron density speeds up the rate of the chain reaction.

reactor core

The central part of the nuclear reactor where the fissile fuel is situated, and where heat is generated as a result of nuclear fission.

rechargeable battery

Colloquial term for secondary *electric cell* or group of such cells.

reciprocating internal combustion engine

A form of prime mover which is an *internal combustion engine* and in which power extraction is by means of pistons whose motion reverses up and down a cylinder. In the form of the *spark ignition*, or 'Otto cycle' engine and the *diesel engine*, this is an important power source for transport.

recoverable oil reserves

The total *oil in place* in a reservoir multiplied by the *recovery factor*. Recoverable reserves depend on a number of factors such as rock permeability, the kind of oil and the reservoir drive mechanism. A coarse, highly permeable sandstone having a high water drive mechanism and containing a light oil will yield a higher percentage of oil than a finer-grained sandstone containing a heavy oil and dependent on a solution gas drive mechanism. As recovery techniques have improved, percentage oil recovery has increased remarkably at the primary recovery stage. *Secondary recovery* is usual practice and can yield almost as much oil as the primary stage.

recovery factor

The percentage of the total *oil in place* in a reservoir which can be recovered by a combination of primary, secondary and tertiary techniques.

rectification

The process whereby an alternating electric current is changed into a direct electric current.

Rectisol process

A process used to purify gas produced as a result of *coal gasification*. It is a gas absorption process employing organic solvents such as *methanol* at temperatures below 0°C. The gas from the coal gasifier is passed through an absorption column equipped with trays (counter-current to the solvent) and acid gases such as hydrogen sulphide and carbon dioxide are absorbed. Spent solvent is regenerated and recycled to the top of the absorption column. The clean *synthesis gas* may then undergo the *shift reaction* and *methanation* if a *pipeline quality gas* is required.

A Rectisol unit is employed by the South African Oil and Gas Corporation in its Sasol coal conversion complex at Sasolburg. In this case the cleaned synthesis gas is subjected to the *Fischer-Tropsch process* whereby it is converted into liquid fuels.

recuperator

A particular form of *heat exchanger* used to recover heat from hot waste gas. Normally this heat is transferred to preheat air for combustion so that the recuperator is a gas to gas heat exchanger. If the heat is used to preheat boiler feedwater a gas to liquid heat exchanger is used, and this is known as an economizer.

See also *boiler*.

reduction

The addition of an electron or an ion to an atom during a chemical process. Also the addition of hydrogen or other electron-donating elements to a substance. Reduction is important in the extraction of many metals from their ores, in *petroleum refining* and in *coal gasification* and *liquifaction* processes.

See also *hydrogenation*.

refinery fuel

Fuel used to provide heat for refinery processes. Products of the refining processes themselves are generally used, in particular *petroleum gases* and residual *fuel oils*. Petroleum refining processes are generally around 90-95 per cent efficient ie for every 100 tonnes of crude processed 90-95 tonnes are made available as refined products, the remaining 5-10 tonnes being flared or used as refinery fuel.

refinery gases

Gases obtained from *petroleum refining* processes consisting chiefly of methane, ethane, propane, and butane. The first two are usually employed as refinery fuel whereas propane and butane are liquified by compression and sold as liquified petroleum gases (LPG).

refining

See *petroleum refining*

reflector

In nuclear terminology, a material placed around the core of a nuclear reactor to reflect the neutrons back towards the centre of the reactor core. The most suitable material for reflectors, as with *moderators*, is one composed of light atoms and so light water, heavy water, or graphite are appropriate choices.

reforming

A general term used to describe a number of secondary refining processes such as *alkylation, dehydrocyclization*, and *isomerization*. Reforming is used to increase the proportion of particular constituents in a distillate, or to introduce a compound that is absent, in order to improve the ignition quality of the distillate or to prepare aromatics for use as chemical feedstock. Both *catalytic reforming* and *thermal reforming* are employed, the former being crucially dependent on the catalyst present, the latter relying on the application of heat and pressure.

refrigeration

The technology whereby cooling of spaces is achieved and maintained. All practical refrigeration processes rely on some kind of *heat pump* cycle to 'pump' heat out of cold spaces against its natural direction of flow.

The most commonly employed refrigeration process is that relying on a mechanical vapour compression, or 'Rankine' cycle. The main elements of this cycle are a mechanically driven compressor, usually powered by an integral electric motor; a condenser where the compressed refrigerant gas gives out heat to the environment; a throttle or expansion valve, through which the liquid expands to low pressure; and an evaporator, in which the low pressure liquid is evaporated by absorption of heat from the space to be cooled. (This cycle can be used for heating, by using ambient heat to warm the evaporator and by using the heat given out by the condenser to warm a space.)

The efficiency of such a refrigerator is measured by the 'coefficient of performance' which is the ratio of the heat absorbed in the evaporator to the mechanical energy expended by the compressor.

In the absorption cycle refrigerator, compression of the refrigerant is achieved by absorption in a liquid, chosen as one for which the refrigerant has high chemical affinity. Such absorption lowers the vapour pressure of the refrigerant, causing it to evaporate from the evaporator and transfer its heat of vaporization to the absorber. The latent heat necessary to vaporize the refrigerant is taken from the space to be cooled. The cycle is completed by the separation of the two substances in the generator, where refrigerant vapour is boiled off from the absorbant/refrigerant mixture. The refrigerant vapour is condensed, throttled to low pressure and evaporated once more. The absorbant is recirculated by a pump to the absorber vessel. Typical chemical pairs used are water (refrigerant) with lithium bromide solution (absorbant) and ammonia with water.

Absorption cycle heat pumps have also been developed for heating purposes.

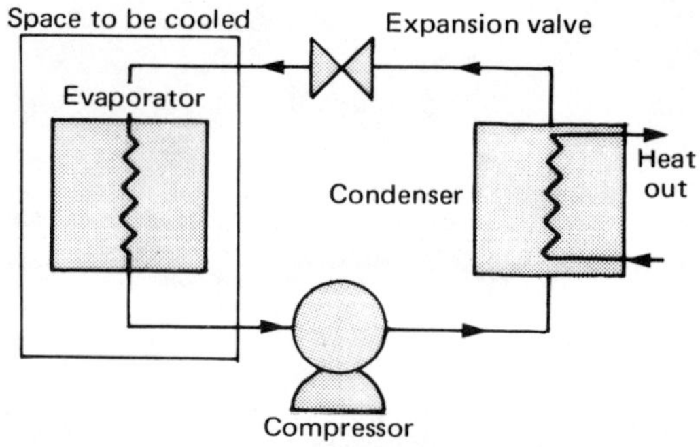

Refrigeration: vapour compression cycle

The coefficient of performance of an absorption cycle refrigerator is defined as the ratio of the heat absorbed at the evaporator to the heat applied to the generator. COP values for such machines are generally well below 2, much lower than corresponding values for Rankine cycle types. It must be remembered, however, that absorption cycle refrigerators (and heat pumps) can operate from relatively low-grade heat, such as that produced by flat-plate solar collectors.

Two other refrigerant cycles deserve mention. The air cycle refrigerator (or air cycle heat pump), used mainly for aircraft air-conditioning, employs air as the working fluid in an open or closed cycle. Air is compressed, cooled and then allowed to expand either directly into the cabin, or through cooling coils.

In the steam jet cycle water is the refrigerant. It is evaporated into a vacuum produced by the steam jet, and subsequently compressed by the jet itself. This type of refrigerator has limited applications.

Cooling can also be obtained by the 'Peltier effect', which is the reverse of the *Seebeck effect.* Power applied across two dissimilar semiconductor junctions causes one to be heated, the other cooled. By connecting such junctions thermally in series significant cooling can be achieved. Again these devices have only limited application at present.

See also *solar cooling.*

re-injection

The re-introduction of gas or water into the neighbourhood of a producing well, carried out in order to maintain pressure and thus a good rate of oil flow from the well. It is also a method of prolonging the life of a producing oil field, since re-injection boosts the naturally occurring drive mechanisms.

rem

Initials standing for rontgen equivalent man, a unit of effective radiation absorbed, one rontgen being equivalent to 0.83 *rad.* It is the unit used to calculate the allowable dose equivalent for any part of the body.

The rem is an approximate measure of amounts of damage caused by different kinds of radiation on living tissue, damage which also varies according to the conditions in which radiation occurs.

Organ irradiated	Dose, rem
Whole body Bone marrow Gonads	5
Skin and bone Thyroid gland	30
Hands and forearms Feet and ankles	75
Other single organs	15

Source: The Flowers Report, HMSO, 1976

ICRP recommendations of allowable radiation dose equivalents for radiation workers, per year

See also *dose limit; maximum permissible dose.*

renewable energy sources

A term applied to energy sources which do not rely on finite reserves of fossil or nuclear fuels. These sources are directly or indiectly due to the sun, (except for *tidal energy*, which arises from the earth's rotation), and include *solar energy* (direct), *wind energy, wave energy* and *hydro-electric power.* These are all likely to be available as long as we require them. *Fusion energy*, since it exploits a very large resource, and, to a lesser extent, *geothermal energy* could be included under this general heading.

See also *ambient energy sources.*

research octane number

The *octane number* of a motor spirit as measured by a standard test in which a single-cylinder engine, whose compression ratio may be altered, is allowed to run until premature detonation or *knocking* occurs.

reserves to production ratio

The ratio of the proven recoverable reserves of an oil or gas field to the production from that field in any given year. The change in the reserves to production ratio for a given field, over a period of time, is a fair guide to the likely life-span of the field and its rate of future production, for the R/P ratio cannot fall below about ten without an imminent drop in production. R/P ratios of 20-30 are considered high, though there are fields in the Middle East where the figure exceeds 100.

reservoir

In the petroleum industry an accumulation of *crude oil* or *natural gas* in a porous rock such as sandstone or limestone and which is trapped because it is overlaid by impermeable rock.

residue

In petroleum refinery terminology, also known as residual crude oil, the constituents of petroleum remaining in the *fractionating tower* after all the gases and distillates have been boiled off. This residue is often used as a fuel oil, or it may be subjected to *vacuum distillation* when some gas oils, fuel oils and lubricating oils are separated out as heavy distillate. The residue remaining after vacuum distillation is usually used as a heavy fuel oil or in bitumen manufacture.

retreat mining

A mining system in which extraction commences at the boundary of a deposit, working backwards towards the point of entry.

ring compounds

Another name for *alicyclic hydrocarbons.*

RON

Abbreviation for *research octane number.*

room and pillar

See *bord and pillar.*

run-up time

In power technology, the time required to start a system, ie to take it from rest to full power output. This time is measured in minutes for *gas turbine generators* and hydroelectric power systems and in hours for steam turbogenerators linked to fossil-fuelled boilers. Nuclear reactors may have even longer run-up times.

Because of their short run-up times, gas turbines and pumped storage hydroelectric systems are used to meet peak demand. When the total capacity of such systems is not sufficient, coal- and oil-fired stations are kept running at part load so as to be able to respond rapidly to a sudden increase in the demand for electric power.

See also *spinning reserve*.

Russel rectifier

See *wave energy*.

S

SAE

Initials for the Society of Automotive Engineers who have laid down widely accepted standards for *lubricating oils.*

salt dome

A dome-shaped formation of rock salt which has been forced upwards through overlying strata until it lies under a layer of impermeable *caprock.* Should petroleum migrate subsequently through the adjacent porous rock layers, the salt dome effectively forms a trap. Such geological structures are regarded as likely sites for petroleum reservoirs and would be investigated by means of the various survey techniques.

Salter's duck

A device for converting the energy of sea waves into electrical energy, developed by Dr Stephen Salter at Edinburgh University.

See also *wave energy.*

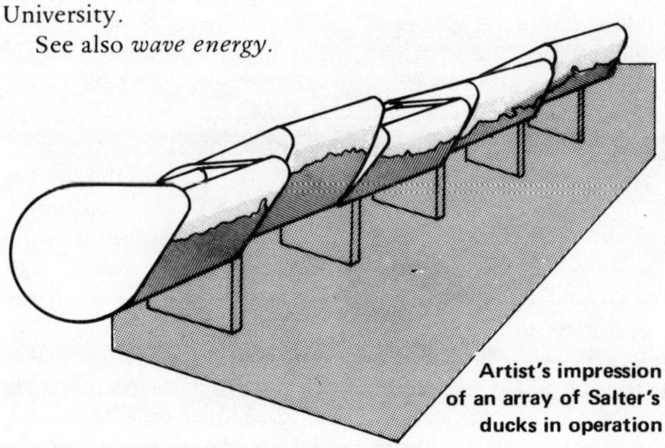

Artist's impression of an array of Salter's ducks in operation

sandstone

A sedimentary rock which has been compacted and cemented to varying degrees by other minerals. Calcareous and siliceous sandstones are two common varieties. Sandstones are porous and permeable and are therefore of great importance to the petroleum industry for they are the reservoir rocks in which most petroleum deposits have been found.

Sasol

Administrative shorthand for South African Coal, Oil and Gas Corporation.

A coal conversion plant producing mainly liquid fuels has been in operation at Sasolburg for some years. Coal with a high ash content is gasified by the *Lurgi process* and then converted to liquid hydrocarbons by the *Fischer-Tropsch process.* A second plant is under construction at Sasolburg which is designed to make South Africa self-sufficient in motor spirt and diesel oil by the mid-1980s.

satellite solar power system

Any of a number of proposed schemes for the use of large orbiting satellites for solar energy conversion. Placed in geosynchronous orbit around the earth, such satellites would be in direct sunlight almost continuously. By conversion of solar energy to electricity (using solar cells or a thermodynamic cycle) and transmission to earth using microwave radiation or lasers, such systems could provide continuous power. The cost would be high, and even proponents of such schemes admit that there are considerable technical problems involved, not least of which is the necessity for low cost rocket transportation to ferry materials and components for their construction.

seaweed

Colloquial name for multicellular algae found growing from the splash zone to deeper sea water. Common species such as wracks are still collected, dried and burned directly as fuel on a small scale, but their possible use as *energy crops* has involved investigation of their suitability for conversion to *methane* gas by anaerobic digestion. Experiments conducted so far have shown that giant kelp Macrocystis pyrifera yields larger quantities of methane than other species. Kelps are easily cultured and harvested and results so far suggest that further study is warranted.

See also *mariculture.*

secondary cell

A rechargeable *electric cell* which in practical applications is colloquially termed *accumulator* or *storage battery.*

secondary fuels

Fuels made by manufacturing or conversion processes from primary sources of energy. Examples include electricity, coke, hydrogen and synthetic natural gas. The primary source may be a *primary fuel*, such as coal or uranium, or an *ambient energy source* such as hydroelectric power or wave energy. Manufacture of secondary fuels invariably involves conversion losses though they are often capable of higher end usage efficiencies as compared with readily available primary competitors; in addition they may be cleaner and more convenient to use. All fuels derived from renewable primary sources are, by definition, secondary fuels, though in order to obtain useful energy from these ambient sources it is not always necessary to manufacture a secondary fuel; an obvious example is the use of solar energy in providing space and water heating.

secondary recovery

The recovery of oil from a reservoir by employing artificial techniques after the natural drive mechanism is exhausted. The methods used include water flooding, when water is injected down an injection hole, and gas drive which uses gas injected into the gas cap above the oil. Great care has to be taken when using water injection that oil is not trapped and left behind in quantity because the permeability of the rock strata may vary considerably. Water injection is most effective where permeability is reasonably uniform. Solvent and thermal techniques have been developed recently which improve the recovery level. Miscible gas drive, for example, is accomplished by adding large amounts of LPG to the natural gas in a gas drive, so making the gas and oil miscible. This helps obtain more oil from the pore spaces in the rock; but the process is expensive, as is that of using polymer additives with the injection water in water drive mechanisms. The injection of steam or hot water is a commonly used thermal technique, and in situ combustion involves burning some of the oil in the reservoir underground, sustaining the combustion by pumping down air which also displaces the oil.

Secondary recovery is becoming particularly important in parts of the world such as the United States where the rate of oil production is peaking out and sufficient new reserves are not being found.

sedimentary rocks

Rocks formed from the breakdown of older rocks and often incorporating the remains of dead and decaying plants and animals. Particles of rock which are the products of weathering or other agencies may be deposited in the sea, in lakes and rivers or on land; salts may be deposited from solution; the hard parts of animals and plants may form deposits; the pressure of the overlying water and/or other sediments, heat and changes within the sediments and the adjacent crustal layers slowly converts the deposits into rock. Sometimes these changes will result in the formation of *coal* seams, petroleum *reservoirs, oil shales* and *tar sands*, all of which have current or potential importance as *energy sources.*

Seebeck effect

A phenomenon demonstrated for the first time by Seebeck in 1821 in which a current is observed to flow in a circuit composed of two dissimilar metals, or semi-conductors, whose junctions are maintained at different temperatures.

The Seebeck effect is the principle upon which the *thermocouple* operates.

See also *thermoelectric effect.*

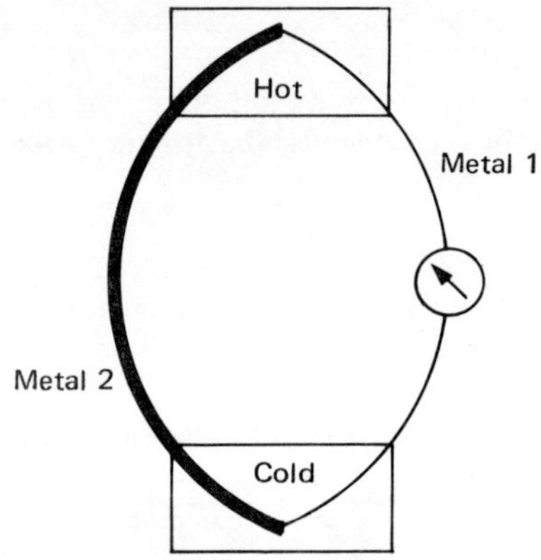

Seebeck effect

seismic surveys

The investigation of rock strata by recording movements induced in the ground by explosions or mechanical devices. The techniques used are derived from the study of earthquakes where seismograph readings show the position of quakes, their intensity and other data. The waves created by ground explosions are reflected or refracted back to the surface, the patterns recorded on magnetic tape representing the subterranean discontinuities between different types of rock, their arrangement and thicknesses. Geophones are arranged at fixed intervals on the surface along a line which may be up to 5,000 metres long for reflection surveys. As the shock waves from the explosive charges travel through the ground the vibrations induce electrical impulses in the geophones and are passed via cables to be recorded on magnetic tape. Photographic monitoring of the shock waves is also carried out as a check. The magnetic recordings are analyzed later. Variations to the techniques include moving the position of the point at which the shock waves are generated along the line of geophones so that several readings are taken of the same underground points from explosives set off in different places. Seismic techniques are also employed during offshore surveys in which case a geophone cable is trailed behind a ship and shock waves from the strata underlying the sea bed are recorded in a similar way to those on land. Data from seismic surveys are used with those obtained by other methods to construct a complete picture of the sedimentary strata likely to yield petroleum and other sought after materials.

See also *gravimetric surveys; magnetic surveys.*

semi-coke

Another name for *coalite.*

semi-water gas

A mixture of carbon monoxide, hydrogen, carbon dioxide, and nitrogen (ie *producer gas* and *water gas*) which is obtained by passing an air and steam mixture through incandescent coke. Semi-water gas is a *low-Btu gas* with a calorific value of around 150 Btu/ft^3 (5.6 MJ/m^3).

sensible heat storage

The containment of thermal energy by transfer to an insulated hot body, without causing any change of phase. For example, the storing of hot water in solar heating systems for use during periods of low insolation is a form of sensible heat storage.

This form of heat storage is limited in terms of storage density, but has the advantage of simplicity over *latent heat storage* or storage in the form of chemical energy. As in the case of latent heat storage, thermal insulation is required to reduce heat losses, especially where long term storage is required. This problem severely hinders the application of sensible heat methods to interseasonal storage.

Materials being used, or under development, for sensible heat storage include water, bricks and refractory materials, rock, liquid sodium metal, molten salts, aluminium, organic liquids, cast iron and various alloys.

separative work units

An indication of the amount of energy required to increase the abundance of the fissile isotope, uranium-235, in an uranium enrichment plant.

Units of separative work per year are the measure used for the throughput capacity of the plant.

sewage

In an energy relevant context, human wastes conveyed in sewers to sewage treatment plants for treatment and recovery of a 'sludge gas' which is rich in *methane.* The solid residues or sludge are a rich source of nitrogen and find use both as a fertilizer (thus saving fertilizers derived from energy intensive chemical processes) and as a soil conditioner. Sewage has been treated by aerobic and anaerobic means; often a mixture of the two is used in the same processing plant, but gas production is usually performed anaerobically in large sludge digestion tanks. The liquid sludge, after initial purification, is passed into the tanks and heated to 26°-37°C (higher temperatures [about 50°C] can be used when thermophilic bacteria are active). It is decomposed by anaerobic bacteria and methane-rich gas is liberated. This has often been used to provide the energy required to operate the sewage plant; gas has even been used to power vehicles; but, especially in the United States where supplies of natural gas are fast dwindling, attempts are being made to increase the output of methane from sewage treatment so that it can be upgraded to gas of pipeline quality. It is estimated that in the USA some 70 billion cubic feet of methane could be produced per year by sewage treatment alone. Larger quantities could be produced from municipal refuse. Anaerobic methods have not progressed much and more attention is being directed towards aerobic digestion techniques. Excessive sludge production does have disposal problems and there are other environmental considerations. Domestic and industrial effluents in sewage are contaminated increasingly with toxic chemicals that block the degradatory activities of the anaerobes. With present digester design and investigative techniques it is difficult to diagnose digester failure quickly, and this can be problematical.

See also *waste materials energy.*

SGHWR

Initial letters for *steam generating heavy water reactor.*

shale

A relatively common sedimentary rock laid down over a number of geological periods. It may contain a high proportion of carbonaceous material, and when a shale contains upwards of 33 per cent organic matter it is called an *oil shale*, from which a crude petroleum called *shale oil* can be obtained.

shale oil

Oil derived from *oil shales*, usually by heating the shale in special retorts to convert the organic matter in it to a crude *petroleum.*

shift conversion

Another name for *shift reaction.*

shift reaction

Also known as shift conversion. The process whereby the carbon monoxide : hydrogen ratio of the product stream emerging from a coal gasifier is adjusted so that it is rendered suitable for *methanation.* The desired ratio is one part carbon monoxide to three parts hydrogen and, if insufficient hydrogen is present, the mixture is subjected to the shift reaction in which the carbon monoxide reacts with water vapour to give carbon dioxide and hydrogen:

$$CO + H_2O \rightarrow CO_2 + H_2$$

In this way, the proportion of hydrogen in the gas stream, in relation to carbon monoxide, is suitably increased. Iron or chromium catalysts may be used and it is important that the gaseous mixture is purified before undergoing shift conversion, otherwise the catalyst may be 'poisoned' by impurities.

show

In petroleum exploration, the indication of oil, gas or water during drilling operations into the strata under investigation. A show does not necessarily mean that commercially exploitable quantities of material have been found; it merely indicates the possibility.

side-stripper

Secondary *fractionating towers* used to refine further the various light and middle distillates collected from the primary *fractionation* process used in *petroleum refining.*

silicon

A non-metallic element, the second most abundant on earth and a constituent of many rocks occurring combined with other elements principally as silicates. Silicon has a chemistry rather similar to that of carbon, and compounds containing both elements have been prepared – the silicones or siloxanes. They are extremely useful as electrical insulators and are used in water repellent films, greases and oils. Silicon is perhaps best known for its use in a pure form in the electronics industry in solid-state devices, such as transistors and more recently in *solar cells.*

symbol	Si
atomic number	14
atomic weight	28.06
melting point	1420°C
boiling point	2600°C

silicon cell

A *photovoltaic cell* which is capable of converting from 10-15 per cent of incident light directly into electricity.

The cell, which was first developed by Chapin and others in 1954, comprises layers (wafers) of single crystal high purity silicon which forms a large, flat semi-conductor diode to which electric contacts are added to the top and bottom surfaces. These contacts are usually metal, vacuum-deposited and sintered but may be plated. Cells with an n-doped sun facing surface and a p-type sub-surface (n-on-p) are more resistant to radiation degradation than p-on-n cells and are hence the predominant type, though p-on-n cells are generally more efficient before irradiation.

As a result of US and Russian space research funding in the 1960s, silicon cells have been engineered into arrays of

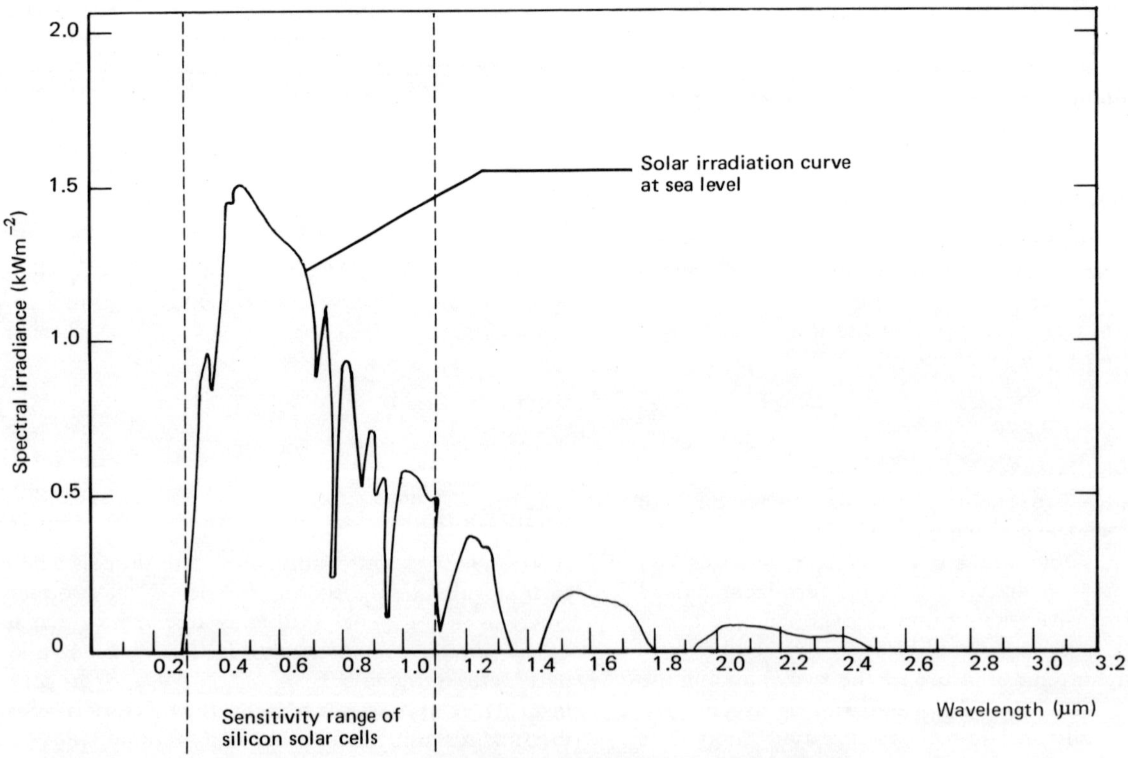

Sensitivity range of silicon solar cells compared with the solar spectrum at ground level

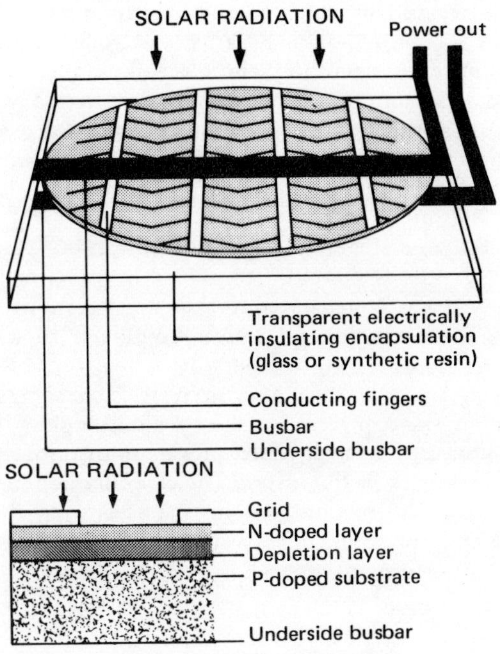

n-on-p silicon cell for terrestrial use

tens of kW and, in 1972, Skylab A of the US programme used 25 kW of solar power from silicon cells.

These cells, though successfully utilizing *solar energy* on some residential and other buildings, have so far been found impracticable for commercial application, for apart from the cost, vast areas of land are required where maximum insolation can be obtained. It has been suggested that solar energy obtained on the ground, in such high insolation areas as Australia and North Africa, could be beamed in microwaves to power relay satellites and then back down into those areas of the world with high energy demands. Silicon cells have also been suggested for use in *satellite solar power systems.*

Silicon cells can deliver between 0.4 and 0.5v at 11mW cm^{-2} representing 11 per cent conversion efficiency for a solar input of 1kW m^{-2} (20 per cent conversion is possible). The limited range of photon energies (corresponding to less than 1100nm) which are compatible with this system accounts for the fairly low efficiency. (Any photon energy in excess of 1.08 electron volts or 1150nm is wasted since one electron is released for each photon absorbed.)

Although silicon cells are still very expensive in terms of cost per installed watt, their cost has fallen from $200-$600 per watt in the 1960s to around $10 per watt now, and further cost reductions are expected. A target cost of $0.50 per watt in 1985 has been set by the US ERDA 'Low Cost Solar Array project'.

Development work is proceeding on polycrystalline silicon cells for terrestrial use. By eliminating the costly crystal-growing process these cells could be made more cheaply, but their efficiencies are of the order of 10 per cent or less.

See also *solar cell.*

silver oxide cell

A primary *electric cell* consisting of zinc and silver oxide electrodes, and an electrolyte of potassium hydroxide paste. Its principal use is in hearing aid devices.

simultaneous maximum demand (SMD)

The maximum demand on an electricity supply system, usually occurring in extreme summer or winter conditions. In the United Kingdom, SMD usually occurs on a very cold midweek winter evening, but in parts of the United States where air-conditioning equipment is the dominant component, SMD is reached during the most humid days in the summer months.

slagging gasifier

A term used to describe the reaction chamber used in some *coal gasification* processes in which non-combustible constituents of coal are converted into a liquid slag which is readily removed from the bottom of the gasifier. In this way gasification of succeeding charges of coal can proceed more efficiently.

slow neutrons

Also known as *thermal neutrons.* Neutrons in a nuclear reactor slowed down by repeated collisions with core material to a speed comparable with that of gaseous molecules. Slow neutrons are far more likely to occasion a fission on colliding with a fissile nucleus and *fast neutrons* are purposely slowed down by *moderator* material to increase the probability of such fissions taking place.

sludge gas

Another name for methane-rich gas produced in sewage disposal plants by the breakdown of *sewage* in sludge digestion tanks. The proportion of methane is usually about 65 per cent and the gas has a calorific value of 650-750 Btu/ft^3.

SNG

Abbreviation for *synthetic natural gas.*

sodium

A metallic element and one of the alkali metals. It is very reactive and therefore does not occur alone in nature but always combined with other elements as the chloride, carbonate, nitrate, etc. In these forms it is widely distributed; as sodium chloride, for example, in the sea and in salt deposits.

Sodium is isolated by the electrolysis of sodium hydroxide (caustic soda) or, of sodium chloride (salt).

Liquid sodium, being a metal, has excellent thermal conductivity and is being developed for heat transport and storage in *solar energy* systems. It is also used as a *coolant* in *liquid metal-cooled fast breeder reactors (LMFBR)*, sometimes with potassium in an alloy commonly called Nak (Na for sodium, K for potassium). Its advantages, besides its high heat-removal abilities, are: it boils at a very high temperature and therefore need not be pressurized; it can be heated to very high temperatures without generating pressure; it tends not to absorb *neutrons* nor to slow them down – important in a breeder reactor which depends on a fast neutron *fission chain reaction*; and because of its strong attraction to other elements, it is likely to retain the dangerously radioactive fission products vented or leaked into it, though not the radioactive gases xenon and krypton which bubble through it.

It also has disadvantages: this characteristic of retaining radioactive products makes it potentially dangerous; also it is likely to become radioactive itself, forming radioactive isotopes which emit gamma radiation.

The sodium-cooled nuclear reactor must have two separate cooling systems. The primary circuit must be confined entirely within the *biological shield* wall, any maintenance being carried out by remote control because the sodium in it, cooling the reactor core, becomes highly radioactive. The second sodium circuit has a heat exchanger inside the biological shielding (but protected against neutrons) to pick up heat and carry it outside to a second heat exchanger where steam is generated. This puts molten sodium in very close proximity with water with which it reacts very violently so the steam generators have to be built to a very high standard.

Because sodium reacts so strongly with water and any other materials its open surfaces in the cooling circuit are covered by inert gases, *helium* or *argon*; which may cause undesirable bubbles in the sodium if swept into its flow. Other aspects of handling sodium coolant are that it must not be allowed to cool beyond its melting point of about 98°C or it solidifies; and should it boil it could be expelled from the coolant channels with a consequent increase in the power level of the reactor. Finally, unlike gas or water coolants, sodium is opaque and thus hides the reactor internals from inspection.

A possible development of sodium as an energy source is the *sodium-sulphur battery*.

symbol	Na
atomic number	11
atomic weight	23.00
density	978 kg m^3 (0.98 g/cm^3)
melting point	97.82°C
boiling point	882.9°C

sodium-sulphur battery

A secondary cell in which a solid beta-alumina electrolyte separates a liquid sodium negative electrode and a liquid sulphur positive electrode. The electrolyte acts as an ion filter so that sodium passes through it to react with the sulphur, to form sodium polysulphide, when the electrodes are connected to an external circuit. On recharging, the polysulphide is broken down to its constituent elements, sodium and sulphur.

The cell operates at high temperature (300°C) and the battery of cells is therefore contained in a thermally insulated box. Heat is generated within the cells during both charge and discharge, and this is employed to keep the battery at its operating temperature when in use. Because of its high *energy density* and the relative cheapness of the constituent raw materials, the sodium-sulphur battery shows promise as a possible power source for battery-powered vehicles in the future. The chief development snag is the difficulty encountered in manufacturing the beta-alumina electrolyte in such a way as to ensure reliable and consistent performance in use.

soft coal

A collective term including brown coal, lignite, and other low calorific value coals. The average calorific value of soft coals is approximately half that of a *hard coal.*

solar battery

See *solar cell.*

solar cell

Any of a number of devices for converting solar radiation directly into electricity. These include (1) thermionic cells, in which a heated metal cathode in a vacuum emits energetic electrons that are collected on a cooled anode. Such a device requires very high temperatures (at least 1200°C) and only *focusing collectors* can achieve these. Efficiencies of up to 20 per cent have been achieved at these temperatures but the inherent complexity and cost of such combined arrays are a severe disadvantage, and no practical power applications of these devices have been developed. (2) Thermoelectric generators, which exploit the *Seebeck effect.* This arises as an emf between heated and cooled junctions between dissimilar metals, or between n- and p-type semi-conductors. As in the case of thermionic devices, high efficiencies can only be obtained at high temperature differentials, (about six per cent at 750°C), but the simplicity of this type of device means that the low efficiencies, about one per cent, of thermoelectric cells connected to *flat-plate collectors* may just be tolerable for some applications, particularly since the diffuse radiation component can be used. Heated by radio-isotope sources, thermoelectric generators have been used in spacecraft, but the expense of the materials makes large scale terrestrial use

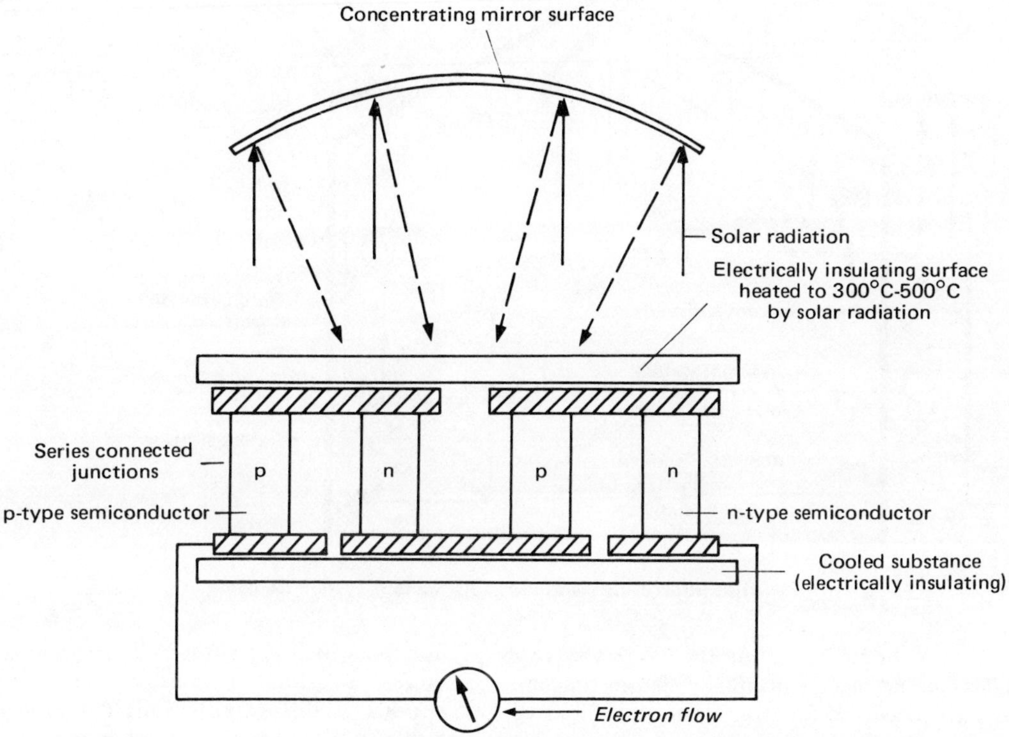

Thermoelectric solar cell

unlikely. (3) Photovoltaic cells. These generate electricity directly from light by the separation of charges in specially treated semi-conductors. The space exploration programmes of the past quarter century stimulated extensive research and development of such cells as the main source of electrical power for spacecraft. They have always been far too expensive for large scale terrestrial use, but research is continuing on ways of reducing costs. *Silicon cells*, *gallium arsenide* and *cadmium sulphide cells* have been put forward as possible contenders for large scale use.

Photovoltaic cells are the most promising of all the cell types and the term solar cell is generally taken to mean this type of cell.

solar collectors

See *flat-plate collectors; focusing collectors; solar energy.*

solar cooling

The application of energy from the sun to effect cooling. The simplest form of so-called solar cooling does not deserve the name since it involves no more than allowing large ponds of water to cool at night (by evaporation and radiation to a clear sky); the chilled water is covered during the day and used to keep a house cool.

True solar cooling involves the use of *solar collectors* to produce heat which is then used to drive a heat-actuated *heat pump.* A heat store may be used to enable the heat pump to run when the sun is not shining; alternatively, cold water can be stored and circulated as required.

Solar cooling installations have the advantage that the energy is usually most abundant when it is most needed. Combined solar heating/cooling installations are under intensive development in the United States.

Typically, solar cooling systems can be driven by flat-plate collectors operating at temperatures below 100°C, feeding hot water to an *absorption cycle heat pump.* (Other heat pump types can be used, namely those based on the Rankine cycle or dessicant systems.) Absorption cycle heat pumps based on the lithium bromide/water cycle and on the water/ammonia cycle have been successfully employed for solar cooling in the US and Australia. Other chemical pairs are being considered for this purpose, particularly where storage is envisaged in conjunction with the absorption cycle.

solar energy

This term may be applied to any form of energy which has its origin in the *sun.* In practical terms it refers to solar radiation falling on the earth's surface and its atmosphere. Almost everything which happens on the surface of the earth depends directly or indirectly on *radiant energy* received from the sun – either now, or in the past. For example, *photosynthesis* in plants occurs in the presence of sunlight. Solar energy plays a vital role, therefore, in plant production and has contributed in the past to the growth of forests which subsequently became the coal fields that we exploit today. Oil, too, owes its origin to solar energy.

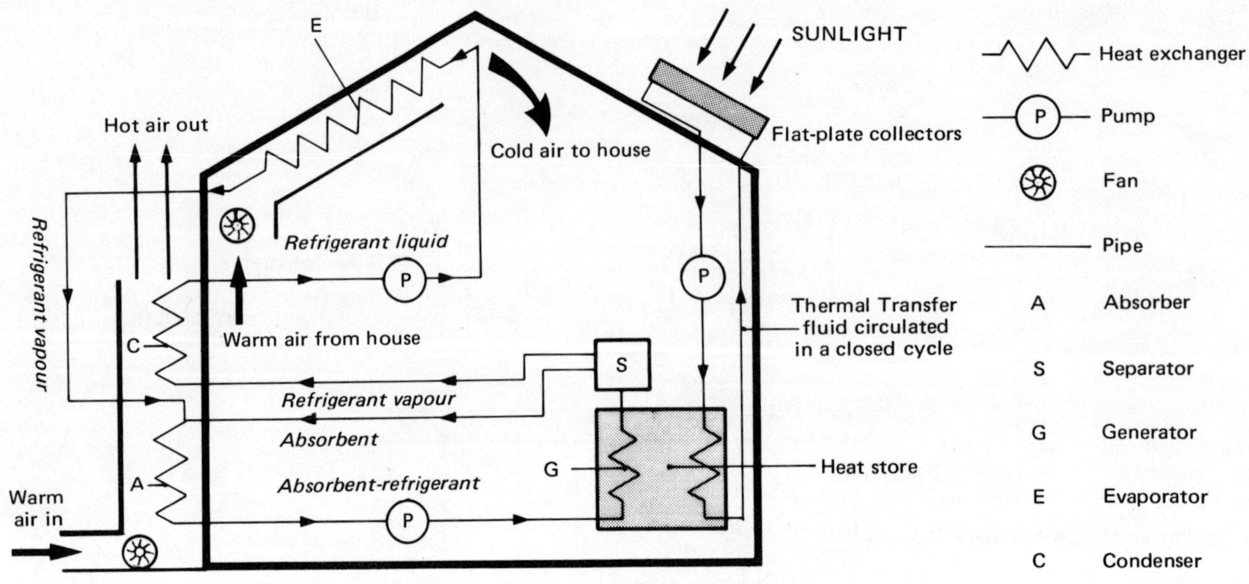

Absorption cycle heat pump solar cooling system (schematic)

Origins. Solar energy originates from the *thermonuclear reactions* that are taking place continually within the sun. Such reactions have occurred for thousands of millions of years. They involve hydrogen nuclei (protons) and the nuclei of other light elements, mainly helium, at a temperature of about 20,000,000°C and convert in each second 4.4×10^6 tonnes of matter into 3.9×10^{26} joules of energy.

Nature and Distribution. The overwhelming bulk of the energy released leaves the surface of the sun as electromagnetic radiation. Although the wavelengths of the radiation range right across the spectrum, visible light (40 per cent) and *infra-red* (heat) radiation (50 per cent) predominate.

Of the 3.9×10^{26} joules of energy radiated in each second by the sun, the earth intercepts only 1.75×10^{17} joules or 1.52×10^{18} kWh per annum. This compares with a total consumption of 7.5×10^{13} kWh in a year. Not all the intercepted solar radiation reaches the earth's surface. The atmosphere reflects back at least 30 per cent – perhaps as much as 50 per cent.

Of the energy retained, the atmosphere, land surfaces and the oceans absorb about two-thirds as heat energy. The hydrological cycle takes up much of the remaining one-third of the radiation in evaporating water. A much smaller proportion (about 0.2 per cent) drives the atmospheric and oceanic convections and circulations and the ocean waves. The energy of the wind and waves is therefore solar in origin. Photosynthesis in plants takes up an even smaller proportion – as little as 0.02 per cent – of the solar energy retained by the earth.

The intensity of radiation outside the earth's atmosphere is 1.4 kW/m^2 (12270 kWh/m^2/annum). In passing through the atmosphere this energy is absorbed, reflected and scattered by atmospheric water, dust and gases. That portion which is transmitted arrives at the earth's surface as direct radiation, or 'sunshine'. The scattered component appears as diffuse radiation. The ratio of direct to diffuse radiation depends, among other factors, on the cloud cover, so that on a cloudy day all the radiation is diffuse. The total radiation is the sum of the direct and diffuse components and is used as an important meteorological measurement. Measurements of sunshine intensity and duration have been carried out for a considerable number of years in many places, so that long-term averages are available for each hour of the year. Such measurements are important in the design and sizing of solar collectors.

The amount of solar energy reaching any part of the earth's surface depends upon latitude. The two arid zones, which encircle the earth between latitudes 15°N and 35°N and 15°S and 35°S, receive the highest proportion of solar radiation (at least 1750 kWh/m^2 per annum). High humidity and frequent cloud in the equatorial belt between 15°N and 15°S reduce the solar energy reaching the earth's surface in this zone. Nevertheless, the amount of energy received per unit area is second only to that reaching the surface in the arid zone.

The next highest solar energy recipients lie between latitudes 35°N and 45°N and 35°S and 45°S. In these latitudes seasonal effects are greater, but the total received in a year is between 1300 and 1750 kWh/m^2. In latitudes more than 45° from the equator, the comparatively low elevation of the sun above the horizon and the short days in the winter months limit the usefulness of solar energy during the months when this energy would be most useful for heating.

Exploitation. Solar energy can be exploited directly, by conversion to thermal, electrical or chemical energy; or development of solar heating. Developments in heat storage

indirectly, by the use of *energy crops*; or by the use of *wind energy, wave energy* or *ocean thermal gradients*, all of which owe their origins to the heating of the atmosphere and oceans by solar radiation.

Direct conversion systems include: (1) *Flat-plate collectors*, in which a fluid such as water is heated to moderate temperatures for heating, or for driving *solar cooling* devices. (2) *Focusing collectors*, which can achieve high temperatures by concentrating direct radiation on to a small area. These can be used for process heating at high temperatures as well as power production, and several prototype systems are being developed. (3) *Solar cells*, which achieve direct conversion of direct and diffuse radiation to electricity by photovoltaic, thermoelectric or thermionic means. (4) Chemical conversion systems, which produce fuel directly from solar radiation. Although a long way from practical exploitation, such systems offer the possibility of storage. The dissociation of water into hydrogen and oxygen, by means of chemical cycles, has been suggested. (5) Biochemical conversion systems, including the splitting of water by the use of chloroplasts; (also known as 'biochemical water splitting'). Although still at the laboratory stage this method shows some potential as a means of converting sunlight to hydrogen fuel.

Advantages and disadvantages. The great advantage of utilizing solar energy is that it is inexhaustible, at least in terms of the lifespan of planet earth. It is also safe in use. By comparison, the supply of fossil fuels is finite, recent estimates suggesting that known reserves of oil will last only 50 years at the present rate of consumption, while coal stocks could last 200-300 years. Because solar energy is widely distributed it can be utilized in a de-centralized way if required, a feature of particular interest to agrarian societies or developing nations. Although in capital terms it is an expensive form of energy to install, compared with the large-scale energy industries of the developed world, a solar system can be built up from small units in a modular fashion, and running costs are low. If necessary local materials and labour can be employed, at least for the simpler thermal systems.

The use of solar energy does not release air or water pollution (though this may not be true for the manufacturing processes required to supply the constructional materials).

Perhaps the greatest disadvantages of solar energy for use in temperate regions lie in its unpredictability and the fact that supply is almost exactly out of phase with the demand for heating. Since heating is the simplest application the latter is an important factor in preventing the

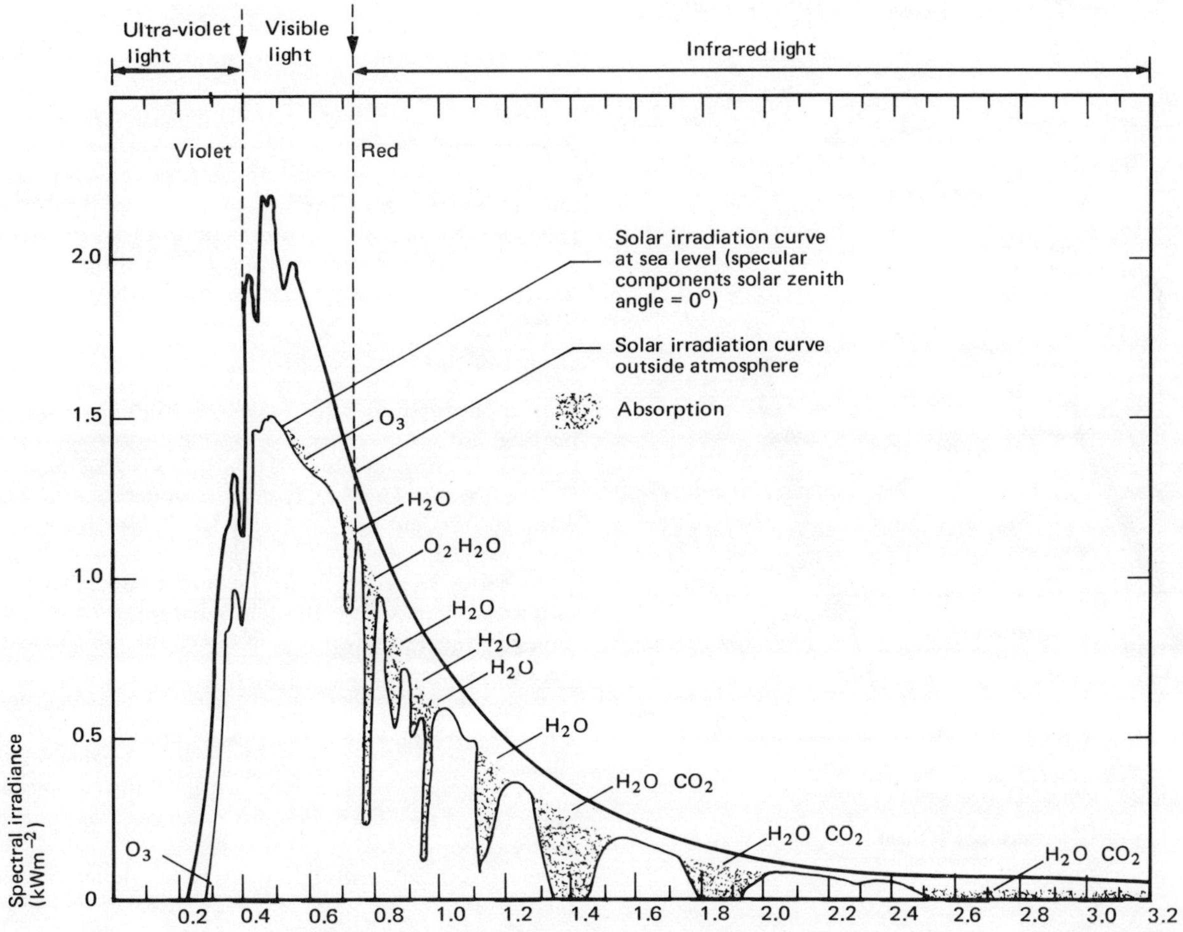

Solar spectrum outside the atmosphere and at ground level

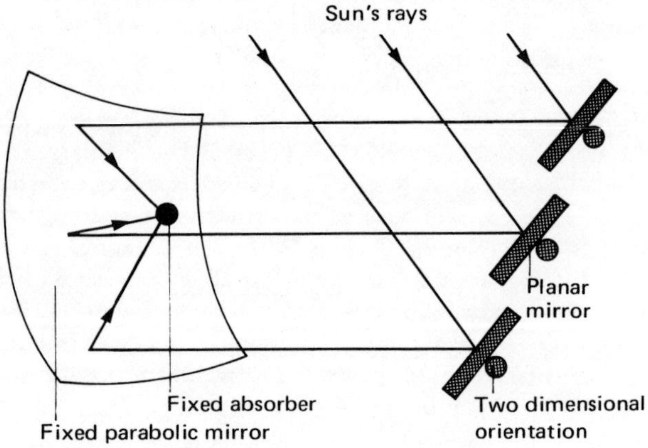

Solar power plant: heliostat type

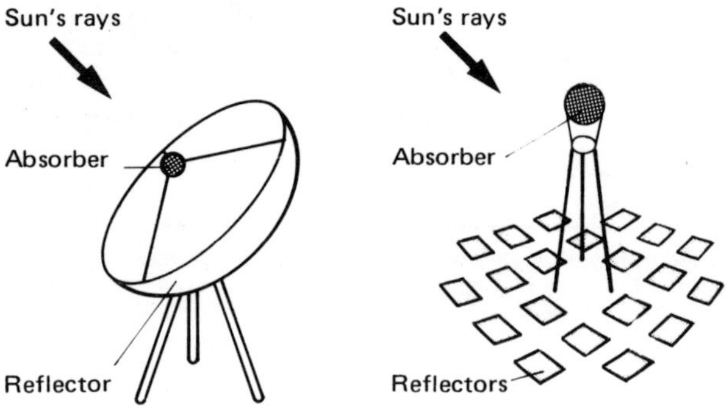

Central receiver collectors
(paraboloid and articulated flat mirrors)

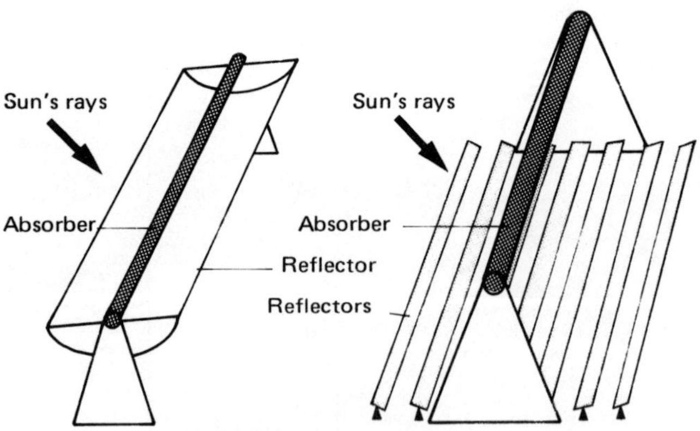

Linear focus collectors
(parabolic and articulated flat mirrors)

**Solar power plant
(geometries)**

development of solar heating. Developments in heat storage may reduce the importance of this problem. Short-term, as opposed to seasonal, variations can be met by smaller-scale, (buffer), storage or by the use of back-up systems using conventional fuels. Furthermore, several of the third world nations lie in those zones which receive the highest insolation. Thus they would benefit greatly by the application of appropriate modern technology to harnessing this resource.

Future Developments. Within the past ten years the major industrial nations have become aware of their dependence on fast diminishing reserves of fossil fuels as a source of energy. In consequence, governments and industrial concerns are now sponsoring many research projects aimed at harnessing solar energy – both directly and via wave and wind power. Theoretically, sufficient energy to meet all mankind's energy needs in the foreseeable future reaches the earth's surface from the sun. The development of solar energy will require significant advances in technology to reduce the cost of collecting and converting this large potential energy supply.

solar gains

In architectural design, these are solar heat inputs resulting from the orientation and configuration of a building. Solar gains can be useful for *passive solar heating*, but can also be a problem in buildings with large glazing areas, causing summer overheating that can only be rectified by the use of energy-expensive air-conditioning. Strategies for the reduction of energy consumption in buildings need to pay a great deal of attention to the maximization of useful solar gains to reduce winter heating bills without at the same time leading to excessive summer overheating problems.

solar heating

The application of radiant energy from the sun to space heating, hot water or process heating purposes.

Solar heating for domestic purposes is generally achieved by the use of *flat-plate collectors* and thousands of solar water heaters are now in use in Israel, North Africa, Japan and Australia.

In those areas which are favourable for sunlight a *solar house* can receive enough solar radiation, even on a winter's day, to provide all heat requirements for the whole 24 hour period.

Focusing collectors can be used to provide the higher temperatures necessary for industrial process heating, but the economics of this method are still marginal in all but a few favourable areas of the world. Simple concentrating collectors have been designed for cooking purposes in sunny regions.

Passive solar heating methods are being increasingly investigated for domestic and office use.

There are many different designs for active solar heating systems, ie those relying on the use of solar collectors,

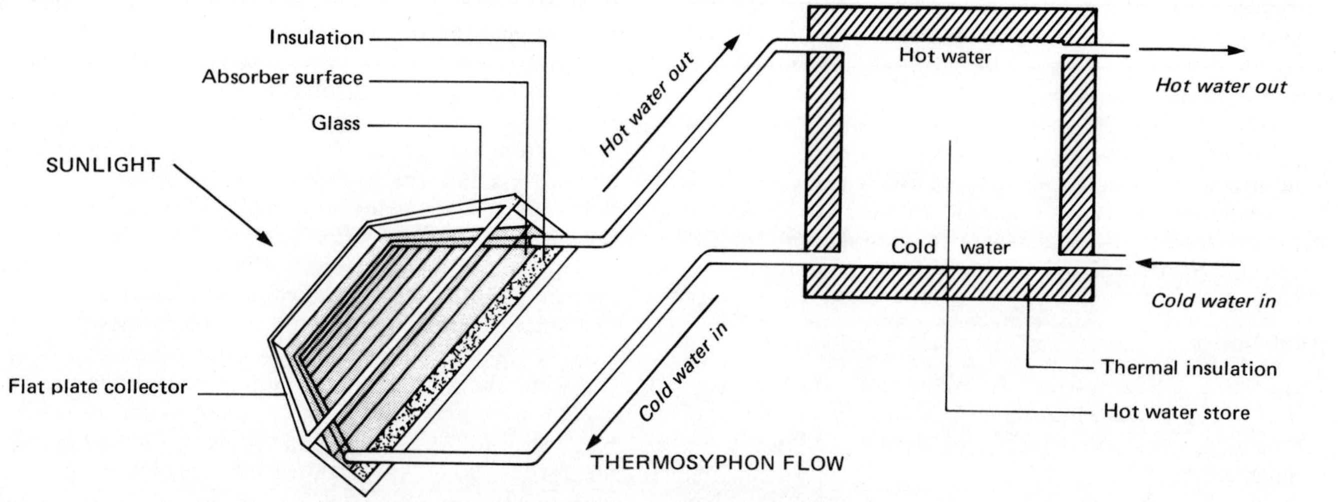

Thermosyphon solar hot-water system

storage tanks and other components additional to the building fabric. These vary considerably in cost, complexity and efficiency.

The simplest means of providing hot water, for direct use or for space heating, exploits the thermo-siphon principle (see diagram). Hot water from the collector rises and is replaced by cold water from the tank, which is arranged to be slightly above the collector. Mixing in the tank must be avoided, and various stratification methods have been tried. (This problem exists in almost all hot water storage systems.) This system is simple and has no moving mechanical parts, but it is not always possible to arrange the collector below the tank.

Most solar heating systems rely on an electric pump to circulate the heat extraction fluid, which may be contained in a closed system to allow the use of anti-freeze agents. Some form of control is often used to prevent circulation of the fluid through the collector when doing so would reduce the temperature of the fluid. An alternative method of avoiding freezing problems is to have a drain pipe on the collector, activated by a temperature sensor.

For use in temperate climates some form of *heat storage* must be incorporated into the system. Although inter-seasonal storage has not yet proved completely practicable because of the large volumes involved, short-term storage, in hot water tanks or pebble beds, has been successfully used.

Because of the storage problem and the effects of economies of scale, solar heating systems are financially more attractive when used for community housing, or for commercial scale use, such as offices.

The application of solar heating to swimming pools is

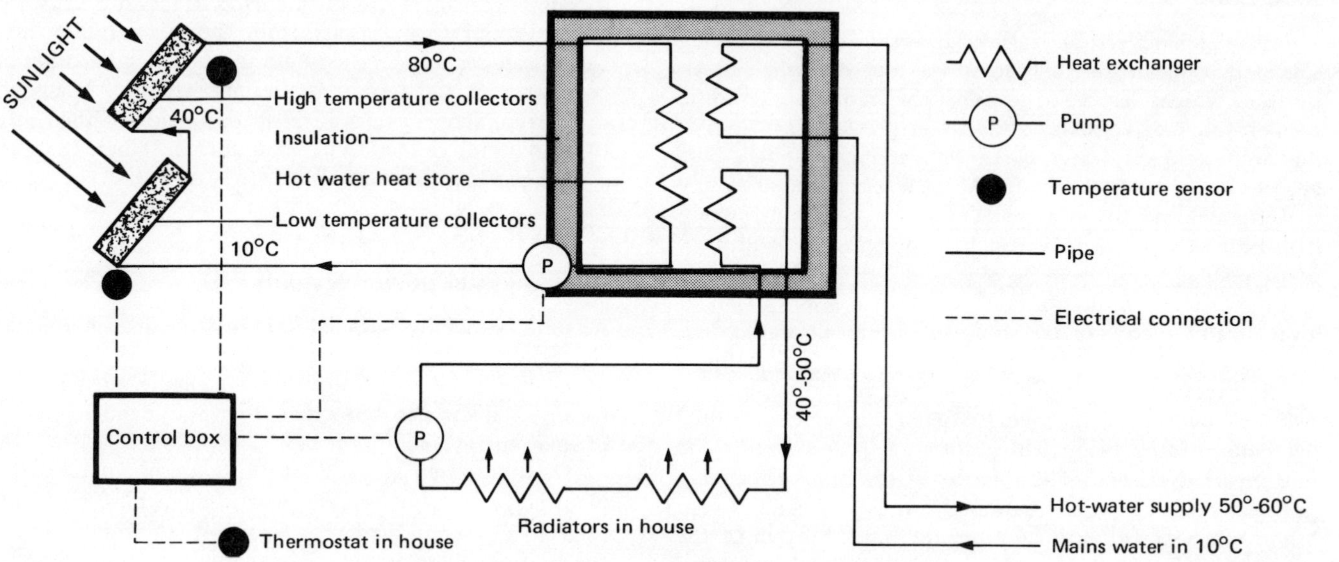

Typical solar heating system incorporating active control

particularly favourable since the low temperature output required leads to higher collector efficiencies at modest cost, and this is now a well established industry in the USA, western Europe and Japan.

Various types of solar house, in which total heating and cooling requirements are met by solar means, have been built around the world and many more are being designed. The incorporation of large storage volume into the house structure is simpler and cheaper if it forms part of the house design, rather than being an add-on component.

solar house

The solar house is designed to utilize solar energy to provide all, or nearly all, heating and cooling requirements. In certain designs electrical power too is supplied, by the use of solar cells.

Typically, solar houses rely on roof-mounted flat plate collectors for hot water and central heating. Cooling may be provided by the use of a heat-actuated heat pump. Heat storage may also be employed to ensure comfort in periods of low insolation, and for winter some form of auxiliary heating may be required.

The earliest solar houses were built in the USA in the 1930s. Since then many more have been built in other countries, including France, England, Japan, Germany, Switzerland, Sweden, Denmark and Norway. (Some of these have been built in latitudes close to the Arctic circle.)

solar panel

The popular name for the *flat-plate collectors* now being put to practical use in converting *solar energy* into heat energy to meet domestic hot water and space heating needs.

solar pond

This is the simplest photothermal system for converting solar radiation into heat. It comprises a pool having a black bottom which serves to absorb the radiation. Adding copper sulphate crystals to the water prevents *algae* growing and so keeps water clear. The copper sulphate also absorbs radiation strongly in the near infra-red, so raising the efficiency of the systems. As water has a comparatively high heat capacity, a solar pond filled with water will store considerable quantities of heat.

Convection, conduction and radiation will all carry heat away from such ponds. However, using a large pond, floating a layer of oil on the water surface or suspending a plastic sheet above the surface will minimize these effects. Another method of improving efficiency involves introducing concentrated brine which forms a layer covering the bottom of the pond. The density of the brine effectively prevents convection, so that the brine layer becomes very hot through absorbing heat from the black bottom of the pond. The cooler water above protects the brine layer against significant heat losses. Pipes carrying heat-transfer fluid pass through the hot brine layer taking the heat to its place of application.

Thus, a solar pond acts as a collector. However, it differs from other collectors in that the pond can also store heat energy. In common with all other collectors, solar ponds suffer from the effects of dust settling out of the atmosphere. Dust and dirt allow water to become heated near its surface. Also, the brine layer will eventually mix with the water.

This means of collecting solar radiation and converting it into heat will operate throughout the year in the tropics and in the summer months only in cooler zones.

Solar ponds can also be used to effect solar cooling. In this case the ponds are left uncovered at night-time, to enable the water to be cooled by evaporation and radiation to the sky, and are covered up during the day by an insulating medium. The circulation of the cooled water so produced can maintain a cool temperature in the house during hot weather.

See also *solar collectors; solar cooling; solar energy; solar house.*

solar still

A device which uses solar energy to distil potable water from brackish water or brine. In parts of Australia, the USA, the USSR, Mediterranean and Third World countries, the only sources of water are contaminated with dissolved salts. Other desalination techniques available either involve using scarce fossil fuels or expensive deionization systems. By contrast, solar stills are simple to construct and they use 'free' energy.

The still comprises a container with a blackened inside surface and a transparent cover – either a raised glass roof, or a plastic film. Solar radiation passing through the transparent cover heats the blackened surface and causes pure water vapour to evaporate from the salt water. (The dissolved salt remains in the liquid water which becomes more concentrated and has to be drained off periodically.) The vapour condenses as droplets on the inside of the roof and runs down for collection in a suitable container.

See also *desalination.*

solar thermal power systems

A term generally applied to any means of generating electrical power from solar energy by using the heat produced in a collector to drive a heat engine, such as a steam turbine. This is also known as 'thermodynamic conversion' of solar energy.

These systems are still at the experimental and pilot-plant stage, but many conceptual designs have been drawn up. In order to achieve reasonable thermodynamic efficiency a high temperature heat input is required and this entails the use of *focusing collectors* of some kind.

In simple systems of this type water is passed along a blackened tube at the focus of a paraboidal mirror, converted to steam and passed to a steam turbine. The mirror must be steered to track the sun to achieve good performance, and this can be carried out by a servo motor linked to a temperature or light sensor. Larger-scale solar thermal power systems are envisaged to consist of a large central tower at the focus of an array of heliostats. The array is able to track the sun by virtue of individual steering of the heliostats. Heat transferred from the tower would be used to generate steam to drive a turbogenerator.

Such systems can only be considered feasible in regions with high insolation over most of the year and low precipitation of rain and dust. Desert sites, especially in the south-western USA, are considered to be the most promising. Advantages include the absence of air pollution and low running costs. Capital costs, however, would undoubtedly be high as a result of the large mirror surfaces required. Land requirements too are large, which is another factor leading to the choice of desert sites. A major problem with such sites, however, is the provision of cooling water for the turbine since the temperature of heat rejection is critical for the thermodynamic efficiency. The use of open-cycle turbines operating on air has been considered. Thermionic, thermoelectric and thermochemical conversion have also been suggested. Energy storage may be required, either as heat or as electrical power, to provide power in periods of low insolation and at night.

See also *solar cell.*

solution gas drive

Also known as depletion drive. A natural drive mechanism operating in a limited or closed reservoir in which gas comes out of solution as the reservoir pressure near the well bore falls to saturation pressure and expands so displacing an increasing quantity of oil from the rock spore spaces. Eventually gas and oil move into the producing wells and the pore spaces become filled with gas. Oil production decreases while that of gas increases; ultimately oil may be left behind and considerable amounts of gas may be wasted. *Gas cap drive* and *water drive* are more effective production mechanisms.

solvent refined coal (SRC) process

A coal refining process designed to produce a clean sulphur-free fuel for use in steam-raising boilers. It was originally studied as a possible method of de-ashing coal prior to use. Coal is crushed and slurried with recycled solvent and fed with hydrogen into a pressurized reactor maintained at a temperature of 450°C. The products emerging from the reactor are filtered to remove char and ash and the resulting liquid fractionated to give an appreciable yield of light liquid products together with the solvent refined coal. Gaseous products are treated to recover sulphur. The refined coal has a low ash and sulphur content and has a relatively low melting point of around 150°C. The process itself requires only small quantities of hydrogen and no catalyst is needed. Pilot plant operation has been successful and a 10,000 ton per day plant has been designed.

solvent refining

The use of solvents to remove impurities from the desired product. Solvents may dissolve the wanted or unwanted substances, or precipitate them. Solvent refining is an extraction technique used extensively in petroleum refining – liquid propane, for instance, is used in the purification of high-viscosity oil to remove asphalt – and in coal liquifaction processes such as the *solvent refined coal process.*

See also *Rectisol process.*

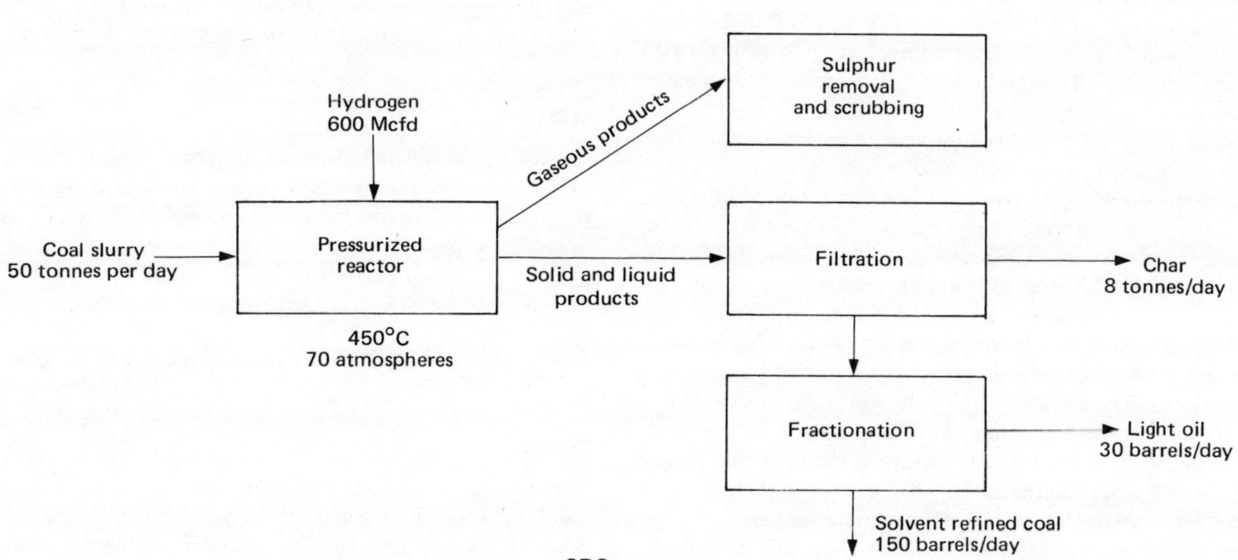

SRC process

sour gas

See *acid gas.*

spark ignition engine

A type of *internal combustion engine* in which ignition of the fuel is achieved by means of a spark generated between the electrodes of a spark plug in the combustion chamber by an electrical discharge. Because such engines use a particular type of fuel, they are also known as 'petrol engines' in the UK and 'gasoline engines' in the USA.

Most spark ignition engines are derivatives of the Otto four-cycle engine, which dates from 1876. During the intake stroke fuel and air are drawn into the cylinder at roughly atmospheric pressure. The mixture is then compressed more or less adiabatically. Towards the end of the compression stroke the mixture is ignited by the spark. The timing of this spark relative to the point of maximum compression is critical. Combustion of the fuel leads to a rapid pressure rise, pushing the piston down (the power stroke). The exhaust stroke completes the sequence, pushing the combustion products out of the cylinder.

A two-cycle variant of the Otto cycle is used for small engines, such as those in motorcycles.

Unlike the *diesel engine*, in which fuel ignition is achieved by the use of rapid compression, spark ignition engines have modest compression ratios (7–11). The tendency of the fuel to ignite when heated by the compression process, whilst a necessary characteristic of diesel fuel, causes detonation or 'knocking' in spark ignition engines. To avoid this, fuels for these engines contain *anti-knock components*, usually lead-based. In addition, high-octane fractions containing high proportions of *aromatic* and *branched-chain hydrocarbons* are blended with the raw gasoline cuts to give motor spirit with the desired *octane-number.*

The reciprocating Otto cycle engine has been developed considerably since its inception and continues to be the dominant engine type for automobile use. A rotary form, the Wankel engine, although now in commercial production, has failed to capture a significant market share, partly because of high fuel consumption and partly because of the huge investment in the conventional reciprocating engine.

This position of dominance has been maintained in spite of serious shortcomings. These include poor part-throttle efficiency and emissions associated with the use of high-octane fuels. Lead-based anti-knock components lead to particulate lead emissions, thought by some authorities to be detrimental to health in urban areas. The use of aromatic blends leads to potentially harmful emissions of benzenoid hydrocarbons. The gasoline engine, however, is considerably cheaper to manufacture than the diesel engine and is less noisy. Further inroads by the diesel engine into the automobile market are hindered by the lack of a widespread distribution system for diesel fuel in many countries.

New forms of the gasoline engine, such as the 'stratified charge' engine, offer the promise of substantial reductions in air pollution. These engines can run on a very weak fuel/air mixture, so reducing emissions of unburnt hydrocarbons and carbon monoxide. In addition, lower combustion temperatures lead to less nitrogen oxides production.

Spartina grass

A species of grass which it has been suggested could be grown as an *energy crop.*

specific gravity

The ratio of the mass of a given volume of a substance at a specified temperature to the mass of an equal volume of water at the same specified temperature, usually 15°C. Specific gravity in relation to calorific value is especially important in fuels for aviation gas turbines where the relationship between energy content and density is critical.

The density of a substance is equal to the product of specific gravity and the density of water in the same units. In the 'cgs system' of units, the density of water is unity, so that specific gravities and densities are numerically identical.

spinning reserve

Reserve generating capacity, present in hydroelectric pumped storage plants or thermal power stations, which is capable of coming into operation rapidly.

The reserve is created by operating the generating plant below full capacity ready to be opened up fully so that the extra generating capacity can be rapidly brought into use in the event of failures in the generating equipment or transmission system, or sudden surges in demand.

standard cell

A form of primary *electric cell*, used for providing standard exact values of *emf* for calibration purposes.

Examples include the *Clark cell* and the *Weston cell.*

starch

A *carbohydrate* polymer of glucose (some 200 glucose molecules comprising each starch molecule). It is an end product of *photosynthesis* and an important energy storage material in plants. Potato tubers, root crops, rice and other grains are rich in starch. When fermented, *ethanol* is produced and this can be added to petrol to increase its octane rating or used as a fuel in its own right.

See also *anaerobic fermentation; photosynthetic energy.*

steam

Water vapour, ie water in its gaseous state. Steam is a widely used working fluid for external combustion engines such as *steam turbines.* As a thermal transfer medium it finds widespread applications in *district heating* schemes, as well as building heating, process heating and power distribution on a factory scale. The specific heat of steam is higher than that of air by a factor of two. Few gases have higher heat capacities (hydrogen is one of them) so that steam, with its advantages of non-flammability and non-toxic nature, is generally the best choice heat transfer fluid.

Superheated steam is a term given to water vapour at a temperature well above the boiling point of water corresponding to the pressure of the steam. Superheated steam behaves as any other gas, but as it is cooled and approaches saturation, departures from ideal gas behaviour become more marked. So-called 'wet steam', which is a mixture of water droplets and steam, behaves quite differently.

Steam occurs naturally from certain geothermal fields and can be tapped to provide power. It is generated in the boilers of thermal power stations and used to drive steam turbogenerators in a closed cycle.

In the fuel and petrochemical industries steam finds application as a hydrogen source, as in the production of hydrogen from hydrocarbons by *steam reforming.* Several coal gasification processes rely on steam to add hydrogen to coal, and adaptations of these processes have been experimentally tested for in-situ gasification of coal or oil shale deposits.

steam generating heavy water reactor (SGHWR)

A type of nuclear reactor in which heavy water is used as *moderator* and heat is extracted by boiling light water. As with *CANDU*, the fuel, uranium oxide 2.3 per cent enriched, is contained in zirconium alloy-clad cans. Each fuel element is enclosed in individual zircaloy pressure tubes. These tubes conduct the light water coolant and act as a pressure vessel. They are arranged horizontally, however, unlike those in CANDU. The heavy water moderator is contained in a *calandria.* Channels through the calandria contain the pressure tubes. Light water coolant boils and the resultant steam drives the turbines directly. Only a 100 MWe experimental plant has been built so far, but this has been operating satisfactorily for some years. Heat output is 309 MWt giving 94.5 MWe with an efficiency of 30.5 per cent.

Plans to build a commercial SGHWR as a prototype for the next generation of nuclear reactors in the UK have been shelved for the time being, further *AGRs* being favoured.

steam generating heavy water reactor (SGHWR) data*	
Peak power density	11.2 kW/l
Heat output	309 MWt
Electrical output	94.5 MWe
Efficiency	30.5%
Type of fuel	Uranium oxide, 2.3% enriched, clad in zirconium
Weight of fuel	21.8 tonnes
Fuel burn-up	21,000 MWD/te
Moderator	heavy water
Coolant	light water
Coolant pressure	63.5 atmospheres
Coolant outlet temperature	282°C
Refuelling	Off load

**This example: Winfrith (UK)*

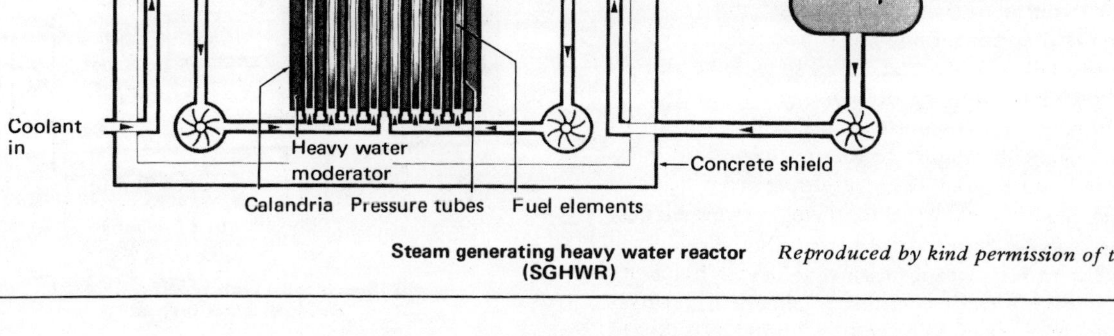

Steam generating heavy water reactor (SGHWR) *Reproduced by kind permission of the UKAEA*

steam reforming

A process in which sulphur-free naphtha is reformed in the presence of steam under conditions of high temperature and pressure (700°C, 30 atmospheres) to give a *synthesis gas* which may then be used in the manufacture of methane (*synthetic natural gas*), *town gas*, or ammonia. The steam reforming of naphtha formed the basis of the gas industry in the 1960s in countries such as the UK where the industry previously had been based on coal-derived gas, and where natural gas had yet to come into widespread use.

See also *catalytic rich gas process*.

steam turbine

A rotary machine in which steam received at high pressure is allowed to expand in a carefully controlled way, with a consequent pressure reduction, to produce mechanical power. Condensation of the steam at low pressure and temperature, followed by the return of condensate to the boiler, completes what is generally known as the 'steam cycle'.

Steam turbines can be classified as impulse or reaction types, according to the principle of their operation. They may also be classified as axial or radial, depending on the direction of flow of the steam releative to the axis of rotation.

In the operation of the reaction steam turbine, steam entering the casing is deflected on to the moving blades, attached to the turbine shaft, by fixed blades attached to the casing. As the steam expands, first through the fixed blades and then through the moving blades, it transfers internal energy to the mechanical motion of the shaft. This type is more correctly known as the 'impulse-reaction' type since expansion of the steam first occurs so as to raise the steam velocity, relative to the fixed blades giving an impulse to the moving blades; the subsequent expansion of the steam within the moving blades imparts a reactive force to the blades. Axial flow impulse-reaction steam turbines are widely used for electrical power generation, in units up to 1300 MW rating.

In the impulse type steam turbine the steam is allowed to expand in a set of fixed nozzles before impinging on the moving blades. All the pressure drop therefore occurs in the fixed nozzles, and this type derives its name from the fact that the moving blades are driven solely by the velocity component, or impulse, of the steam.

The designer of a steam turbine has considerable freedom in the choice of blade profiles and angles so as to be able to extract the maximum amount of work out of any given quantity of steam. Furthermore, since the process of expansion can be made almost isothermic (constant temperature) throughout by the use of multiple stages (ie sets of fixed blades together with the following moving blades) steam turbines can be designed so as to follow closely the theoretical cycle. In fact, steam turbines achieve efficiencies closer to the thermodynamic maximum than any other heat engines. For an inlet steam temperature of 565°C (close to present technical limits) and a condensing temperature of 35°C, the ideal, Carnot efficiency can be calculated to be 63 per cent. Typical operating efficiencies for the best modern multi-stage turbines lie in the range 42-48 per cent.

Multiple stage expansion is achieved by taking steam out between stages and reheating in special heat exchangers within the boiler. Reheating also serves the function of reducing the onset of condensation – water present in the steam as droplets can significantly shorten turbine blade life.

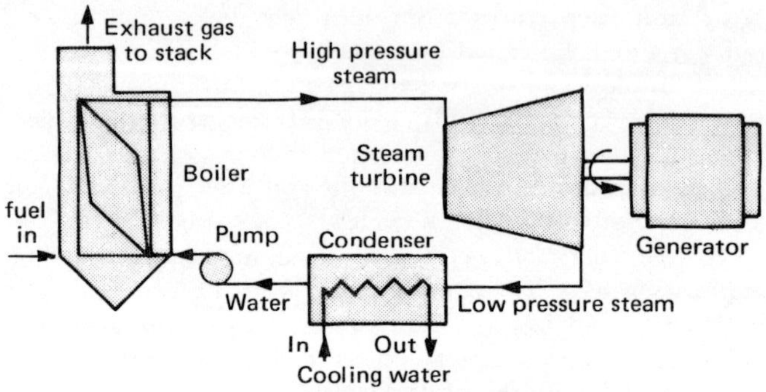

1. Schematic diagram of the basic steam cycle for power generation

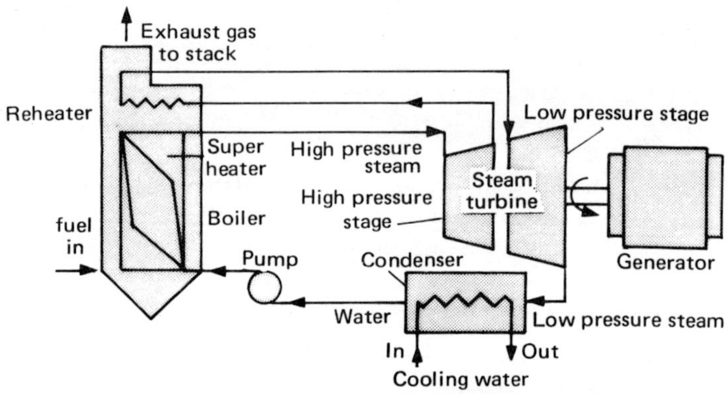

2. Steam cycle with superheat and inter-stage reheat

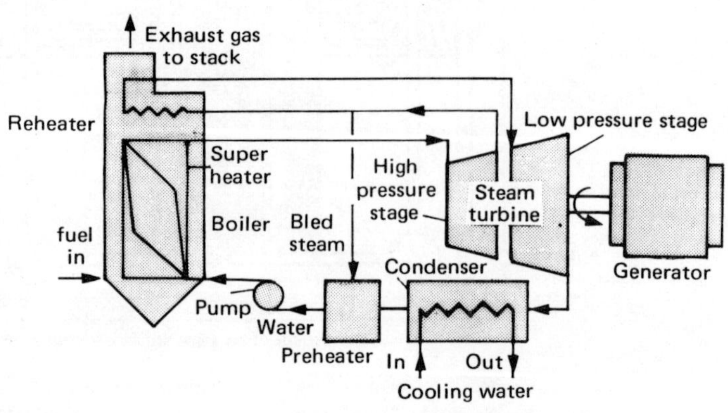

3. Steam cycle with superheat, reheat, and feed water preheating

Since reheating has the effect of using the same volume of steam several times over, albeit with decreasing work output, a greater output of shaft power is obtained per unit mass of steam.

Further increases in the steam cycle efficiency can be obtained by feedwater preheating. This is done by bleeding some of the steam out of the turbine and using it to heat the boiler feedwater, either directly, by mixing, or indirectly via a heat exchanger. Further increases in efficiency can be obtained by using several preheaters, each fed by steam from successive stages of the turbine.

Although achieving high efficiencies, the modern power station turbine rejects considerable quantities of waste heat at temperatures too low to be useful. By accepting a loss in electrical generating efficiency it is possible to increase the temperature of heat rejection to a point where it can be distributed for heating purposes. This is the basis for combined heat and power (CHP) systems, which are in widespread use, both on a factory scale and for district heating schemes.

Steam turbines may be driven by steam from a variety of sources including boilers heated by fossil fuels, nuclear fission or waste exhaust gases at high temperature, geothermal steam sources, and in the so-called combined cycle, by steam generated from the exhaust heat of *gas turbines.* Possible future applications for steam turbines include *solar thermal power systems* for the generation of electricity from solar heat and open cycle *ocean thermal energy conversion.* Power stations using steam turbines are generally known as 'thermal power stations' to distinguish them from hydroelectric and tidal stations in which heat is not deliberately produced.

Although electricity generation constitutes the major application for steam turbines, they are also widely used for ship propulsion and for industrial compressors, pumps and machinery drive.

step-out well

Another name for *appraisal well.*

straight chain hydrocarbon

An *aliphatic hydrocarbon* in which all the carbon atoms are connected in a single chain without branching. Examples include n-butane (short for normal-butane) as opposed to iso-butane (a *branched chain hydrocarbon*) and n-pentane (normal-pentane) as opposed to iso-pentane. The prefix 'n-' generally denotes a straight chain hydrocarbon. Fuel for spark-ignition engines is specially blended so that it contains a high proportion of branched chain hydrocarbons but *diesel fuel*, used in compression ignition engines, benefits from a high proportion of straight chain constituents.

straight run

In petroleum refining terminology, distillate fuels such as gasolines or kerosines that have not been subjected to any further processing following primary fractionation. Further treating and upgrading is usually, but not always, necessary.

strip mining

Another name for *open-cast mining.*

sub-criticality

In a *nuclear reactor*, the situation where the *fission chain reaction* dies out because of an insufficient supply of *neutrons* or *fissile fuel*; when each neutron that causes *fission* is replaced by less than one that does likewise, that is, when the reproduction factor is less than one.

substitute natural gas

Another name for *synthetic natural gas.*

sulphur (sulfur)

A light yellow solid element occurring in a number of allotropic forms in the solid, liquid and gaseous phases. It is found in nature as free sulphur in parts of Europe and the United States and chemically combined in sulphides (eg iron pyrites) and sulphates (eg gypsum). Sulphur often occurs in petroleum and natural gas deposits, from which it needs to be removed before oil or gas products are used as fuels, and *desulphurization* techniques are widely practised in the petroleum industry. Sulphur is also found in coal deposits, and in this instance is far less easy to remove prior to combustion. The burning of sulphur-rich coal has led to widespread *air pollution* as a result of *sulphur emissions.*

The principal use of sulphur is in the manufacture of sulphuric acid; it is also used in the rubber, paper and fertilizer industries and may find an important application in the *sodium-sulphur battery* if the problems associated with fabrication and reliable operation are overcome.

symbol	S
atomic number	16
atomic weight	32
melting point	112.8°C (rhombic form), 119.0°C (monoclinic form)
boiling point	444.7°C
density	2000 kg m^{-3} (2.0 g/cm^3)

sulphur emissions

Air pollution by energy production and conversion processes, consisting mainly of sulphur dioxide and related sulphur oxides, usually abbreviated to SO_x to denote a mixture of molecular species.

Sulphur emissions are one of several major components of *air pollution*, being produced when any sulphur-containing fuels are burnt. The health problems associated with atmospheric sulphur oxides, especially when associated with particulate emissions (smoke), are severe and include such debilitating and lethal conditions as bronchitis and coronary disease. In addition, exposure to sulphur pollution associated with particulate emissions tends to kill those already weakened by such diseases.

Because of the strong synergistic effect between smoke and sulphur pollution, and probably also between these and other pollutants such as tars, hydrocarbons and radionuclides, it is extremely difficult to predict the effect of increasing or reducing any one component of air pollution on mortality and illness (morbidity). It is abundantly clear, however, that the improved air standards now in force in almost all industrial countries have led to a marked decrease in the incidence of severe air pollution catastrophes, such as the 1952 London 'smog', and in the numbers of deaths associated with these. Despite this, it is difficult, if not impossible, to set 'threshold values' for pollutants, below which no effects can be expected.

Although the bulk of sulphur emissions are due to man-made sources, natural outputs of these pollutants exist in the form of volcanoes and geothermal fields. The exploitation of *geothermal energy* can lead to sulphur-containing gaseous emissions, but these are not usually associated with smoke. Not all man-made sulphur pollution arises from the energy industries. Severe sulphur emissions occur from smelters manufacturing lead, copper and zinc. In fact the energy industries have a better record than most.

Sources of sulphur emissions. Coal is the main offender as a source of airborne sulphur compounds since most coals contain appreciable quantities of the element. The proportion of sulphur oxides in the stack gas of a coal burning process is directly related to the sulphur content of the coal, which can vary from 0.2 to 7 per cent by weight, depending on source. Because of their huge consumption, coal fired power stations constitute a major source of airborne sulphur.

Petroleum is generally lower in sulphur content than coal, lying in the range 0.05 to 2 per cent. The average sulphur content of the heavy fuel oils used in power generation is higher, however, since the heavy sulphur compounds tend to have relatively high boiling points.

Most natural gases are relatively free of sulphur and can be burnt without producing appreciable quantities of SO_x compounds in the flue gases. Certain natural gases, however, contain appreciable quantities of hydrogen sulphide, which when burnt releases sulphur oxides.

Geothermal steam exploitation is another source of hydrogen sulphide. Some geothermal fields produce steam containing a considerable proportion of the gas, which in the absence of any form of control will be released without oxidation. Since even low concentrations of hydrogen sulphide are unacceptable due to the odour, and are potentially lethal, emissions of the gas must be prevented.

Control of sulphur emissions. The most satisfactory means of reducing the sulphur pollution associated with burning a fuel is to 'clean' the fuel, ie remove the sulphur, before combustion. Unfortunately this is not always easy, especially in the case of coal where the sulphur is chemically combined with the coal structure. Coal cleaning technology is, however, being developed. Such techniques as *solvent refining* are under investigation. In addition a number of processes for converting coal into clean gaseous and liquid fuels (ie fuels low in sulphur and other undesirables) are being developed. Since the sulphur must be removed at some point, such plants generally include some sort of recovery system to extract the sulphur as a pure element, in which form it is a valuable by-product.

Sulphur removal from petroleum fractions can be accomplished by the process known as *hydrodesulphurization*, by which the feedstock sulphur content is converted to hydrogen sulphide by catalytic reaction with hydrogen at high pressures and temperatures. The hydrogen sulphide is then removed, leaving a clean fuel.

An alternative approach to sulphur control is the use of a fluidized-bed combustion process. Sulphur oxides generated by the combustion process can be successfully removed by the incorporation of limestone (or dolomite) particles into the fluidized-bed.

However, given the massive capital investment in conventional pulverized fuel boilers, a change to new combustion processes would take many years to complete. In the interim, short lead time solutions must be found. Stack gas cleaning equipment to remove sulphur can be added to existing plant at relatively low capital cost. Catalytic processes, by which the sulphur dioxide content is either oxidized to sulphuric acid, or to elemental sulphur, are under trial. Sale of sulphur or of the acid would offset the cleaning costs. Absorption in alkaline scrubbing solutions is another process available.

The use of tall stacks, whilst reducing the concentration of sulphur oxides in the immediate neighbourhood of the power station, disperses the gases over a wide region, leading to international 'export' of air pollution. Sweden and Norway, for example, accuse the UK and Germany of inadvertently dumping large amounts of sulphur oxides from power stations on their territories.

See also *desulphurization*.

sun

A yellow star of average size and brightness, the sun consists mainly of *hydrogen, helium* and other light elements, all of which occur on earth. The specific gravity of the centre of the sun is about 96 (8.5 times that of lead) and at its centre the sun comprises bare atomic nuclei and free electrons undergoing the most violent thermal collisions. This extremely dense, elementary material is known as plasma. The collisions between the nuclei are so violent that the nuclei penetrate one another and the ensuing thermonuclear reactions liberate large quantities of energy.

The temperature and density of the sun are greatest at its centre and both fall sharply with distance from the centre. The tempeature at the surface is a mere 6000K or

0.0003 times that at the centre, while the specific gravity is much less than that of atmospheric air.

The sun consumes about 4.4 x 10^6 tonnes of matter per second in producing 3.9 x 10^{26} joules of energy. The overwhelming bulk of this energy leaves the surface of the sun as electromagnetic radiation whose wavelengths range right across the spectrum. As the graph of intensity of solar radiation outside the earth's atmosphere against wavelength shows (see page 161), the most intense radiation from the sun has a wavelength of about 0.47 x 10^{-6} metres which is within the waveband of visible light.

See diagram under *solar energy.*

superconductivity

The property that many metals have of possessing no resistance to the flow of electrical current at extremely low temperatures. Feasible temperatures are presently no higher than 20K but the discovery of new superconductors is expected to raise this. Superconductivity could have many energy applications in electricity generation, transmission and even in storage.

See also *croygenics.*

survey techniques

Methods of investigating the surface and subsurface geology with a view to finding *petroleum, coal* and mineral resources. The principal types of survey used are geophysical, including *magnetic surveys, gravimetric surveys, seismic surveys* and those involving radioactive and electrical readings. Geochemical and photographic methods are also used and drilling too is of great importance with analysis of core samples obtained providing a great deal of information on rock types, fossils and sequences in the geological timescale.

sweet gas

Natural gas which contains no sulphur compounds, unlike *acid gas* or sour gas which contains hydrogen sulphide, for example.

syncline

The opposite of *anticline.* A geological structure in which the rock strata are arranged in a bowl-shaped configuration.

syncrude

An abbreviation for *synthetic crude oil.*

synthane process

A *coal gasification* process, at pilot plant stage, in which coal is treated with a steam/oxygen mixture in a pressurized fluidized-bed to give a *synthesis gas* with a high methane content suitable for up-grading to *pipeline quality gas.* Unconverted char is removed from the fluidized-bed and used for steam generation. One potential advantage of this process is that it may be used to gasify a wide range of coals with different coking characteristics.

synthesis gas

The name given to the mixture of carbon monoxide and hydrogen produced as the result of the first stages of oil or coal conversion processes. The product stream is also likely to contain a variety of other liquid and gaseous products, and impurities, and it is usually necessary to purify synthesis gas before it can be up-graded into useful liquid or gaseous products.

synthetic crude oil

Crude oil made from coal by liquifaction and to a smaller extent as a by-product of *coal gasification.* A number of *coal liquifaction* processes are being developed at present with a view to large-scale manufacture of synthetic crude oil from coal and these are listed under the coal liquifaction entry.

synthetic natural gas

Also called *substitute natural gas.* A *high Btu gas* made from fossil fuels which can be substituted for or used with *natural gas.* Its methane content is usually higher than 85 per cent and its calorific value above 900 Btu/ft^3 (33.4 MJ/m^3). Government bodies lay down minimum standard specifications such as delivery pressure, minimum calorific value, maximum sulphur and carbon monoxide content, and water content.

Supplies of natural gas are dwindling in many parts of the world and energy-intensive economies such as the USA are developing SNG processes.

Oil shales and *coal* would seem to be the most likely feedstocks for large scale long-term SNG production. In the short term *naphtha* and LPG feedstocks are being used.

See also *coal gasification; oil gasification; steam reforming.*

synthoil process

A *coal liquifaction* process being developed in the USA in which coal is converted to fuel oil by high pressure hydrogenation in a fixed-bed catalytic reactor. Coal which has been crushed and dried is slurried in recycled process oil and passed into a fixed-bed reactor maintained at a temperature of 450°C under a pressure of 150-300 atmospheres. Ninety per cent of the coal is converted to a heavy fuel oil, and the sulphur in the coal removed by hydrogenation. It appears that the most efficient way of separating unreacted coal and ash from liquid products is by centrifuging, though on a larger scale this may prove to be a problem.

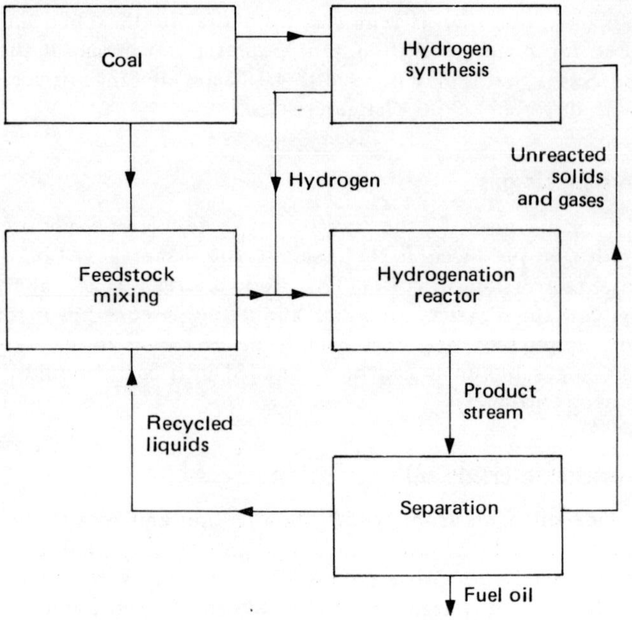

Synthoil process — schematic representation

system load factor

The average demand for power on an electricity supply system divided by the simultaneous maximum demand on that system given as a percentage. Average power demand is given by total sales (in, say, millions of kWh) divided by the number of hours for which that electricity is supplied so, for a period of one year:

$$\text{system load factor} = \frac{\text{Total electricity supplied}}{\text{No of hours in one year}} \times \frac{100}{\text{SMD}} \%$$

Système International (SI) d'Unités

A unified system of scientific measurement based on the fundamental units, the kilogramme (kg), the metre (m) and the second (s). Derived units of importance in energy measurement include the *joule*, the *watt* and multiples such as kJ, kW, MW.

T

tar

1. The heavy, dark liquid residues obtained from *petroleum refining*, also called *bitumen*.
2. The black viscous liquid formed by the destructive distillation of coal and wood, also called *coal tar*.
3. The hydrocarbon deposits in *tar sands*.
See also *Athabasca tar sands; tar sands*.

tar sands

Sedimentary formations of loose grained rock material bonded by heavy bituminous tar (see *bitumen*). Processing separates the tar as an artificial crude oil.

There are enormous quantities of *oil in place*, notably the Canadian *Athabasca tar sands*. However, present technology can recover only a small proportion. Extraction and processing costs are extremely high as huge amounts of overburden and tar sands have to be mined for reasonable oil yields.

Origins and Nature. It is not known whether the organic and inorganic materials in tar sands were deposited together or are the result of a separate migration path on the part of the bituminous materials. The inorganic materials – sand, sandstone and siltstone – are bonded by a bituminous (asphaltic) material, which constitutes up to 12 per cent of the mixture. This asphaltic material is lacking in the volatile matter common to conventional *petroleum*.

Processing. Tar sands are found at varying depths. Overburden is removed and generally the sands are extracted by *strip-mining*. As with *oil shale* extraction, the quantities of material to be mined and eventually disposed of are huge. (One Canadian operation digs out 250,000 tonnes per day and disposes of the same amount in waste.) Mining costs are therefore extremely high.

Oil is separated out by direct heating (at temperatures of around 80°C) or by passing steam through the tar sands. Bituminous material flows quite freely at this temperature and is washed clean of inorganic material. Two tonnes of mined material yield an estimated one barrel of oil.

In situ mining, in this case by pumping high-pressure steam through boreholes and extracting the separated oil through conventional wells, could dramatically decrease mining costs and disposal problems. Although such methods are still at the development stage, their usage would open up vast areas of otherwise unreachable deposits.

The extracted oil is upgraded by further processing (for example *hydrogenation* methods) and used in conventional refineries. Oil from tar sands, when finally processed, is employed in most of the uses of conventional petroleum products.

Reserves. The world's largest deposits are in the Athabasca area, some 34,000 km^2, in Alberta, Canada. Small scale commercial operations have been under way since 1967. Although initial losses were heavy, there was sufficient optimism in the future of oil from tar sands to encourage new operations in 1973 (*Syncrude*). The Alberta authorities have also taken a direct interest in the tar sands. Government subsidies may well be the only way to ensure the future of the tar sands industry. A forecast for the two Canadian plants is that production of ten million tonnes per year should be reached by 1980. (Compare this with Canada's forecasted total oil consumption of approximately 120 million tonnes per year, for 1980. Oil from tar sands is, therefore, not expected to make a large contribution to Canada's oil requirements.) Weather conditions around the Athabasca deposits are extremely severe, both on men and machinery, adding to the high cost of production.

Venezuela has large tar sand deposits in the Orinoco basin region, some 2,300 km^2. China also has reportedly large reserves. One estimate, working from a 30 per cent recovery rate, puts an upper limit of around 300 thousand million barrels in reserves for the Athabasca deposits, and 50 thousand million barrels for the Venezuelan deposits. No other known deposits come near these quantities.

As mentioned, the future of oil from tar sands depends greatly on the development of efficient 'in situ' mining techniques and the competitive cost of other fuels. Oil from tar sands is less difficult to extract, and possibly more accessible in its raw state, than oil from shale. Because of the operations in progress oil from tar sands has attracted more commercial interest than oil from shale. At present artificial oil from *coal liquifaction* techniques receives more attention and finance than oil from shale or tar sands. But the latter have the advantage in that new sources are exploited thereby adding to the overall world energy supply.

In the near future, however, oil from tar sands will affect the world energy supply very little.

tectonic plates

The rigid plates of which the earth's crust is thought to be composed. They move independently and where they come into contact at their edges the crust is stressed and crinkled or stretched. These areas along the edges of the plates are the world's earthquake zones and the places where geothermal activity is closest to the surface.

See also *geothermal energy.*

TEL

Abbreviation for *tetraethyl lead.*

tertiary recovery

Extraction of crude petroleum from underground reservoirs using techniques more sophisticated than those usually associated with *secondary recovery.* Examples of tertiary recovery methods include steam injection, the use of solvents and in situ combustion to aid the flow of viscous deposits. Such techniques are certain to become more widely used as petroleum becomes scarcer in the future and have been suggested as feasible methods of recovering large accumulations of very heavy oil and tar sands occurring in parts of Canada and in the Orinoco basin of Venezuela.

tetraethyl lead

A colourless liquid made from ethyl chloride by the action of lead-sodium alloy. It is used to increase the *octane number* of motor spirit.

tetramethyl lead

A colourless liquid related chemically to tetraethyl lead which is used as an anti-knock additive in motor spirit.

therm

A *heat* unit, equivalent to 100,000 British Thermal Units (Btu's) and often used when *calorific values* of fuels are quoted, usually in the form therms per tonne. It is widely used in the gas industry, and 100 cubic feet of natural gas has a calorific value of approximately one therm.

See *Appendix 2; Tables 2 and 3.*

thermal cracking

Formerly an important process in *petroleum refining* by which *gasoline* and *gas oil* was produced from heavier fractions. This was achieved by subjecting heavy distillate and residues to high temperatures (500°C) and pressures (up to 25 atmospheres). Thermal cracking has been largely replaced by *catalytic cracking*, which tends to give a higher quality product. However, *visbreaking* and *delayed coking* are thermal cracking processes still used today.

thermal efficiency

A term often applied to the generation of electricity from fossil fuels, which indicates the ratio between the amount of energy obtained as the result of a conversion process to the amount of energy put in to that process.

The thermal efficiency of electricity generation at power stations, governed as it is by the laws of thermodynamics, is comparatively low, around 35 per cent at the most efficient stations, though average thermal efficiency has increased over the years as the graph indicates:

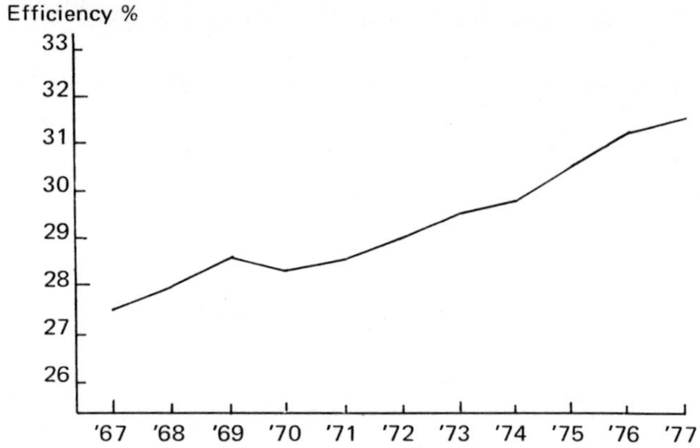

Source: CEGB Statistical Yearbook 1976-1977

CEGB system average thermal efficiency 1967-1977

Advocates of *combined heat and power* generation point out that the overall thermal efficiency of such a scheme would be much higher, perhaps 70 per cent or more; although this would require the efficiency of electricity generation, taken separately, to fall to around 25 per cent. However, the *low grade heat* normally going to waste at conventional power stations would in this case be at a sufficiently high temperature for it to be useful for *district heating* purposes.

See also *efficiency; electricity generation.*

thermal energy

The *energy* possessed by a body as a result of the random motion of its constituent atoms or molecules. The random nature of thermal, or 'heat energy' is important, since it is this that limits the efficiency with which this form of energy can be converted into mechanical work, which is a form of non-random, or ordered, motion. The usefulness of thermal energy is directly related to the maximum amount of work that can be obtained from it, and this depends on the temperature at which the heat can be supplied compared with the temperature of the surroundings. The greater this temperature difference, the more efficiently the heat can be converted to mechanical work, and the more useful is the heat. Thus the total thermal energy of the oceans is vast, but it is mostly useless since no equally large low-temperature sink can practicably be found. However, the temperature difference between the upper, warm layers of the tropical oceans and the deeper, cold waters can be exploited to provide power; (see *ocean thermal energy conversion*). Similarly heat emanating from the earth as *geothermal energy* can be exploited because it is hotter than our surroundings.

thermal energy storage

See *heat storage.*

thermal neutrons

See *slow neutrons.*

thermal pollution

A term generally used to describe the effect of large-scale discharges of heat to the environment from the activities of man.

Thermal pollution is a necessary end product of any chain of energy conversion processes since all such processes involve losses, and these losses must eventually appear as low-temperature heat rejected to the surroundings. It is therefore inevitable that all the energy contained in fuels will ultimately be degraded to low-grade thermal energy at the approximate temperature of the earth's surface. The average temperature of the earth must therefore rise until the extra input of thermal energy is balanced by the increased radiation losses of the earth to space. It is this line of reasoning that leads to the postulate of a heat limit for the earth.

In the immediate future, the problem of thermal pollution will probably arise where inherently inefficient processes are carried out on a large scale. Power stations are an obvious example; the thermal discharges of large stations have been under scrutiny for some time.

Discharges to rivers may have strong influences on the local ecological balance while atmospheric heat releases may alter local weather patterns. Large conurbations may also lead to measurable micro-climatic effects.

Certain energy technologies are able to extract power from naturally occurring sources in such a way as leads to no net input of heat to the environment. Such *ambient energy sources* include *wind energy, wave energy* and *hydroelectric power.*

See also *environmental effects.*

thermal power stations

Plant in which heat produced by the combustion of fossil fuels, or in a nuclear reactor core, is used to generate steam for driving turbogenerators thereby producing electricity.

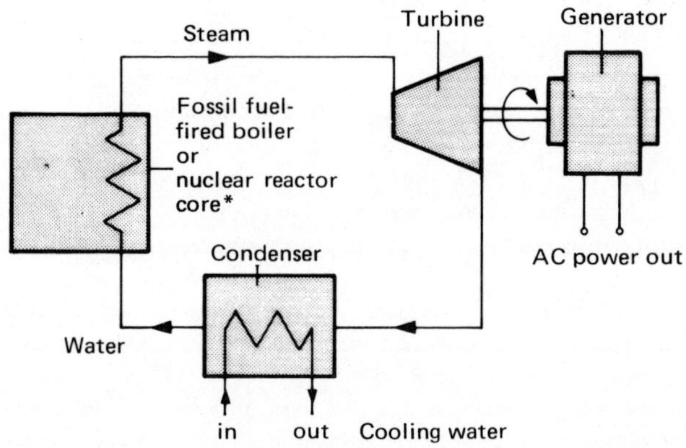

*in the case of a nuclear reactor, the heat may be extracted from the core by the use of a primary coolant circuit, which transfers heat to the secondary steam circuit

Basic elements of a thermal power station

Increases in *thermal efficiency* over the years has been largely due to the installation of large generating sets designed to run as *base load plant* and in future it is envisaged that nuclear power stations and large coal-fired plant will fulfil this role. Peak demand will be met increasingly by *gas turbine generators* and *pumped storage schemes.*

The siting of thermal power stations is usually dictated by the necessity of large quantities of water for cooling purposes, so rivers and coastal sites are favoured. Modern units are invariably built close to coalfields or oil refineries to minimize fuel distribution costs. Stations providing both heat and electricity (*combined heat and power* plants) tend to be smaller units sited close to centres of demand in order to render feasible the distribution of heat.

See also *electricity generation.*

thermal reactors

Nuclear reactors which are designed so that fission is induced by slow 'thermal' neutrons. They are also known as burner reactors because the amount of fissile plutonium created in the fission chain reaction is less than the quantity of fissile uranium fuel used up. The *conversion ratio* is less than 1.

thermal reforming

The reforming of *petroleum fractions* by the application of heat (550°C) and pressure (100 atmospheres), similar in many respects to *thermal cracking* though the feedstock used is a straight run gasoline. The purpose of thermal reforming was the up-grading of gasoline to give a fuel of better ignition quality, and the process has now largely been replaced by *catalytic reforming.*

thermal transfer media

Gases or liquids used to extract heat from where it is produced to where it is required for use, or for rejection, to the atmosphere. The characteristics of a satisfactory thermal transfer medium are: a good conductor of heat; a high heat capacity; low resistance to flow; chemical stability at the temperatures of use; and low coefficient of thermal expansion; non-corrosive to the heat exchangers, pumps and pipes; low cost. For use in nuclear reactors, thermal transfer fluids should additionally have low neutron absorption coefficients and should retain as little induced radioactivity from the core as possible.

Water, in both liquid and vapour form, is an important heat transfer agent, performing well under most conditions for many heat extraction purposes. When higher temperatures than 100°C are required, without resorting to pressurized equipment, organic oils, or such gases as carbon dioxide, helium, or even air, can be used. In general, however, liquids are better heat transfer agents since they have higher heat capacities. For the stringent conditions of the *liquid metal-cooled fast breeder reactor,* liquid sodium and potassium metals are used since these combine high boiling points with excellent heat transfer and low corrosion of plant.

thermal water

Underground water which is at a higher temperature than that of normal ground water. It is found chiefly at great depths and occurs widely in volcanic regions and in *fault* and block mountain areas. Such water has long had therapeutic uses where it comes to the surface as springs.

See also *geothermal energy.*

thermionic effect

A phenomenon discovered by Edison in 1883 in which a current is observed to flow between two electrodes in a vacuum when the electrodes are at different temperatures.

See *thermionic generator.*

thermionic generator

A device for converting heat into electrical energy. A thermionic generator consists of two electrodes, an emitter and a collector contained in a vacuum. When heat is supplied to the emitter, electrons flow through the vacuum to the collector electrode. If cathode and anode (the emitter and collector respectively) are connected to an external circuit, an electric current, capable of doing work, will flow. Thermionic generators can operate with a range of heat sources eg radioactive isotopes, solar energy, conventional burners, though the only important development to date has been in radioactive-powered devices.

See also *nuclear cell; solar cell.*

thermocouple

A comparatively inefficient device, which operates on the principle of the *Seebeck effect,* to convert heat energy directly into electrical energy. A thermocouple comprises a pair of dissimilar, conducting materials joined by a hot and a cold junction. If the two junctions formed between these materials are kept at different temperatures, an electric current flows in the circuit.

A collection of two or more thermocouples connected together in series or parallel is known as a thermopile.

thermoelectric effect

A phenomenon discovered by Seebeck in 1821 in which thermal energy is converted directly into electrical energy. If two dissimilar metals or semi-conductors are joined, and a temperature difference maintained between the two junctions, a current is observed to flow. This is referred to as the *Seebeck effect.*

There are two other significant phenomena in thermoelectricity, the Peltier effect, which is the Seebeck effect in reverse, and the Thomson effect which relates the magnitude of these effects to the direction of flow of the electric current.

Thermoelectric generators are unlikely to ever be used in the generation of large amounts of electric power but small-scale developments seem feasible.

thermoelectricity

Electrical energy derived directly from thermal energy by means of the *Seebeck effect.*

thermonuclear reactor

See *fusion reactor.*

thorianite

A mineral of thorium oxide, ThO_2, which is isomorphous with *uraninite.* The mineral consists of black cubic crystals which can be found in pegmatites and gem gravel washings; it often contains rare earths and uranium and is strongly radioactive. It occurs in Ceylon and in the Malagasy Republic.

See also *thorium.*

thorite

A mineral which is a silicate of thorium, with the formula $ThSiO_4$. It consists of black tetragonal crystals. Other varieties are the orange-yellow orangite and uranothorite which contain uranium oxide. It occurs in syenites and syenitic pegmatites. Thorite occurs in Norway, thorite and orangite in Sweden, orangite and uranothorite in the Malagasy Republic and uranothorite in Ontario.

See also *thorium.*

thorium

A dark grey radioactive metal of the actinide series. The naturally occurring isotope is thorium-232; other isotopes include thorium-233 and -234. Thorium is chemically reactive so it is never found as the free metal but as derivatives of the oxidized state. The reserves of thorium are large and easily accessible, though there is very little in the sea. It is found as the mineral *thorite,* which contains the silicate, and *thorianite,* which contains uranium and thorium oxides. These minerals are found associated with acid and basic igneous rocks and their derivatives. Thorium also occurs in beach sands as *monazite* in Brazil, India and North and South Carolina. This ore contains 3-9 per cent oxide.

Although thorium has been relatively ignored as a possible source of *fission energy,* it is a potentially extremely useful nuclear fuel as the reserves of economically available thorium easily outstrip those of uranium. Thorium-232 is analogous to uranium-238 in that it is a *fertile material;* on taking up a neutron and losing two beta particles it becomes uranium-233, which is fissile. Indeed a mixture of thorium and highly enriched uranium may be used in a number of prototype designs of the *high temperature gas-cooled reactor* (HTR) from which it is hoped to extract and reprocess significant and useful amounts of fissile uranium-233 for further reactor use. This is sometimes referred to as the thorium/uranium fuel cycle.

Attractive as this idea may appear to be, particularly to those anxious to postpone any commitment to the *fast breeder reactor,* it should be added that there is as yet little or no operating experience with this fuel cycle and it may be some time before reliable fabrication and reprocessing techniques are developed.

See also *nuclear fuel cycle.*

symbol	Th
atomic number	90
atomic weight	232.04
melting point	1700°C
boiling point	4500°C
density	11.725 kg m^{-3} (11.7 g/cm^3)

thorium-232

The naturally occurring isotope of thorium, it has a half-life of 1.39×10^{10} years. It is used as the fertile fuel in the high temperature gas cooled reactor where it is converted to the *fissile fuel* uranium-233.

See also *thorium.*

thorium/uranium fuel cycle

See *thorium; nuclear fuel cycle.*

three-phase supply

The supply of electrical energy through a three-phase system. Most electricity supplied for industrial and domestic use is in the form of alternating current. This is at a frequency of 60 cycles per second in the USA and 50 cycles in the UK. In a three-phase *electric generator* there are three pairs of field coils arranged so that there is an angle of 120°C between each coil. The three-phase supply uses three live wires and one neutral to carry away current. More power can be transmitted in this way than in a single or *two-phase supply* system and relatively less conducting material is required.

tidal energy

Energy derived from the diurnal rise and fall in the level of coastal waters. The twice daily ebb and flow of the tides is due to the gravitational pull of the moon, and to a lesser extent the sun, acting on the oceans as they rotate with the earth. The periodicity is approximately 12 hours and 25 minutes. The largest tides (spring tides) are due to the combined influences of the moon and sun when they are in line with the earth. When at right angles to each other lower tides, so-called neap tides, result. The rise and fall in the ocean levels is accompanied by lateral movements of the water; these are obstructed by the land masses, in which indentations produce many local variations in tide ranges and currents. The largest tidal ranges are to be found in such areas as Canada's Bay of Fundy (16 metres), the UK's Severn Estuary (13-14 metres), the Rance Estuary in France (12 metres) and Secure Bay, Australia (11 metres).

The use of tidal energy dates back to the Middle Ages in Europe when tidal mills operated. An enclosed pool was filled through gates by the rising tide. These closed when the tide dropped below the level of the pool and the contained water was released at low tide to turn the wheel of the water mill. It was not until the twentieth century that large-scale application of tidal power was considered. Between 1928 and 1940 tidal projects were proposed in France, the USA, Britain and the USSR. However, none of these was built at the time, and the French installation at Rance was the first large tidal barrage.

At present there are only two tidal power stations in operation. The Rance unit in France, completed in 1967, has a gross production of 555,000 GWh; the smaller, experimental plant at Kislayaguba near Murmansk in the Soviet Union was completed in 1969.

Tidal energy systems. Present and proposed tidal power schemes rely on the use of estuaries or basins that can be economically enclosed by a barrage to form a tidal reservoir. In the simplest system, analogous to the early tidal mills, the reservoir is filled at high tide and emptied at low tide, and power is generated during both processes. Electrical output is therefore produced in two short bursts in every tidal period of 12 hours and 25 minutes. These high peak powers necessitate the use of expensive turbines and lead to problems of matching to demand. Peak tidal power output will not necessarily occur at times of peak electrical demand, and the output periods are 50 minutes later each day.

By starting to fill the reservoir before high tide, and emptying before low tide, a more continuous power output is obtained, but this is at the expense of efficiency since a lower average head of water is used. There are still periods of zero power output, however. Essentially, this is the system used at Rance. The French electricity system, by virtue of a large hydroelectric pumped storage component, is able to absorb the remaining tidal power peaks.

Smoothing by means of pumped storage is already an established technology but, as with tidal power, the number of sites available is limited. Other means of power storage, such as compressed air, have been suggested in connection with tidal power and other natural energy sources. An alternative approach is the use of a double reservoir, with a high pool filled at high tide, and a low pool discharged at low tides. The head between the two pools would be used to generate power. Such a scheme is proposed at Passamaquoddy Bay in the Bay of Fundy on the United States/ Canada border.

Worldwide tidal sites. The Bay of Fundy, on the eastern seaboard of North America, is one of the world's best tidal sites. Numerous proposals for tidal schemes here have been put forward. As well as the Passamaquoddy Bay project already mentioned, attention has been given to the Chignecto Bay area and the Annapolis Basin.

The Severn estuary in Britain is another potentially useful tidal site. It is estimated that a barrage project there could provide approximately six per cent of the UK electrical demand. Sites in Korea, Australia, Alaska, the USSR and India are currently under investigation.

Turbines for tidal power. Since the exploitation of tidal power involves relatively low heads, huge volumes of water must be used relative to power generated. Turbine dimensions and inlet channels must therefore be large. In addition, tidal turbines are required to operate with reversible flow and with varying heads. For system reasons, it is advantageous to over-fill the basin at high tide, and over-empty it at low tide. This requires turbines capable of doubling as low-head pumps.

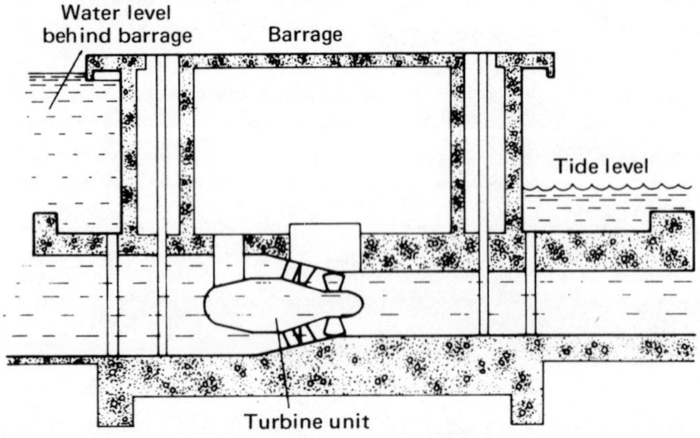

Tidal energy (La Rance unit)

The design developed for Rance and finally installed there was a horizontal axis bulb turbine with a variable pitch propeller runner. The generator is encased in a 'bulb' on the reservoir side of the runner and was specially developed so as to be compact enough to fit in the bulb casing without making this so large as to disrupt the water flow. Bulb-type turbines are now in use for low-head river hydro installations.

At the Bay of Fundy site tube turbines are proposed. Each unit consists of an axial flow-turbine mounted in the water duct and connected to a generator unit at the periphery of and external to the duct by an inclined shaft. These slant-axis turbines can be connected to compressors so that large quantities of compressed air can be produced for storage underground. This can be used to drive gas turbines at peak demand periods so eliminating the intermediate generation of electricity for use in pumped storage.

Straight-flow turbines have also been designed. These have the alternator mounted peripherally in a rim at the tips of the turbine blades.

Prospects, advantages and disadvantages. Tidal energy can never make more than a limited contribution to world energy demand. There exist only a limited number of sites where there is both an adequate tidal range and a marginal basin that can be enclosed by a barrage to provide sufficient power. Its potential role should not be underestimated, however. The exploitation of tidal power produces no air pollution or thermal pollution and other environmental effects should be minimal. The tides are also a highly predictable and effectively inexhaustible resource.

Financially, tidal schemes suffer from the disadvantage that they cannot be built up piecemeal but must be fully constructed before any power output, and hence financial return, is obtained. Combined with high capital costs, and long construction periods, this leads to very high interest charges on the capital required to build them. For these reasons electricity generated from tidal barrages will probably be more expensive than that from other sources.

Around 40 per cent of the capital cost of the typical tidal power unit proposal is for electrical and mechanical equipment. Most of the remainder is for civil engineering expenditure on barrage construction etc. Many of the running problems suffered by the Rance installation have been overcome and corrosion minimized by cathodic protection. Much has also been learned from developments in the offshore oil and gas industry and from land reclamation developments.

TML

Abbreviation for *tetramethyl lead.*

toluene

An aromatic hydrocarbon chemically related to benzene and which is produced in petroleum refining and coal tar distillation. Also called methylbenzene it is a colourless liquid which is added to gasoline fractions in the blending of *motor spirit.* It is also converted to benzene for the same purpose.

topped crude

Crude oil which has been distilled to remove some of the more volatile constituents.

total energy systems

The use of a single fuel, or a single plant, to provide the total heat and power needed for an area or industrial site. A synonymous term that might be used is integrated energy supply.

Most equipment used to generate electrical power depends on the use of heat generated from fossil fuels or nuclear fuels. A proportion of this heat is eventually wasted – rejected to the atmosphere in cooling towers of power stations and the like.

In the concept of total energy utilization the rejected heat is used as a form of direct heating for domestic or industrial purposes. In addition, the design of the generator plant is such that a specified proportion of the heat input is deliberately used to produce direct heating, rather than to generate as much electric power as possible, it being recognized that in the heat-electricity conversion process there is energy loss.

Electricity generating stations use the heat provided by fossil or nuclear fuels to raise the temperature of water in the boilers to high pressure steam that is used to activate the turbines. The hot steam expands and cools, delivering mechanical energy that turns the turbine blades in the process. The rejected steam is cooled in condensers in the cooling towers, the resulting liquid water being returned to the boilers for re-use.

In total energy systems the heat that is normally dissipated in the cooling system is used to supply heating for purposes in the locality of the power station.

This principle is exemplified in the design of *back-pressure steam turbine* plant. In the traditional steam turbine steam is heated under high pressure to a temperature of about 550°C and eventually brought down to a rejection temperature of 15°-20°C. This latter temperature is obviously too low for use in domestic and other heating.

However, if the turbine is designed (at a higher initial cost) to operate at higher steam pressures, the rejection temperature is raised to 80°-90°C and the rejected steam and water can be used for heating purposes. This advantage is gained at the cost of lower power-producing efficiency but this is more than compensated for by the amount of heat energy that is made available. The overall efficiency of the system rises from 35 per cent to over 70 per cent and there is a further saving because large and expensive cooling towers are not needed.

In the *gas turbine generator* the rejection temperature of the exhaust gases is high – about 500°C. The rejected heat energy can be used to save fuel by preheating the input fuel, or for district heating purposes. The ratio of useful heat output to electrical power output can be varied by controlling the proportion of exhaust gases that are fed back for preheating purposes. The overall efficiency of the system is high but the electrical power generating efficiency is low.

In reciprocating internal combustion engines the power generating efficiency is low – 30-40 per cent. The last 60-70 per cent is either rejected as exhaust gas to be utilized in a waste heat boiler or is passed through a jacket cooler where at a temperature of approximately 90°C it can be utilized for district heating or other low temperature heating purposes.

Applications. Total energy systems are widely used in urban heating systems, combined with power systems. For example, back pressure steam plant accounts for 20 per cent of the power requirements of British industry.

town gas

Another name for *coal gas*, or coal gas substitute, now largely replaced by *synthetic natural gas* and *natural gas.*

transformer

A device in which electrical energy received at one voltage is converted to electrical energy at a different voltage. Transformers are used in the electricity transmission system to this end and may be of the step-up type when they increase the voltage, or the step-down type when they reduce it. Electricity can be transmitted over long distances more efficiently at higher voltage and lower current.

transformer oil

Oil used in transformers as a cooling agent and for electrical insulation. It is a low viscosity oil and has a high resistance to oxidation.

transmission losses

Also known as distribution losses. Losses of electrical energy between the point of generation and the point of use. Electricity is usually distributed at very high voltage in order to keep transmission losses to a minimum, but even so these losses can be quite appreciable. For instance, in the United Kingdom losses on the national transmission system account for approximately seven per cent of electricity generated at power stations. Gas transmission systems also suffer from distribution losses due to small leakages, and these amount to one or two per cent in the UK.

See also *electricity transmission.*

transmutation

In nuclear terminology, the conversion of one element into another by radioactive decay or neutron irradiation. For instance, uranium-238 is said to have undergone transmutation when it is converted by neutron bombardment and β-particle emission to plutonium-239.

tritium

A heavy isotope of *hydrogen*, of atomic mass 3. Tritium is a component of the hydrogen bomb and is an important potential fuel in *fusion reactors*. Tritium is produced by *neutron* irradiation of *lithium* in *fission reactors* and there are large resources of lithium available as future sources of tritium.

turbine

A machine that converts the energy of a flowing stream of liquid or gas to rotary mechanical power. The main turbine types are *gas turbine, steam turbine* and *hydraulic turbine*. The first two are *heat engines* whilst the third is a purely mechanical device, converting fluid motion to shaft power. The steam turbine can be considered to be a particular form of *expansion turbine*.

turbine oil

A lubricant used in steam turbines and prepared from heavy distillate and residue by vacuum distillation and solvent refining. Turbine oils often contain anti-oxidants and anti-rust compounds and require a high resistance to emulsification.

turbogenerator

An integral unit comprising a *turbine* and *generator* in which high pressure steam from a power station boiler is converted into mechanical energy in the turbine and thence into electrical energy by the generator.

See also *electricity generation*.

two-phase supply

An electricity supply system in which alternating current is generated in two phases. In a two-phase alternator the pair of field coils is arranged so that each makes an angle of 90° with the other. Current is conducted away by two live wires and one neutral. A two-phase system is more effective than a single-phase one but not so efficient as a *three-phase supply*.

U

ULCC

The initial letters for ultra large crude carrier, a description given to the world's largest crude oil tankers, those exceeding 350,000 dwt. Tankers of 500,000 dwt are being built at present.

ultraviolet radiation

A form of electromagnetic radiation whose wavelength range lies just outside the visible spectrum. The ultraviolet range is usually reckoned to be radiation with wavelength between the bounds 10^{-8} and 4×10^{-7} metres. Ultraviolet radiation is emitted by most light sources, and is a significant component of the electromagnetic radiation given out by the *sun*. However, because of absorption by ozone in the upper layers of the atmosphere, only the near ultraviolet (radiation with wavelength of greater than 3×10^{-7}m) penetrates through to the earth's surface. The ozone layer acts in some respects as a protective blanket, for far ultraviolet radiation (wavelength $<10^{-7}$m) can be harmful to plant and animal tissue.

UMASS system

A system for closed cycle *ocean thermal energy conversion* that uses either *ammonia* or *propane* as its working fluid, proposed at the University of Massachusetts.

underground coal gasification

See in *situ coal gasification*.

uraninite

See *uranium*.

uranium

A hard grey radioactive metal which occurs in a number of isotopic forms. As found in nature uranium consists of three isotopes of mass numbers 238, 235, 234 with relative abundances 99.28 per cent, 0.71 per cent and 0.006 per cent respectively. The most important isotopes with respect to the production of *nuclear energy* are uranium-238, uranium-235 and uranium-233. These decay in a series of stages to an isotope of lead. Uranium-238 with a half-life of 4.51×10^9 years becomes lead-206; uranium-235 with a half-life of 7.1×10^6 years becomes lead-207; uranium-233 (formed from thorium-232 decay) has a half-life of 1.62×10^5 years.

Uranium metal exists in three allotropic forms with transition temperatures of 660°C-770°C. In transition from the alpha to the beta phase, the metal expands considerably so that metallic uranium fuel elements in nuclear reactors must not be allowed to reach temperatures above 660°C.

symbol	U
atomic number	92
atomic weight	238.0
density	19,050 kg m^{-3} (19.05g/cm^3)
melting point	1135°C
boiling point	4000°C

Geology. Uranium is of widespread occurrence in the earth's crust. It never occurs as the free metal but as primary and secondary minerals which are present as ores in acid and basic igneous rocks, sandstones, shales, limestone and in sea-water.

The content of uranium in the earth is thought to be 1×10^{14} tons, down to 15 miles, with 3×10^9 tons in the sea at a concentration of 1 to 3 micrograms per litre.

There are over one hundred uranium minerals known, the most important primary mineral being *uraninite* which is approximately the composition of UO_2 (uranium dioxide) but contains, besides the higher oxide of uranium UO_3 and oxides of lead, *thorium* and rare earths. When in a microcrystalline form it is called *pitchblende.* It also contains a minute amount of radium.

Secondary minerals are present with deposits of primary minerals from which they may have been formed; they are usually brightly coloured and are found dispersed in sedimentary rocks, sandstone and limestone.

Exploration. Uranium exploration begins by the detection of the radioactivity given off by the decay products of uranium. Instruments mounted either on vehicles or low flying aircraft look for the gamma rays which are evidence of radioactivity. More evidence is assembled on the ground by studying likely geological formations, by detection of trace elements of the ore in soil, stream and lake sediments, surface water, and groundwater, followed ultimately by explorative drilling.

Extraction. This takes place by two basic processes: an acid leach employing sulphuric acid, or an alkaline leach employing a mixture of sodium carbonate and bicarbonate. The final product, *yellow cake,* a mixture of uranium oxides, contains 85-88 per cent by weight of uranium.

Uses. Uranium is used as a source of fission energy. For use in nuclear reactors the products from the uranium ore must be in a state of high purity. The refined product is converted into the metal, dioxide or carbide, which may be used with or without cladding in suitable metallic or other containers, according to the method of heat transfer used. These forms of uranium may consist of any of the following kinds:

natural uranium (0.72% U-235, 0.006% U-234, 99.27% U-238)

depleted uranium (less than 0.72% U-235)

enriched uranium (more than 0.72% U-235)

weapons grade uranium (over 90% U-235)

If the uranium is to be enriched it must first be converted into *uranium hexafluoride* and then undergo one of several enrichment processes. The selection of a given form of fuel for a nuclear reactor is dependent on a number of economic and technical factors. The advantage of using natural uranium is that no expensive isotope separation plant is required, and there is no dependence on enriched uranium. In breeder reactors, uranium-238 is used to produce the fissile plutonium-239 by absorption of neutrons.

The complete fission of all the nuclei in one pound of uranium releases the same energy as the burning of 1250 tons of coal.

See also *nuclear fuel cycle.*

uranium-233

This isotope does not occur naturally but is formed by the decay of thorium-232 in the high temperature gas cooled reactor.

See also *nuclear fuel cycle.*

uranium-234

A natural isotope which occurs only in very small quantities, and which has a half-life of 2.45 x 10^5 years.

uranium-235

A fissile fuel which occurs in natural uranium to an extent of 0.7 per cent. It has a half-life of 7.1 x 10^6 years and a low radioactivity level. It is important in uranium enrichment where the uranium-235 content of the nuclear fuel is increased over that in natural uranium.

See also *uranium.*

uranium-238

The most abundant isotope of uranium, it is a *fertile* fuel which is used in *breeder reactors* where it transmutes to plutonium-239.

See also *uranium.*

uranium dioxide

A fine black powder, present in nature in *uraninite*, which is produced when ammonium diuranate is heated in air to give U_3O_8 which is then further reduced in hydrogen at about 800°C to give the dioxide, UO_2. The dioxide must be cooled under hydrogen or carbon dioxide to prevent reoxidation. The pure dioxide is pyrophoric and readily catches fire. It is used as a fuel in *nuclear reactors* in the form of pellets.

See also *nuclear fuel cycle.*

uranium enrichment

See *nuclear fuel cycle.*

uranium hexafluoride

A yellow solid, UF_6 which forms a vapour at temperatures above 56°C. It is produced from uranium tetrafluoride, UF_4, by heating in a stream of fluorine. On cooling the vapour, the solid is collected; this is reactive towards water and must therefore be kept entirely dry. It is used in uranium enrichment processes.

See also *nuclear fuel cycle.*

uranium tetrafluoride

A dark green powder, UF_4, produced by heating uranium oxide in a stream of hydrogen fluoride.

See also *nuclear fuel cycle; uranium.*

useful energy

The quantity of energy required to perform a task, irrespective of the form in which the energy is supplied. The concept of useful energy is necessary to evaluate the energy demand of an economic system in such a way as to be independent of the mix of fuel supplies. Only in this way can the effects of changes in the fuel supply mix on the primary energy demand of the economy be measured.

For example, the energy required to run a water heater will depend on the type of fuel supplied; electricity can be converted to heat at 100 per cent efficiency, whereas the use of natural gas involves flue losses. However, the useful energy requirement can be calculated from the mass of water to be heated and the required temperature rise. Knowing the efficiencies with which the various fuels can be used to perform this heating, the energy that must be supplied to the heater can be calculated. The process can be applied to the house as a whole, to a factory, or more important, to an entire energy economy. In the latter role, the method is particularly relevant to strategies for fossil fuel replacement by ambient energy sources.

U-value

A measure of the thermal conductance of a building element or structure, such as a wall or roof, calculated as the rate of heat flow through unit area per unit temperature gradient. U-values are usually measured in units of $\mathrm{W\,m^{-2}\,deg^{-1}C}$. For a given material, x, and ignoring heat transfer by radiation, the U-value is related to the so-called 'k-value' (which is a measure of specific thermal conductivity) by the equation:

$$U_x = \frac{k_x}{t_x}$$

(where t_x is the thickness of the material, x used to make the building element to which U_x relates).

The U-value is an important measure of the effectiveness of thermal insulation and can be used to calculate the expected heat loss of a building, under stated weather conditions, at the design stage. In this way the building heating (and cooling) requirements may be estimated.

As can be seen from the relationship between the k-value and the U-value, the latter is a function of the thickness. Since a low U-value implies a good thermal insulator, one of the simplest ways of reducing building heat loads is by increasing the thickness of insulation in the exterior walls and roof spaces.

V

vacuum distillation

When applied to petroleum refining, a distillation process carried out at pressures lower than atmospheric to separate out the constituents of heavy distillate and petroleum residues. Sufficient vacuum is obtained by means of vacuum pumps or by using a steam ejector. Vacuum distillates so obtained, which consist of heavy gas oils, fuel oils and lubricating oils, may be further *side-stripped* to obtain a higher degree of separation. The residue left behind after vacuum distillation is sometimes referred to as vacuum residue and consists of the heaviest fuel oils and bitumens.

vaporizing oil

Another name for *power kerosine.*

vapour cycle turbine

See *expansion turbine.*

visbreaking

Short for viscosity breaking. A *thermal cracking* process, operating at 450°-500°C, in which petroleum residues are cracked to give a lower viscosity fuel oil as the principal product. Small percentages of petroleum gases and gasolines are produced as by-products.

viscosity

The property of a fluid which determines its flow characteristics. Viscosity varies inversely with temperature. It is an important characteristic of crude petroleum, *lubricating oils* and fuel oils, and also of drilling fluids.

See also *pour point; viscosity index.*

viscosity index

A measure of the change in *viscosity* of an oil with temperature change, that is, the extent to which an oil thins as its temperature increases. A typical automobile engine oil must be thin enough for the engine to start easily from cold, but as the engine temperature increases its tendency to thin must not be so great that it ceases to lubricate. An oil that fulfils these requirements has a high viscosity index (VI). Such multigrade oils are classified according to a method introduced by the American Society of Automotive Engineers over fifty years ago. Oils are given an SAE number and this is an indication of viscosity level. An SAE 20W/50 multigrade oil has the viscosity at −18°C of an SAE 20W oil and at 100°C that of an SAE 50 oil. The W is used for oils which are thin at low winter temperatures.

See also *lubricating oils.*

vitrification

A process under development which may not be in commercial operation for many years. It involves turning liquid high-level waste from *nuclear reactors* into glass blocks in a reprocessing plant. It is proposed that a solution containing

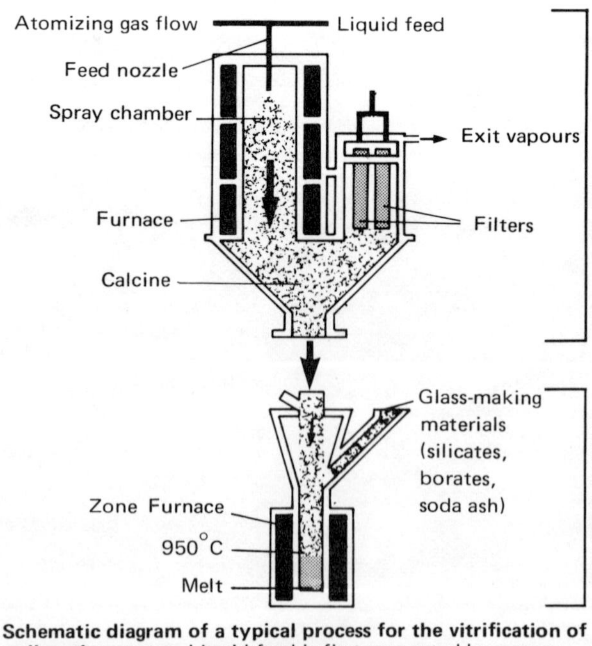

Schematic diagram of a typical process for the vitrification of radioactive wastes. Liquid feed is first converted by a spray process to a fine-grained calcine. This is then mixed into glass-making components and vitrified in a zone furnace.

the waste products is boiled for a period in order to reduce the volume and the concentrated waste would then be solidified by incorporation into glass. Safety hazards would be reduced by vitrification in that the solid waste would not be liable to leakage or vaporization. Though it is intended that the glass blocks will be resistant to leaching by water they must still be disposed of securely. The alternatives under consideration include underground burial in stable geological formations.

See also *nuclear fuel cycle*.

VLCC

Initial letters for very large crude carrier, until recently the largest crude oil tankers in use, but now overtaken by the ultra large crude carriers. Vlcc's are normally classed as tankers over 150,000 dwt.

volatility

The propensity that a substance has to change into a vapour. Generally speaking, low boiling point compounds are more volatile than those of a higher boiling point. Volatility is an important feature in *gasolines* and is carefully controlled and checked, otherwise unsatisfactory ignition qualities will be introduced.

W

waste heat boiler

A heat recovery device used to generate steam or hot water from exhaust gas streams.

A waste heat boiler has advantages over a gas to gas heat exchanger, arising from the high heat transfer rates. These include compact size, low capital cost, controllability and relative resistance to corrosion in exhaust gases (due to the cooling of the boiler tubes).

Like any other boilers, waste heat boilers can be classified as water tube or fire tube. Water tube types can produce higher steam pressures reliably but are less useful for handling high pressure or contaminated exhaust flows. Fire tube types can operate only at moderate steam pressures but are less susceptible to gas fouling. Superheaters, regenerators and economizers may be fitted to the water tube types to increase efficiency.

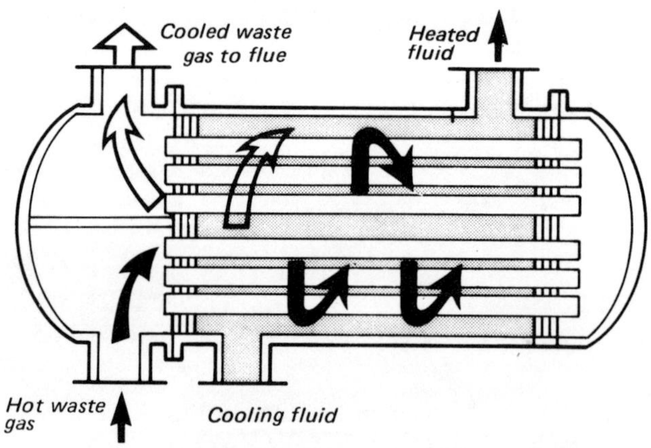

Typical waste heat boiler (tube type)

For heat recovery from internal combustion engines a packaged boiler is used, taking heat from both the engine coolant and exhaust.

For use with gas turbines, whose typical efficiency is 25 per cent, waste heat boilers producing low pressure steam can lead to an overall efficiency of 80 per cent; if high pressure steam is required an overall efficiency of 70 per cent is possible.

See also *boiler; heat recovery.*

waste heat recovery

See *heat recovery.*

waste materials energy

Energy derived from materials that would otherwise be wasted. Until recently energy has been considered plentiful, but the implications of the energy crisis are such that all possible alternative energy sources are being looked at to help conserve finite fossil fuel resources, and the utilization of waste materials and waste heat will no doubt make an increasing contribution.

Waste materials from agricultural, industrial and domestic sources have to be collected, transported and disposed of. If the waste can be regarded as an energy source and the energy extracted, then not only is the disposal problem solved but energy is both saved and made available for powering machinery, or heating and lighting.

Refuse, crop remains, animal wastes and industrial wastes can provide heat or fuels through biological and chemical conversion processes. Alternatively, they can be composted with a consequent saving in artificial fertilizers which are very energy intensive in manufacture. Agricultural wastes, such as straw, can also be used for manufacturing building or insulating materials.

On the assumption that only about 50 per cent of available waste can be collected, it has been estimated that in the United States waste production could provide up to four per cent of the country's total energy requirements. This does not make it a massive energy source, but on a local basis it could be significant.

Waste can be used directly as a combustion fuel by incineration, or may be used as a feedstock for the production of gaseous or liquid fuels.

The three main techniques of fuel production are *pyrolysis, hydrogenation* and *anaerobic fermentation.* Pyrolysis is a destructive distillation process in which the shredded organic materials are heated to temperatures of 500°C or higher in a closed vessel. Gas, oil and char can be produced, and integrated systems enable ferrous metals and glass to be salvaged as well as oil and char.

Hydrogenation, which involves the chemical reduction

of the organic materials heated to over 300°C in high pressure hydrogen, produces oil of a heavy paraffinic nature. The process is at an early stage of development and is unlikely to become commercial for some time.

Anaerobic fermentation is a well proven method of decomposing organic wastes. In sewage disposal works, for example, methane is produced by the digestion of activated sludge. The sludge gas is used in many plants to produce power for running the sewage works. Schemes to utilize farm manures are also in operation. Pig and cattle manure is readily obtained because of the intensive farming methods used. Apart from producing useful quantities of methane for powering farm equipment, the manure is disposed of in a convenient way. Solid waste residue is used as fertilizer. Small units are now available commercially for farm use and are easy to operate and maintain.

See also *alcohol; biogas; cellulose; energy conservation; heat recovery; incineration of wastes; methane; methanol; photosynthetic energy; total energy systems; waste heat boilers.*

water drive

An essential primary drive mechanism for enabling appreciable quantities of oil or gas to be obtained from an underground reservoir. As oil and gas is forced out of the porous rocks, water displaces it and applies more upward pressure on the hydrocarbons remaining. Water drive is also employed as a *secondary recovery* technique.

water gas

Also known as blue water gas. A relatively clean, low grade *fuel gas* employed mainly in industry for heating materials and in furnaces. The gas is produced by passing superheated steam through a bed of red hot coke. The steam partially oxidizes the carbon to yield carbon monoxide and is itself reduced to hydrogen:

$$C + H_2O \rightarrow CO + H_2$$

Depending upon conditions, the principal constituents of water gas are carbon monoxide (40 to 45 per cent) and hydrogen (45 to 51 per cent). Water gas also contains some carbon dioxide, nitrogen and traces of methane. Both the carbon monoxide and hydrogen are flammable, but the calorific value of the mixture is not very high, approximately 350 Btu/ft^3 (13.0 MJ/m^3).

The reaction between steam and coke absorbs heat and causes the temperature of the coke bed to drop rapidly. As a consequence the bed has to be reheated by means of a blast of air. This makes the process intermittent – approximately one minute of air blast (blow stage) to reheat the coke followed by up to three minutes steaming.

The gases produced during the blow stages are often discharged to waste through a stack. However, provided the coke bed is thick enough, the blow stage yields a mixture of carbon monoxide and nitrogen known as *producer gas.* Because the carbon monixide is also combustible, it can be mixed with the water gas – the resultant mixture of principally carbon monoxide, hydrogen and nitrogen is known as *semi-water gas.*

water pollution

Contamination of surface run-off, river or ocean waters arises from several energy-related activities.

Coal mining. The reaction of rain water with the sulphur content of abandoned mine workings, whether surface or underground, produces acidic water that is harmful to vegetation and river ecology. This is a serious problem in the Appalachians of the USA where 80 per cent of the spoil tips surveyed in a recent government study were so acidic as to be unable to support more than a few plant species. Such problems can be avoided by reclamation of the land, storing the topsoil and subsoil for backfill. Experiments have also been carried out to determine whether neutralization by coal ashes, followed by fertilization and seeding, can eliminate the problem by stabilizing the soil and reducing run-off.

Oil production, transport and refining. The production of oil from offshore fields, such as those in the Persian Gulf, Lake Maracaibo and the North Sea, can lead to surface pollution of the sea by spilled oil. The transport of oil by ship has already led to two major ecological disasters: the 'Torrey Canyon' off the coast of England in 1967, and the 'Amoco Cadiz' off the French coast in 1978. Such losses (about 100,000 tons of oil in the Torrey Canyon incident) are small compared with the estimated 3.5 million tons of oil discharged into the seas of the world annually. Most of these discharges arise from routine cleaning operations and minor spillages, but their cumulative effect is only now beginning to be appreciated. The surface film of oil on the ocean inhibits oxygenation and is toxic to sea life. Dispersants used to clear oil from beaches may also be harmful to fish and plankton.

Oil refining produces liquid wastes having a high hydrocarbon content. These include surface run-off from the refinery area, process water which has been in direct contact with oil, cooling water and tanker ballast. Refineries use considerable quantities of water and care must be taken to ensure that oil fouling of effluents is limited. Treatment methods include physical separation by the use of storage tanks, where the oil wastes float to the surface and are removed; biological cleaning by the use of specially cultivated bacteria or algae; flocculation by means of chemical agents, followed by mechanical separation.

Nuclear fuel cycle. Running releases from nuclear reactors to rivers and seas are generally less significant than gaseous discharges and arise from coolant contamination (in those reactors using water as coolant), routine maintenance and refuelling procedures, and from infiltration of fission products into cooling ponds through faulty or corroded fuel elements. After a certain retention time in holding tanks such wastes are generally diluted with the coolant outlet stream of the turbine condensers.

Radioactive releases in the form of liquid discharges to the environment arise from nuclear fuel reprocessing. These releases are carefully monitored and regulated by national authorities. Some experts, however, have expressed concern about the level of such discharges and the biological concentration of such isotopes as ruthenium-106, strontium-90, caesium-137 and cerium-144 in fish. Actinides such as plutonium and americium are also released in liquid effluents. Continued monitoring is required of the environmental pathways by which such contaminants may reach man's environment. In general, as far as liquid wastes are concerned, releases from fuel reprocessing activities are considerably more important than running releases from reactors. Gaseous emissions, particularly those of krypton-85, argon-41, iodine-131 and hydrogen-3 (tritium) may be more important than either.

Contamination of groundwater can occur through leakages of high-level or medium-level storage tanks. This may become a pressing problem in the future as the volume of such wastes grows, but techniques such as vitrification will, it is hoped, alleviate this situation to some extent.

General sources of water pollution. *Thermal pollution*, arising from large-scale discharges of heat to rivers and the oceans, can cause local problems and may exert a limit on energy production in the long term.

Chemical pollution results from cooling water discharges containing biocidal agents such as chlorine and anti-corrosion additives. The use of dry cooling towers, however, eliminates this problem, but at considerable expense.

water power

Power obtained by using the movement of water or heat flow within it. Such power is derived from two main sources: river flow and tidal movement. Also under investigation is the extraction of the energy of the movement of water particles within waves. Geothermal water is becoming increasingly important on a local scale as a heat source and the injection of water into *hot rocks* for steam generation is a likely further source of geothermal heat. The exploitation of *ocean thermal gradients* is a possible means of adding to energy supplies. Some authorities have suggested that ocean currents might be exploitable, but present indications are that this is not economically feasible.

See also *geothermal energy; hydraulic turbines; hydro-electric power; pumped storage; steam; tidal energy; water wheel; wave energy.*

water wheel

A primitive type of prime mover using the energy of water falling from a height to a lower level. The turning wheel is used to drive machinery, as in a mill. Though largely superseded by turbines and powered machinery, water wheels were once widely used.

In the 'undershot wheel' the axis of the wheel is usually placed at about the same height of the water moving towards the wheel. The water hits the buckets and shoots under the wheel turning it round. To prevent water loss from the buckets when they are full they are covered by a close-fitting channel so that as great a weight of water as possible is turning the wheel.

In the 'overshot' type the diameter of the wheel is generally less than the head of water. Water is directed onto the top of the wheel by a chute – the headrace – and fills the buckets set round the circumference of the wheel, as the wheel turns. As the buckets on one side fill, the weight of water turns the wheel and the empty buckets on the other side move up to be filled from the headrace.

watt

A unit of *power*, defined as one *joule* per second. For practical applications, multiples of this unit are usually used, namely kilowatts (kW, 10^3 watts), megawatts (MW, 10^6 watts), gigawatts (GW, 10^9 watts) and terawatts (TW, 10^{12} watts). For instance, domestic electrical appliances have *power ratings* of up to a few kilowatts, and installed capacity of a large generating system may be of the order of tens of gigawatts.

watt-hour

A unit of work, equivalent to 3600 joules. Because this unit is rather small for practical purposes, it is usually used in the form kilowatt-hour (kWh, 10^3 watt-hours) or a higher multiple such as the terawatt-hour (TWh).

wave contouring rafts

See *contouring rafts; wave energy.*

wave energy

Energy derived from the motion of waves, usually ocean waves close to land. The oceans effectively act as large collectors of wind energy since surface waves of long period can travel hundreds of miles with small losses. Data from weather ships shows that typical wave heights throughout the year lie between two and three metres. In the western Atlantic this corresponds to an average power flow of around 70 kW per metre of wave front. Along the 1500 km coastline of the British Isles, this corresponds to an average energy flow of over 100 GW, more than twice the present peak power demand of the UK. With a conversion efficiency of 50 per cent, a one km stretch of ocean adjacent to the coast could yield as much as 300 GWh per annum. Whilst it is improbable that very large schemes for wave energy conversion could be justified, as a form of renewable energy with few environmental effects, wave energy is receiving some attention worldwide, particularly in the UK and Japan. The Central Electricity Generating Board (CEGB) of England and Wales has shown interest, and several pilot-scale developments, sponsored by the UK Department of Energy, are under way.

As far as an electricity network is concerned, an advantage in temperate regions is that wave power is at its peak in winter when electrical demand is at its highest level. However, the variable nature of wave-generated power would make it difficult to integrate into the network without the use of some large-scale forms of storage. Since the generators will be a long way offshore, the problem arises as to how best to transmit power into the electricity supply system.

The major problems associated with wave power concern the difficulty of maintaining complicated large-scale devices in an extremely hostile and corrosive environment. The construction of equipment that can survive storm conditions and the anchoring or mooring of the plant would be extremely difficult.

Wave energy without storage could probably only be considered as a fuel saver since an installed capacity of standby plant equal to the wave power capacity would be required. The standby plant would only be used during calm periods, but its capital costs would represent a considerable financial burden on any wave energy programme. Clearly there is an optimal level of storage at which the capital cost is minimized. This level would probably require some capacity of standby plant.

The availability factor of a wave power system is dependent on the rating and efficiency of the conversion devices. A unit tailored to operate most efficiently with large waves will have a lower availability factor than a similar unit designed to produce maximum power from smaller waves, and to reject the energy of larger waves.

Wave energy conversion devices. Wave dynamics are such that several characteristics can be exploited: the oscillating vertical motion of the wave; the circular motion of the water particles within each wave; the varying distance between the water surface and the sea floor and the associated pressure changes, and the breaking of waves on the shore or breakwaters. Devices can be designed to exploit any one of these motions.

Oscillating movements. Most ideas have been put forward in this area. There are many design variations, but each device contains a float connected to levers or cables and the linear motion produced is converted to electrical energy. The linkage may be between the float and the sea bed or the shore where the movement of the float itself may actuate the armature of a linear generator; the reciprocating movement of water within an inverted box or tube compresses air and this is used to drive an air turbine connected to a generator; the movement of a series of inter-connected rafts or mats operates pumps or hydraulic motors which compress air or fluid to drive an air or water turbine connected to a generator.

All structures using these systems are prone to damage from the sea. A seemingly attractive design is that based on an idea by the Japanese engineer Yoshio Masuda. When investigating floating breakwaters he discovered that wave height could be reduced if the breakwater was arranged as an inverted box with openings at the top in and out of which the air in the box could flow. The development of what is known as an *air pressure ring buoy* has been to use the movement of the water and air to operate an air turbine. Such devices are now in commercial use as self-powered marine buoy lights, rated at about 70W. Although it requires mooring cables to keep it floating freely on station these do not take any power strain and it is therefore less likely to be damaged than some other devices. This device can also accept waves from any direction, ie it is 'omnidirectional'.

Another wave motion device is the wave contouring raft of Sir Christopher Cockerell, inventor of the hovercraft. Like the 'ducks' of Dr Stephen Salter described below, they are modular, which has advantages in construction and allows the system to be added to very easily.

The wave contouring rafts are aligned at right angles to the wave motion. Each wave passes under the raft which moves up and down in sympathy with the surface profile of the wave. Each unit in the raft is linked somewhat like the links in a bicycle chain and the relative movement of the parts results in energy being transferred via hydraulic compressors to the generators. Trials with scale models in the UK have been successful, and tests with larger models are planned. The contouring raft is not omnidirectional and would need to be steered to face oncoming waves.

The 'HRS rectifier' or 'Russel rectifier' developed at the UK Hydraulics Research Station by Robert Russel, consists of a single structure divided into two reservoirs, an upper and a lower. As a wave peak flows past the floating structure water is allowed to enter the upper reservoir via non-return valves and is then allowed to flow into the lower reservoir by way of a turbine. The lower reservoir is emptied via another set of non-return valves to the wave trough. In this way a small water head is maintained. Suitable turbines for use in this device are under development.

Circular water particle motion. Stephen Salter's 'nodding ducks' are so-named because of their shape and the way they nod up and down in the water. Each unit is a floating vane which oscillates with the particle motion of each wave and contains an electrical generator. The shape of each duck is such that maximum energy is extracted from the waves, its front surface being flat. The rear surface is smoothly rounded so that the nodding movement of the duck does not displace enough water to generate waves there. Power is generated by tangential hydraulic compressors and transmitted to the generators. Numerous ducks can be arranged side by side in a line and tests have been completed with 12 in a string. Further trials are to take place with a string of 20 ducks, each one metre in diameter (1/10 scale). Power output will be 10 kW. It is anticipated that suitably sited strings of ducks, each string 1 km long, could generate 45 MW. This device has achieved high conversion efficiency in test conditions but there are obviously many engineering problems to be overcome.

Pressure changes under the surface. Because the surface is moving up and down as the waves pass, the distance between the surface and the sea floor is repeatedly changing. Such changes create varying pressures which can be utilized. One suggestion is for flexible rubber tubes containing hydraulic fluid to be anchored to the sea floor inshore in concrete troughs. Pressure changes in the fluid are transmitted along a tube to a hydraulic accumulator and motor assembly on shore, the motor driving a generator. Conversion efficiency is claimed to be high.

Other devices have included buoys moored by tight lines. Pressure fluctuations act on a diaphragm which behaves as a hydraulic piston.

Breaking effect of waves. One scheme employing this merely allows the waves to break over a suitably inclined impounding wall into a reservoir. From here it passes through a low-head turbine back into the sea. By generating at low tide some of the tidal energy could be obtained using this system. Such a system would present fewer problems than all other wave energy proposals. Since the plant is sea-based the risk of possible damage is eliminated. The generating plant is conventional and in use throughout the world so that no new technology is needed in this respect or in sea wall construction.

Prospects for wave power. Although a 1 kW wave energy generating device was used in France, near Bordeaux, in the early part of this century, the immense technical problems and the abundance of other energy supplies has not provided a sufficient stimulus for its subsequent development. But the energy crisis of the early 1970s, precipitated by the raising of petroleum prices, and the problems facing the highly industrialized countries such as the United States, whose energy use has increased to the extent that she is now a net importer of energy supplies, have meant a resurgence of interest in tapping a vast energy source. The proliferation of wave energy devices is evidence of this.

wax

See *paraffin wax.*

Weston cell

One of the two principal *standard cells.* (The other being the Clark cell.) There are two types of Weston standard cell.
1. The normal cell which contains a saturated cadmium sulphate solution. This cell is used as the basic standard for calibrating cell voltages, having an emf of 1.0183 volts.
2. A working standard in which the solution is less than saturated above 4°C. This cell is useful because of its negligible temperature coefficient. The emf may be taken to be 1.0186 volts for all practical purposes.

wet gas

Natural gas that contains appreciable quantities of *ethane, propane, butane* and the *pentane-plus fraction. Associated gas* is usually wet; non-associated gas is usually dry.

white damp

Carbon monoxide produced in a coal mine by fire or gas explosions. It is colourless and poisonous.

white spirit

A specially refined light distillate boiling in the range 150°-200°C, widely used as a paint thinner and in dry cleaning. Occasionally known as turpentine substitute in the United Kingdom, and commonly called petroleum spirits in the United States.

wide-cut gasoline

Also known as aviation turbine gasoline. A petroleum fraction overlapping the gasoline and kerosine ranges, it is used as an *aviation turbine fuel* in jet aircraft. It has a broader boiling point range (30°-260°C) than aviation kerosine (150°-250°C) which is also used as a fuel for jet engines. Wide-cut gasoline is little used in the United Kingdom nowadays.

wide range distillate

A term used to describe a petroleum fraction containing hydrocarbons from a broad boiling point range. An example is *wide-cut gasoline.*

wildcat

A well drilled speculatively in the hope of finding petroleum deposits and without adequate surveying.

wind energy

The kinetic energy associated with the movement of large masses of air over the surface of the earth. Such motion results from the uneven heating effect of solar radiation on the seas and land masses and varies with season and location.

Air in contact with the warmer parts of the earth's surface is heated and, being less dense than cooler surrounding air, rises and is replaced by the cooler air. The primary directions of air motion are modified by the rotation of the earth on its axis to produce the familiar 'trade winds'. Local climatic and topographical variations lead to alterations in wind direction and velocity over a short time scale. Because of the frictional losses due to irregularities in the earth's surface, wind speeds are observed to increase with altitude.

Of the solar radiation absorbed by the earth approximately 20 per cent is converted to heat by atmospheric absorption; only a small proportion of this appears as the kinetic energy of the winds. The total power of the winds is estimated to be 2 x 10^{10} kW, or about three times the world energy consumption rate. Only a small proportion of this power can be converted to useful energy, so that the potential of wind energy is limited, though not insignificant.

History and Development. For many centuries man has taken advantage of the kinetic energy which the wind possesses. Extant records show that the Chinese operated windmills some 2,000 years ago. The first accounts of windmills in England and France date from the early 12th century. For centuries wind energy provided one of the major sources of power for grinding, pumping, forging and milling purposes.

From 1890 wind-driven devices linked to *generators* started providing electricity on a small scale, mainly in the agricultural sector. The gradual change towards fewer large scale power supplies diverted attention away from the relatively diffuse energy sources, including wind energy, towards fossil fuels and later, nuclear sources. However, Denmark, Sweden, the UK, USA and USSR experimented with various wind-powered projects for providing electricity on a larger scale. For example, in 1931, the Russians set up a wind-powered system at Balaclava and for several years it generated approximately 250,000 kWh/yr for the Yalta electricity grid. Earlier this century Denmark records some 33,000 wind-powered generators each with a capacity of around 35 kW of electricity.

The world's largest wind turbine operated between 1941-45 at Grandpa's Knob in Vermont (USA). Designed by Palmer Cosslett Putnam, it consisted of a 175 ft twin-bladed propeller connected to a 1.25 MW generator, and in 1942 it produced approximately 180 MWh of electricity. (It went out of service in 1945 after one of the 8 ton blades snapped off.)

World War II shortages temporarily encouraged research into alternative energy sources, including wind energy. However, the majority of wind projects were abandoned through lack of interest and financial backing. The recent general reappraisal of world energy supplies has again switched attention and resources to alternative energy schemes. Significant R & D programmes aimed at utilizing wind energy are under way in the USA and smaller programmes exist in Europe and USSR. (At Gedser, in Denmark, for example, a fully automated 200 kW wind turbine produced approximately 400,000 kWh/yr.)

Theory of wind exploitation. The energy in the wind increases with the cube of wind velocity. In North America and the UK, in particular, wind movements are at their strongest during the winter months, which corresponds to the peak demand period for energy supplies.

Theoretically a *wind turbine* can recover 59 per cent of the kinetic energy in the wind. Mechanical losses through conversion reduce this to around 40 per cent. (These figures relate, of course, only to the flow of air through the area covered by the sweep of the rotor.) Smaller units tend to have higher efficiency rates and are usually geared to less than maximum wind velocities to maintain a more constant flow.

Even-flowing wind movements over gradual hill slopes yield substantially more power than turbulent movements over rough country. The positioning of wind-driven devices is extremely important as wind strength and continuity may vary dramatically within short distances. Thorough surveying is a vital preliminary to choosing sites.

Storage. Storage facilities are necessary to reduce the problems caused through the intermittent nature of wind movements. If the windmills are linked to generators then surplus electricity may be used for storage purposes. One method entails storing electricity in *lead-acid accumulators.* However, this method is expensive and the battery charge is comparatively short-lived. A second method could utilize excess electricity in pumping water to a high reservoir. Releasing the water later could produce power in hydroelectric generators (see *pumped storage schemes*).

Several schemes envisage using depleted or empty natural gas reservoirs to store compressed air. In times of strong wind, surplus electrical energy from the windmills could compress air into the reservoirs. Then, on windless days, the air pressure could reverse the process through the generators to produce electricity again.

A further method could involve electrolyzing water into hydrogen and oxygen and storing the two gases for later use in fuel cells. (This final method would suffer from large conversion losses at each stage and capital costs would be extremely high. However, one advantage is the potential application to offshore windmills.)

Advantages and limitations. Most authorities concede that the present day economics of wind-generated electricity are marginal and that wind energy will only be developed on a large scale when energy prices have risen by a considerable amount. The capital costs, though subject to debate, seem to be comparable with those pertaining to the next generation of nuclear reactors but considerably greater than the installed costs of present fossil-fuelled power stations. It should be remembered, however, that the wind is not a firm power source and that availability factors for wind generators will generally be lower than those achieved by nuclear reactors. Hence the often considerable costs of storage must be added to the installed capital costs of a wind power system.

There exists considerable potential for cost reductions in wind power systems. These arise mainly through the economies of scale associated with the use of large units, but there are also possibilities for the development of cheaper components and materials of construction. The newer designs, such as the Darrieus rotor, or modifications of it, offer the promise of simple, high reliability generators at moderate cost. Considerable research on such designs is still required, however.

The main advantage of wind generated power is that it exploits a clean, renewable resource and does not cause wind turbine

problems of *thermal pollution*. Even if used purely as a fuel saver, supplying power only when available without storage, the use of wind energy will conserve fossil fuel resources and reduce atmospheric pollution. It remains to be seen whether this is sufficient justification for its use.

Prospects for wind energy. A project under study in the USA, the Heronemus Scheme, envisages a series of 300,000 wind-driven devices spread across the Great Plains. These would possess an accumulative generating capacity of 189,000 MW. (Compare this to the 1970 US total installed capacity of 360,000 MW.) Such a large scheme would represent a major contribution to a nation's energy supply. A land-based scheme of this magnitude has little application in western Europe as a prerequisite is for vast areas of land with low population density.

The world's oceans and seas could provide possible locations for wind energy schemes. Indeed the shallow waters of the southern North Sea have been suggested recently as the site for a scheme involving clusters of floating windmills. However problems relating to storage, navigational hazards and interference with fishing may restrict the viability of such sea-based schemes.

wind turbine

A device used to convert the kinetic energy of the wind to mechanical power, generally in a rotating shaft which may be linked to an electrical generator (hence the term wind generator or 'aerogenerator').

The kinetic energy of a moving volume of air is proportional to the product of the mass of that air and the square of the velocity. Since the mass of air moving through an area is itself directly related to the velocity, the power density of the wind, ie the power available per unit area, can readily be seen to be directly proportional to the cube of the wind velocity:

$$\text{power density available} = \tfrac{1}{2}\, p v^3 \quad (\text{W m}^{-2})$$
$$\text{where } p = \text{density of air (kg m}^{-3})$$
$$v = \text{wind velocity (m sec}^{-1})$$

Since p is approximately 1.2 kg m^{-3}, for a wind speed of 5 m sec^{-1}, the available power is 75 W per square metre.

The rapid increase of wind power with speed leads to a serious problem in wind turbine design. At speeds above the maximum rated wind speed the turbine cannot extract any extra useful energy from the wind; indeed, the blades may have to be braked to a halt. Winds even slightly less than maximum produce considerably less power: a reduction in wind speed by 21 per cent leads to a halving of the available power in the wind. The choice of maximum rated wind speed is therefore a critical design parameter. If set too low the turbine will be unable to make use of the most powerful winds; if set too high the availability factor of the wind generator will be low, ie it will not be a firm enough source of power. In this context an accurate knowledge of wind strengths and their variability is an important part of the initial appraisal of a wind power scheme.

It is not possible to extract all of the energy in the wind since to do so would imply bringing the air completely to rest, resulting in an accumulation of air at the turbine. It

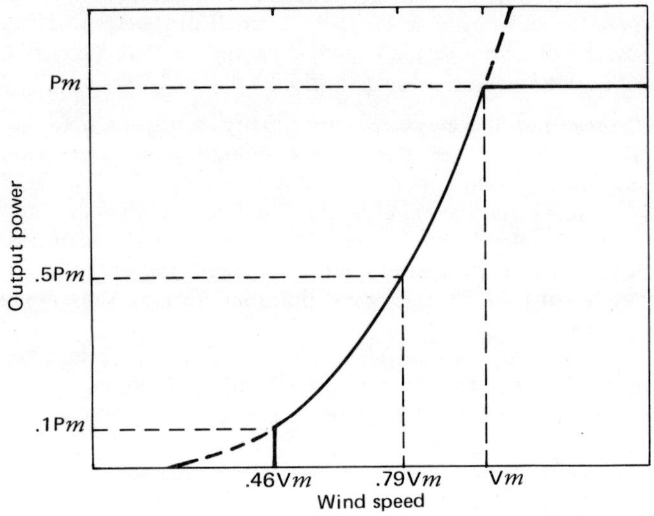

Power output as a function of velocity for a typical wind turbine

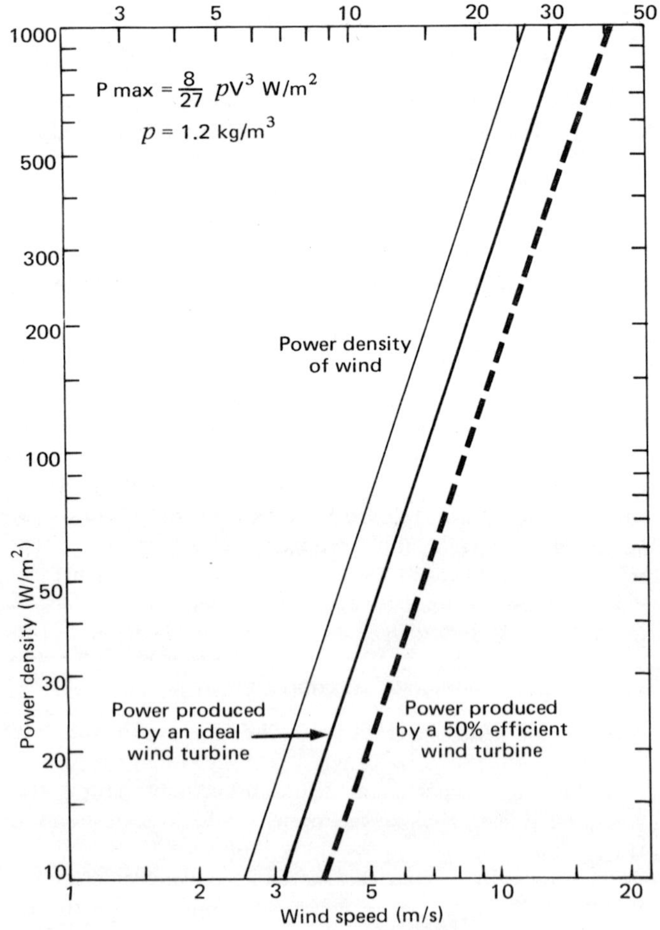

Power density : wind speed

wind turbine

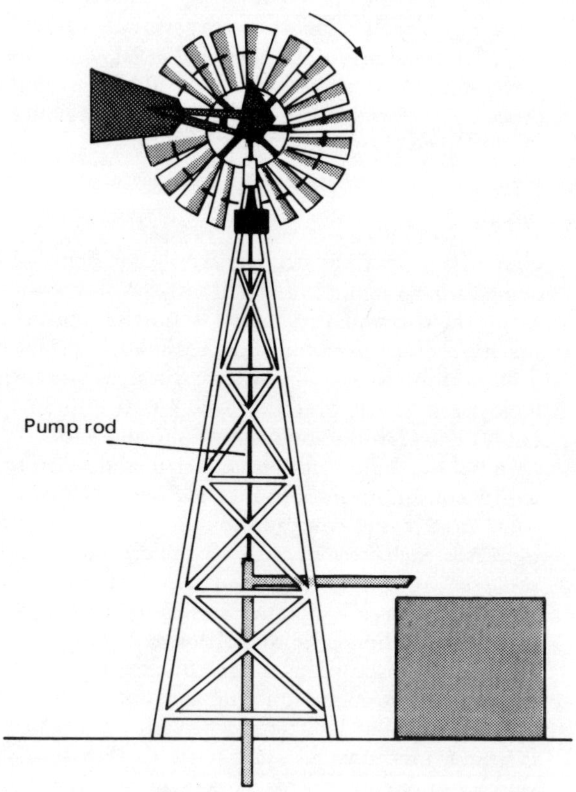

Typical farm wind pump

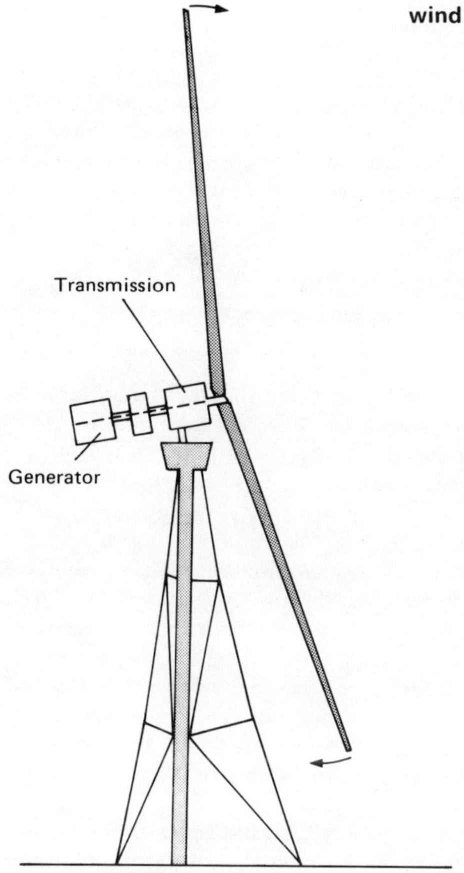

Putnam wind turbine

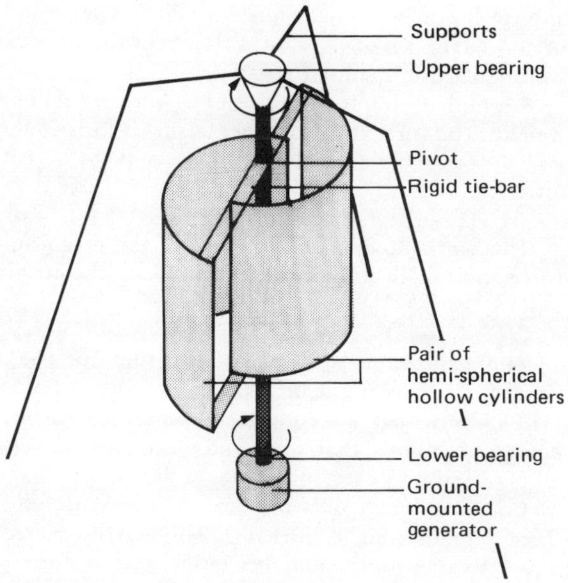

Savonius rotor: a vertical axis wind turbine

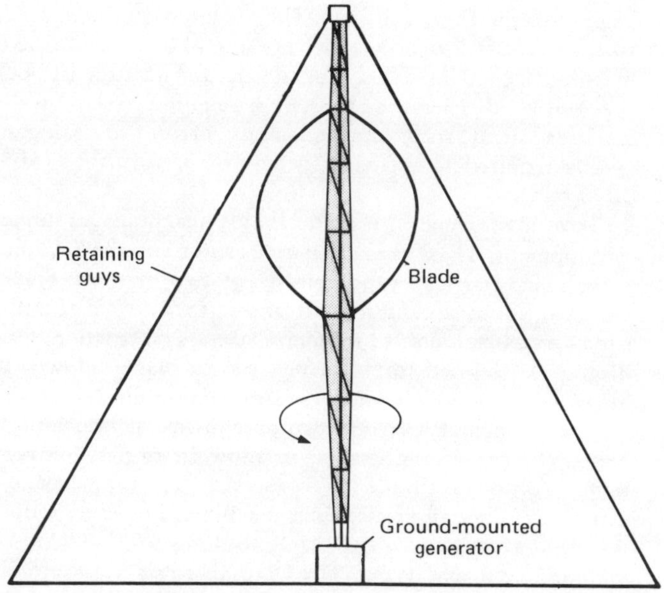

Darrieus-type vertical axis wind turbine

can be shown that the maximum proportion that can be extracted is 16/27, or 59 per cent, of the power of the wind. Efficient wind turbines have efficiencies of 50 to 75 per cent, measured as the ratio of power extracted per unit swept area of rotor to the maximum of 59 per cent of the power in the wind.

The graph shows plots of total wind energy as a function of wind speed, together with the amounts of energy extractable by an ideal wind turbine, and by a 50 per cent efficient one.

The wind speed at which a particular wind generator starts to rotate and produce output power is known as the 'cut-in speed'. At high wind speeds it is necessary to shut down the machine – the lowest wind speed for which this is necessary is known as the 'furling speed'.

Windmills, used for pumping water and grinding corn, typically had a small number of large canvas sails rotating on a horizontal shaft. Wind pumps, as developed in the USA and Australia, were built with a large number of much smaller metal blades, giving the high torque, low speed characteristics needed to pump water. Electric generators, however, were usually driven by wind turbines having only two or three blades. Modern designs of aerogenerator follow this pattern, with aerofoil blades rotating about a horizontal axis. These types of machine, exemplified by most of the prototype wind generators built in this century, are known as horizontal axis types, to distinguish them from the various forms of vertical axis machine now being developed. Representative horizontal and vertical axis designs are shown in the diagrams.

In modern horizontal axis aerogenerators the generator proper is typically mounted in one of two locations: coaxially with the rotor at the top of the tower; or at the foot of the tower, linked to the rotor by means of a right-angle drive and vertical shaft. The latter arrangement leads to easier maintenance but higher capital cost. Vertical axis machines generally use a ground based generator. In both types of wind generator a gearbox is required to step up the low speed of the rotor (approximately 50 rpm) to the higher speeds required for efficient generator performance (500-3000 rpm).

Two main types of vertical axis machine are under development. These are the Darrieus rotor and the Savonius rotor. The Darrieus type consists of two or more blades arranged about a central shaft and shaped like a rotating rope, or 'tropaskien', to minimize stresses on rotation. (See diagram.) Modified forms having straight blades allowed to hinge outward in high winds are also being studied.

The elementary Savonius rotor consists of two hollow hemi-cylinders arranged so as to allow air to pass between them. Although simple to construct, this type has low efficiency. Like all vertical axis machines, however, it does not need a mechanism to keep it pointing into the wind, as do horizontal axis types. The Darrieus rotor is potentially highly efficient, but has the disadvantage that it is not self-starting, requiring an initial impulse before it can rotate under its own power.

The majority of large-scale wind turbine research is directed at the two-blade, high speed horizontal axis machine, usually incorporating variable pitch blades to maintain constant rotational speed in varying wind strengths. Development work is also proceeding on generators and their control systems with a view to attaining automatic switching and control.

wood

Plant tissue that has become woody or lignified due to the deposition of lignin in the cell walls. Wood is widely used as a fuel in the underdeveloped countries and precise figures for its use are unobtainable. Estimates of well over 2,000 million tonnes per annum for wood consumption as fuel have been made which is over 1,000 million tonnes coal equivalent (1 tonne of coal is equivalent to 1.77 tonnes of wood). In some south-east Asian and African countries wood consumption accounts for over 90 per cent of their total annual fuel consumption. Although forest production increases each year because of the growth of existing trees the area of forest is depleted each year as felling outstrips planting. The use of special wood burning stoves could undoubtedly increase the efficiency with which wood is currently consumed as a fuel but because many countries do not take proper economic account of their use of wood at present there is little incentive for them to adopt more efficient methods of using it or to investigate alternative sources of energy.

See also *forest resources* and *Maps 19* and *20.*

work

In mechanics, work is done when a force moves its point of application. When a system is described as performing work, or producing a work output, this means that it is capable of causing motion against a resistance.

A simple example of performing work would be raising a weight. The force resisting such raising is the product of the gravitational acceleration and the mass of the object:

$$F = mg$$

The work done is the product of the magnitude of the force and the distance moved, in this case the height, h:

$$\text{work done} = mgh$$

This expression is identical with that for the *potential energy* of a mass m at height h. Since the potential energy can, in principal, be converted completely to *mechanical energy*, it follows that work and mechanical energy are, for all practical purposes, interchangeable.

Common units of work are 'newton metres' (SI) and 'foot pounds weight' (British). The newton metre (Nm) is equal to one *joule*, and this latter unit is more generally used.

XYZ

xenon poisoning

See *poisoning*.

xylenes

Aromatic hydrocarbons, existing in three isomeric forms and used in the blending of *motor spirit* and as a petrochemical feedstock. Xylenes are generally prepared by the dehydrogenation of naphthenes.

yellow cake

A mixture of uranium oxides, often denoted by the chemical formula U_3O_8. It is produced from uranium ore by chemical separation, treatment, and purification and may be converted directly into nuclear fuel consisting of naturally-occurring uranium, or be further converted to uranium hexafluoride, the compound of uranium most suited to the various enrichment processes.

zinc

A grey-white metal of lustrous appearance, though tarnishing quickly in moist air. It is widely distributed in nature in a number of ores and is usually isolated by conversion to the oxide followed by reduction with coke, or by electrolysis of zinc sulphate solution. It has been used for many centuries in alloys and is nowadays used in galvanizing iron to prevent corrosion. The oxide in particular has a multitude of present day application, eg in paints, pharmaceuticals, printing inks and electrical equipment. Zinc metal is an important component of the dry *Leclanché cell* and may in the future find an important application in zinc-air batteries if research and early development work proves promising.

symbol	Zn
atomic number	30
atomic weight	65.38
melting point	419.6°C
boiling point	907°C
density	7133 kg m^{-3} (7.1 g/cm^3)

zircaloy

An alloy containing 98 per cent zirconium, 1.5 per cent tin, 0.35 per cent iron chromium and nickel and 0.15 per cent oxygen. It has good corrosion resistance, a very low neutron capture cross section and is strong enough to be used for cladding uranium fuel in some designs of nuclear reactor.

zirconium

A grey-white lustrous metal obtained from the ores baddleyite or zircon by a reduction process. Zirconium metal is very highly corrosion-resistant, even withstanding the environment of nuclear reactors where as *zircaloy* it is used in pressure tubing. Because zirconium has a low absorption cross-section for neutrons it is also widely used as fuel cladding material.

symbol	Zr
atomic number	40
atomic weight	91.22
melting point	1852°C
boiling point	4377°C
density	6506 kg m^{-3} (6.5 g/cm^3)

Energy Resources Atlas

Note:
Some differences occur between figures where different sources have been used to describe a particular type of energy in separate tables and on different maps.

Map 1: World Energy Consumption in Relation to Population Distribution

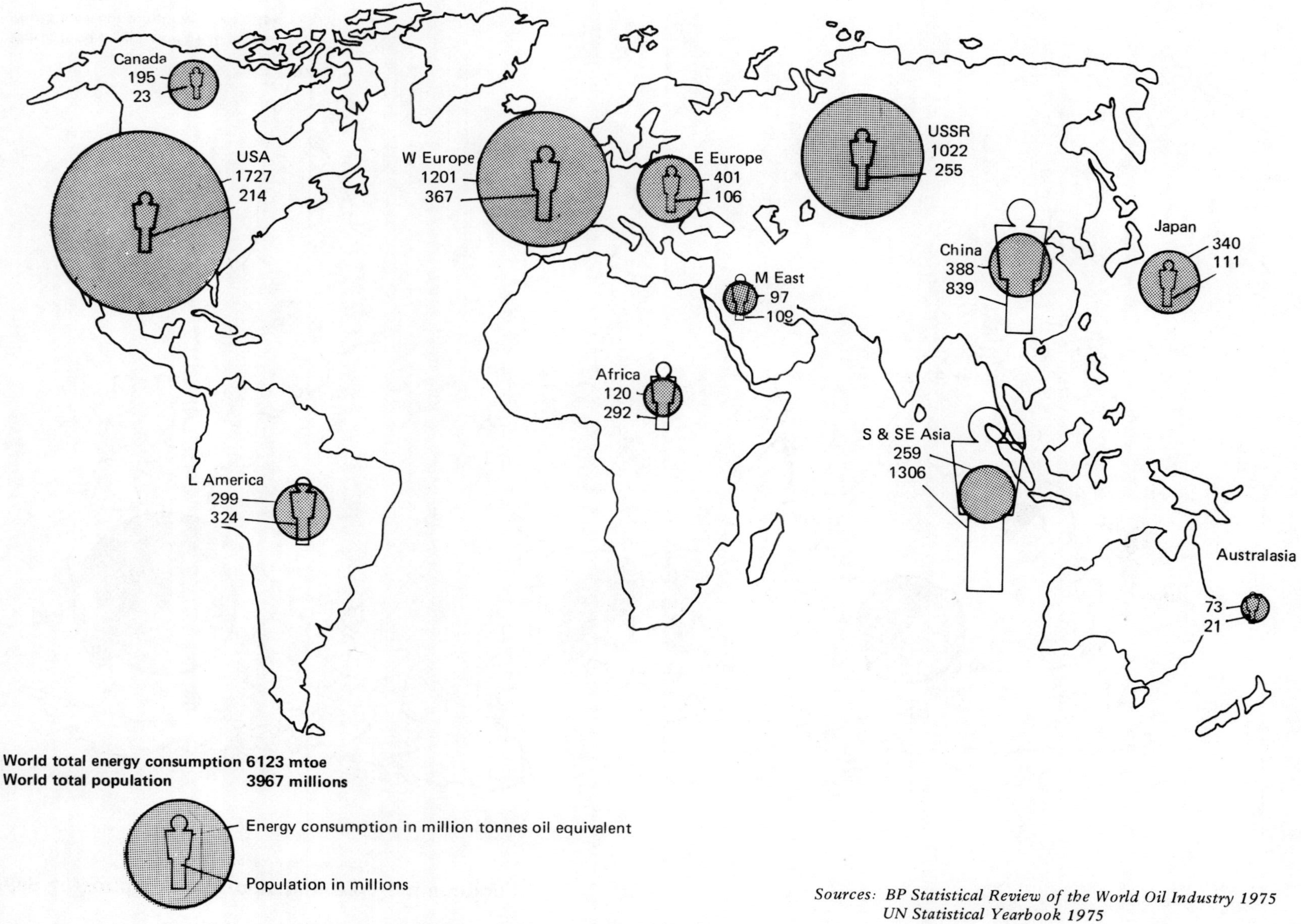

Sources: *BP Statistical Review of the World Oil Industry 1975*
UN Statistical Yearbook 1975

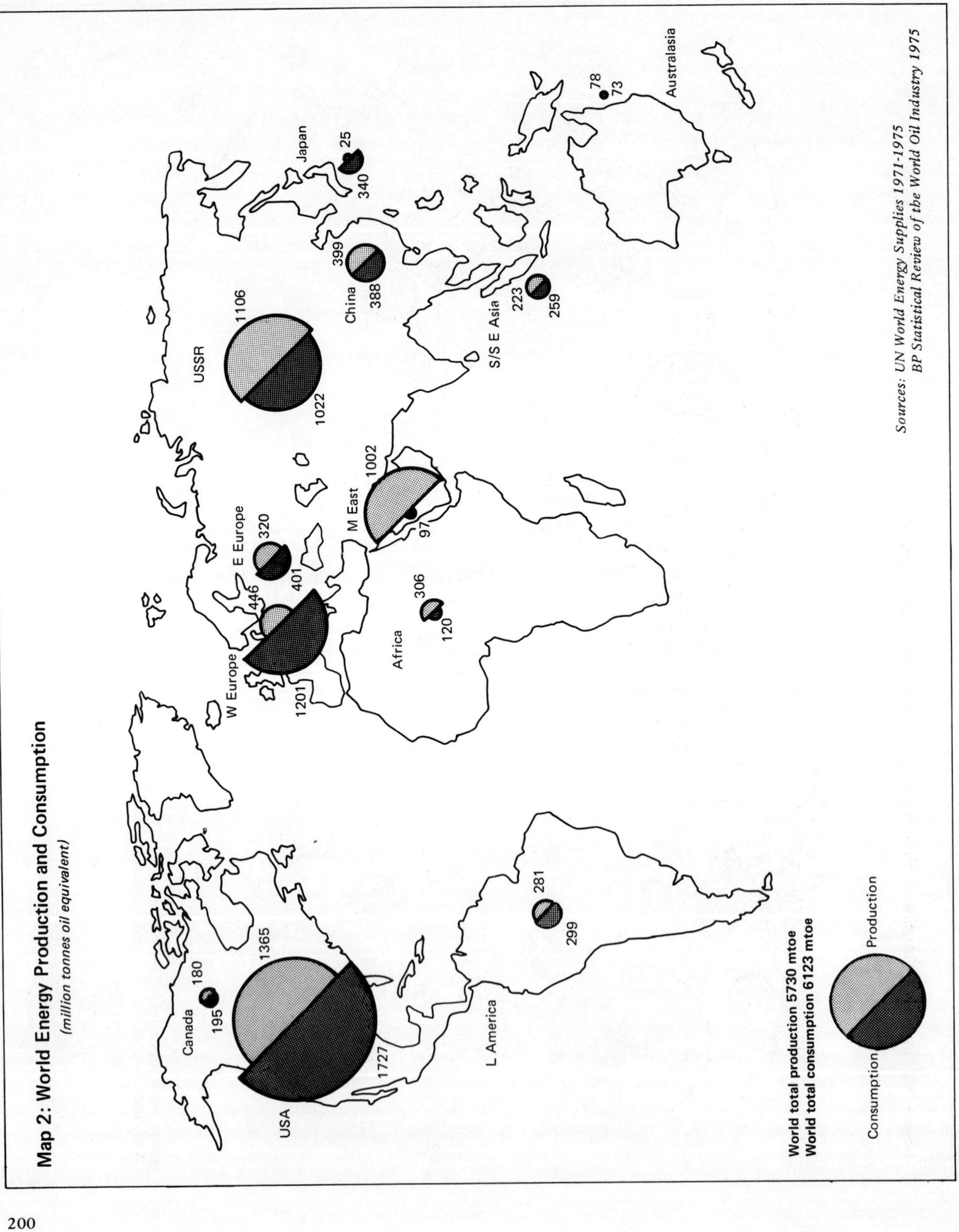

Map 2: World Energy Production and Consumption
(million tonnes oil equivalent)

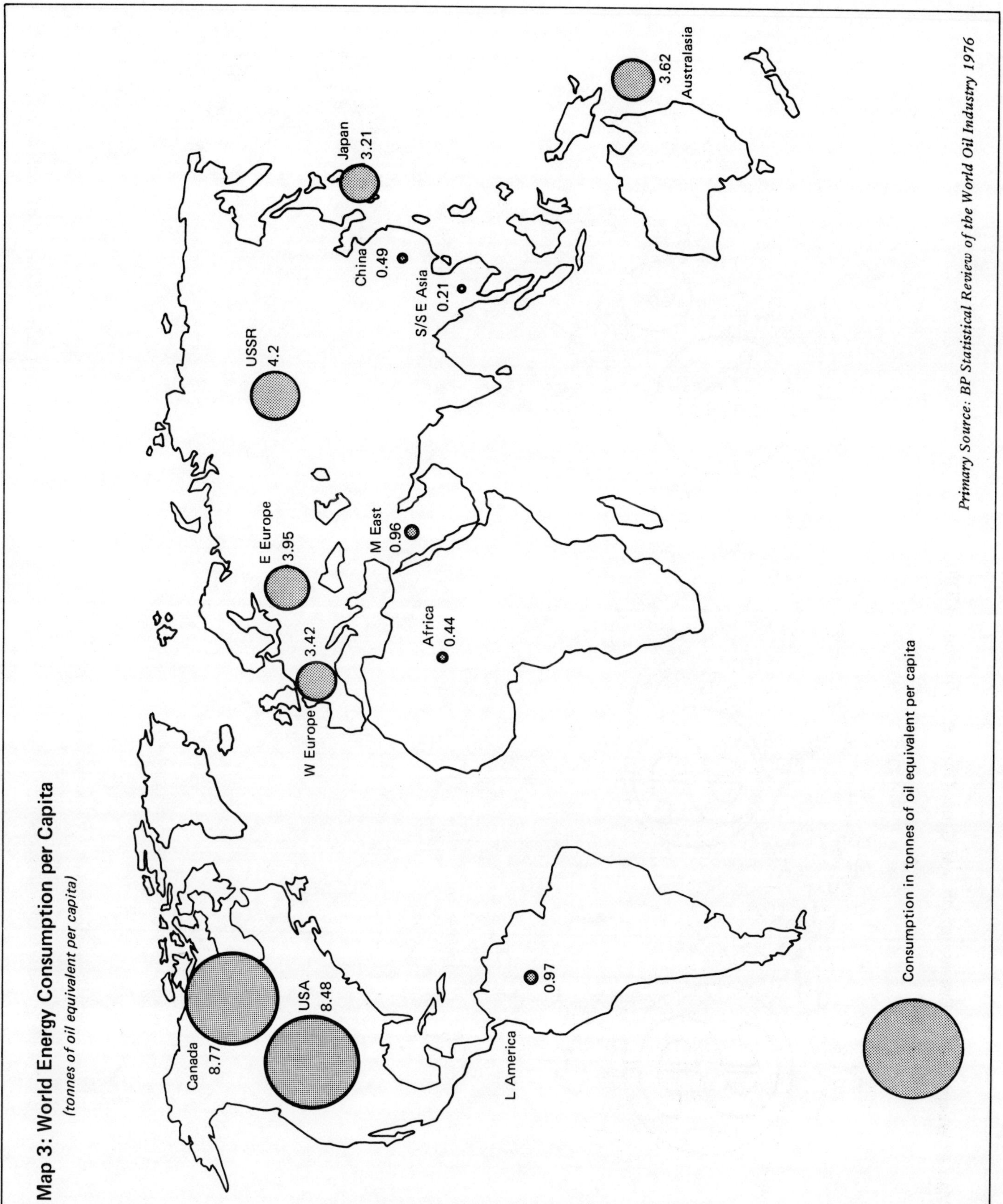

Map 3: World Energy Consumption per Capita

(tonnes of oil equivalent per capita)

Primary Source: BP Statistical Review of the World Oil Industry 1976

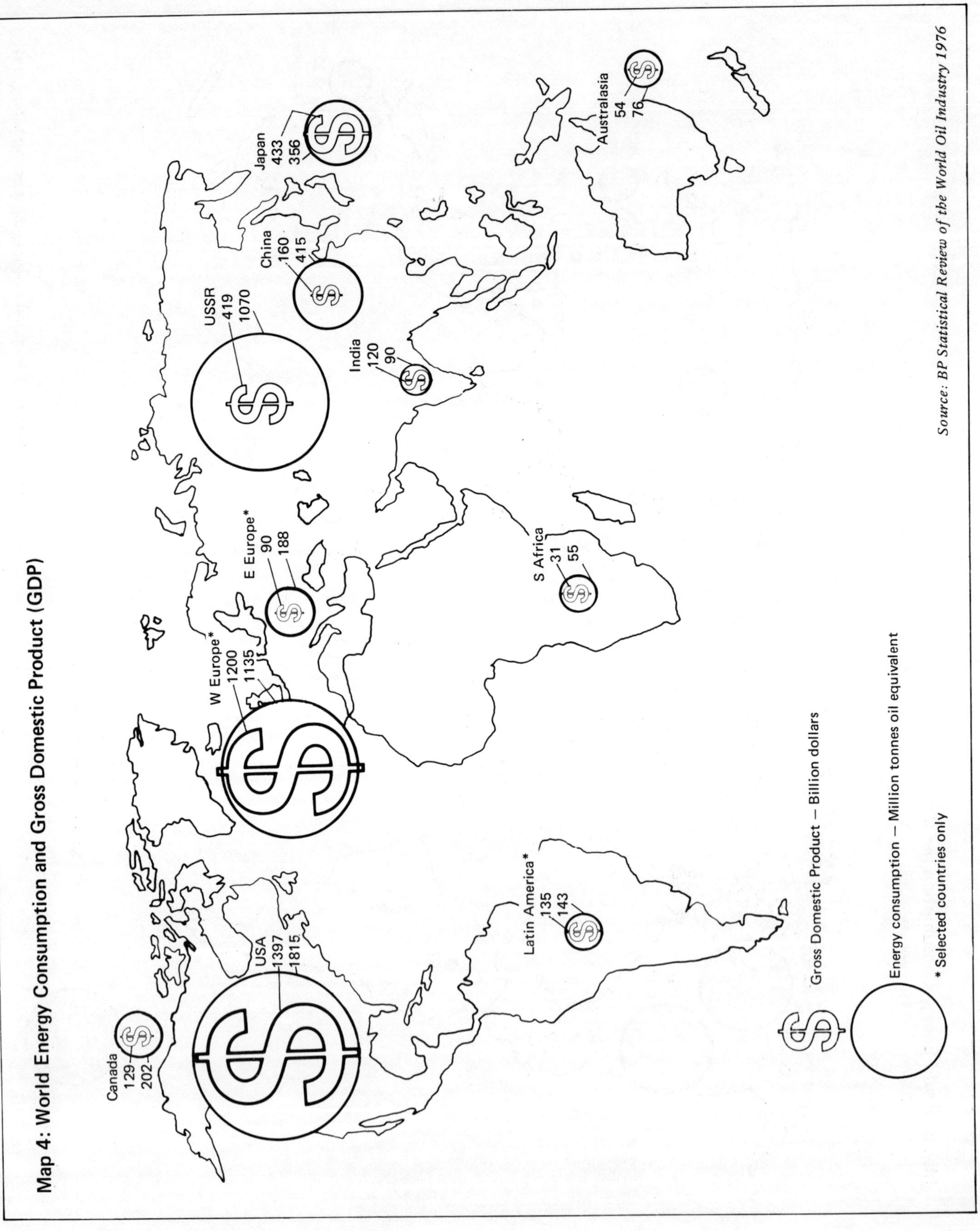
Map 4: World Energy Consumption and Gross Domestic Product (GDP)
Canada
129
202
USA
1397
1815
Latin America*
135
143
W Europe*
1200
1135
E Europe*
90
188
S Africa
31
55
USSR
419
1070
India
120
90
China
160
415
Japan
433
356
Australasia
54
76
Gross Domestic Product – Billion dollars
Energy consumption – Million tonnes oil equivalent
* Selected countries only
Source: BP Statistical Review of the World Oil Industry 1976

Map 5: Changes in Energy Consumption since 1950

(million tonnes oil equivalent)

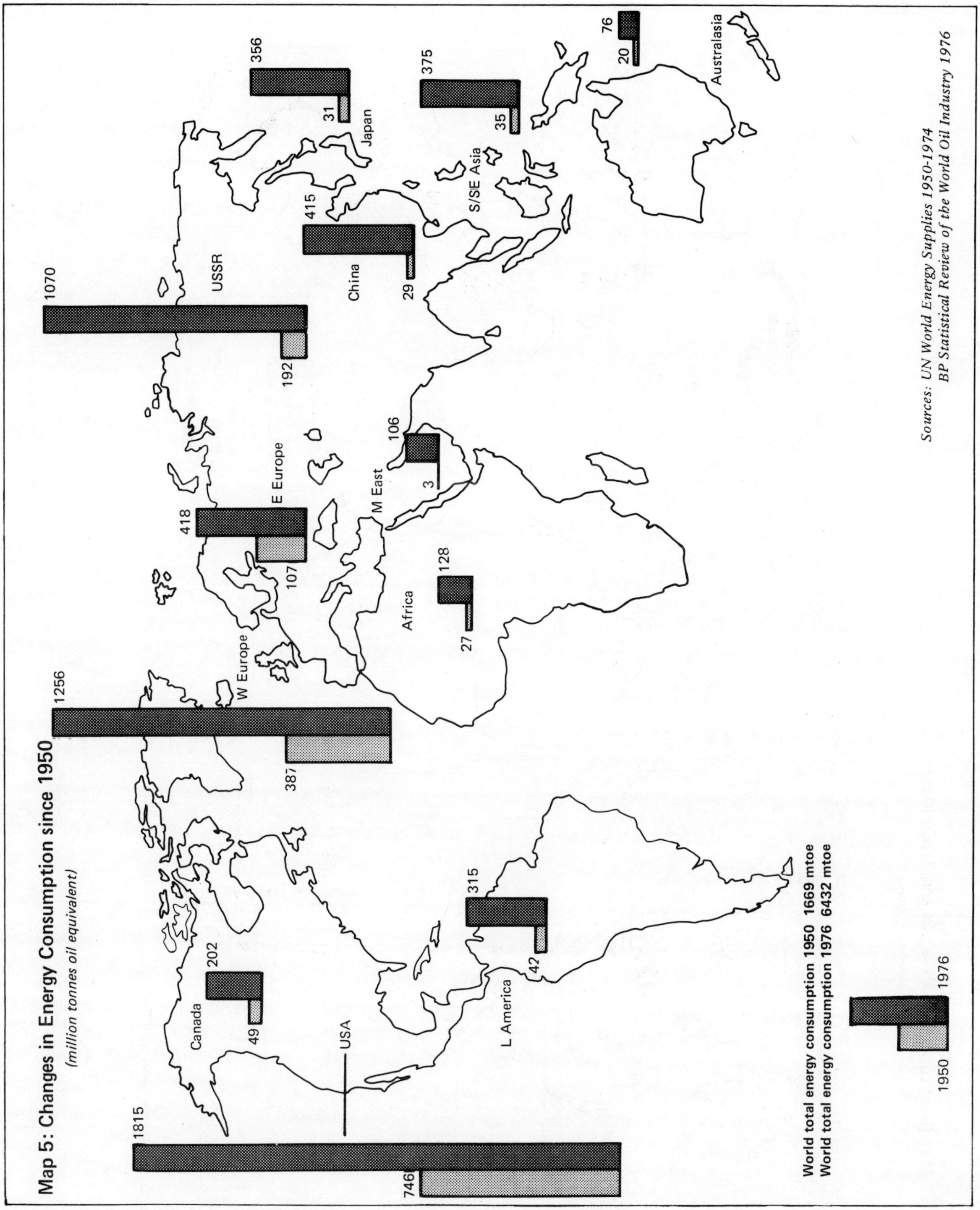

Sources: UN World Energy Supplies 1950-1974
BP Statistical Review of the World Oil Industry 1976

Map 6A: World Coal Reserves and Trade Movements

(million tonnes oil equivalent)

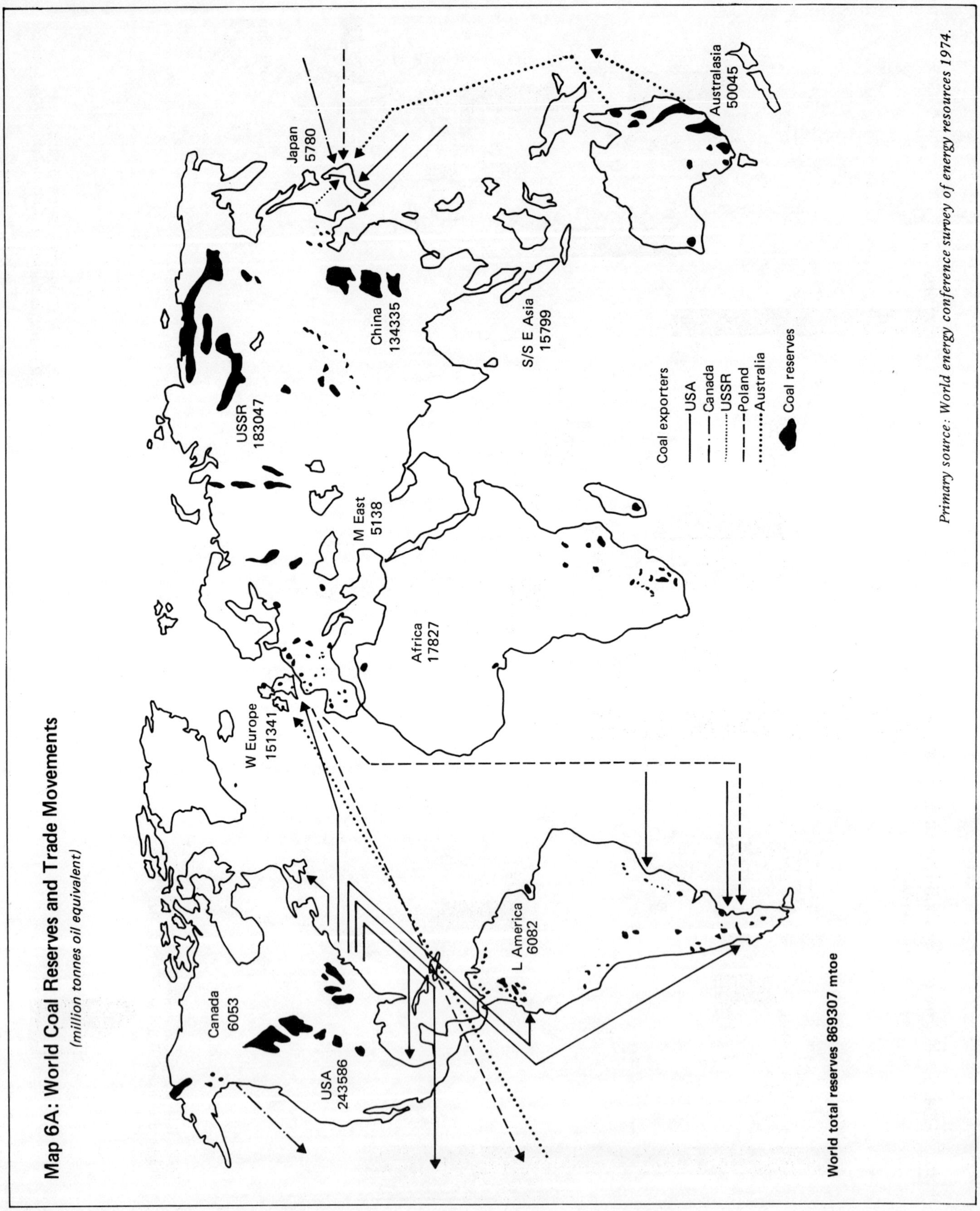

Primary source: World energy conference survey of energy resources 1974.

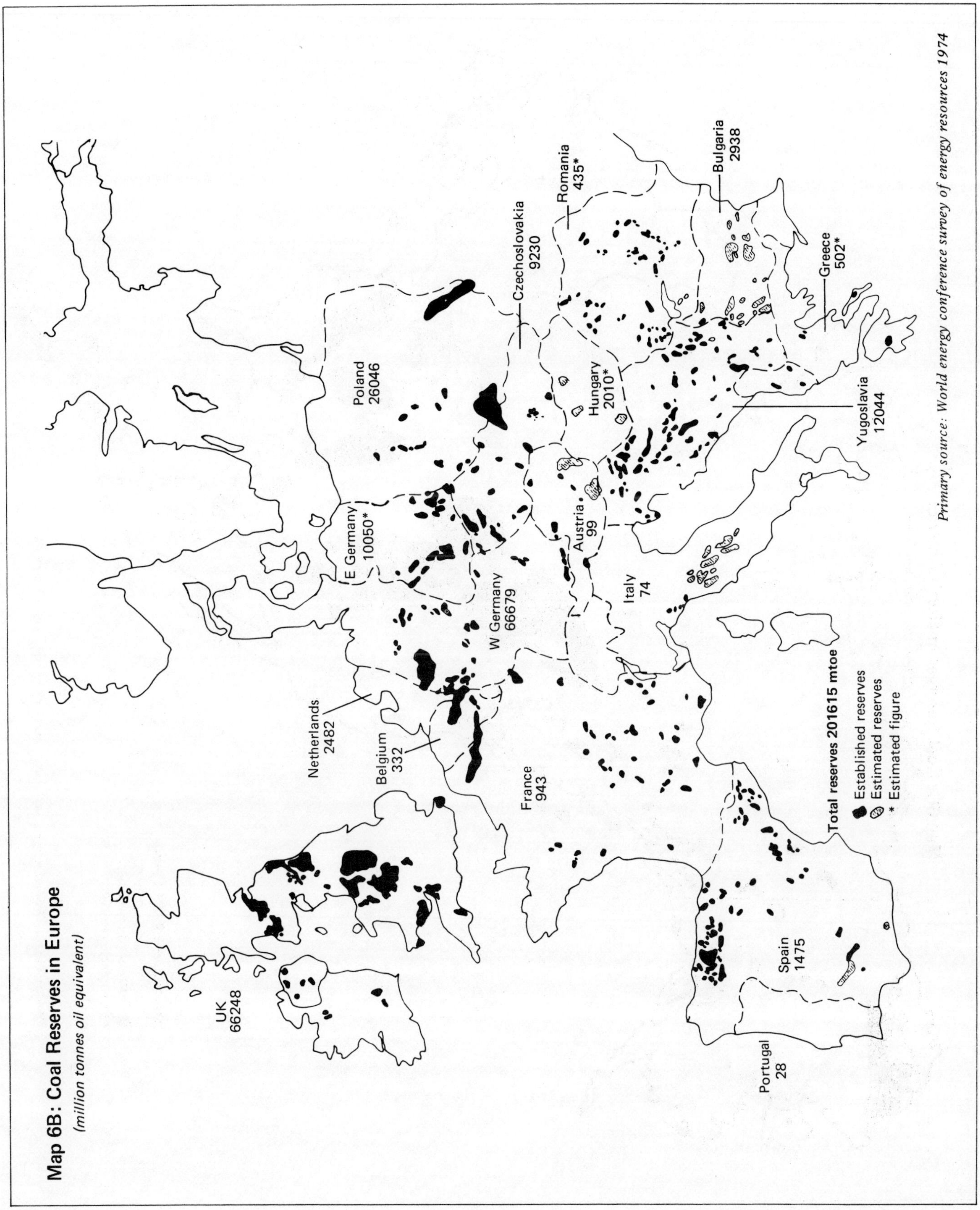

Map 6B: Coal Reserves in Europe
(million tonnes oil equivalent)

Primary source: World energy conference survey of energy resources 1974

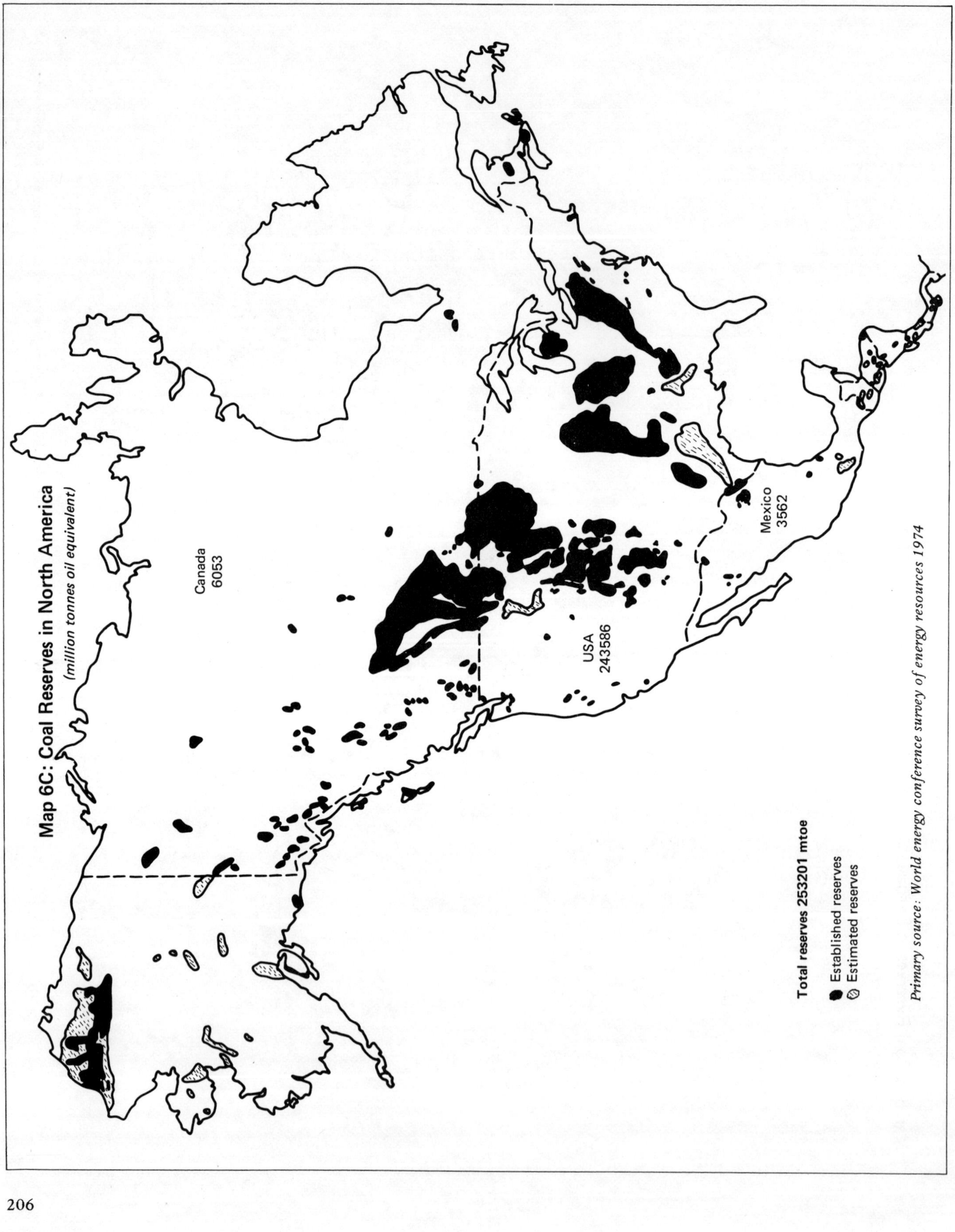
Map 6C: Coal Reserves in North America
(million tonnes oil equivalent)
Canada
6053
USA
243586
Mexico
3562
Total reserves 253201 mtoe
Established reserves
Estimated reserves
Primary source: World energy conference survey of energy resources 1974

Map 6D: Coal Reserves in the Middle East, USSR and Asia
(million tonnes oil equivalent)

Primary source: World energy conference survey of energy resources 1974

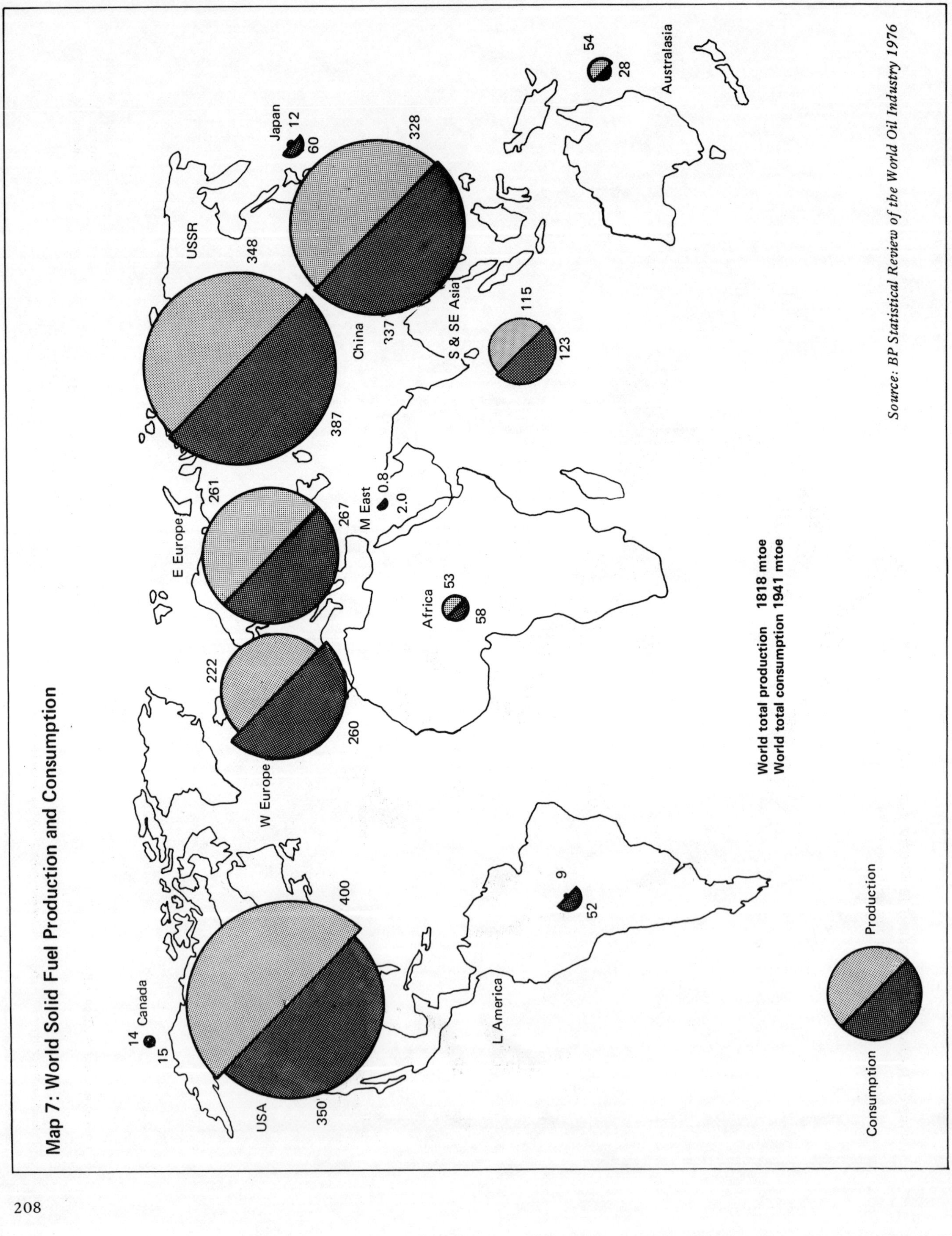

Map 7: World Solid Fuel Production and Consumption

Source: BP Statistical Review of the World Oil Industry 1976

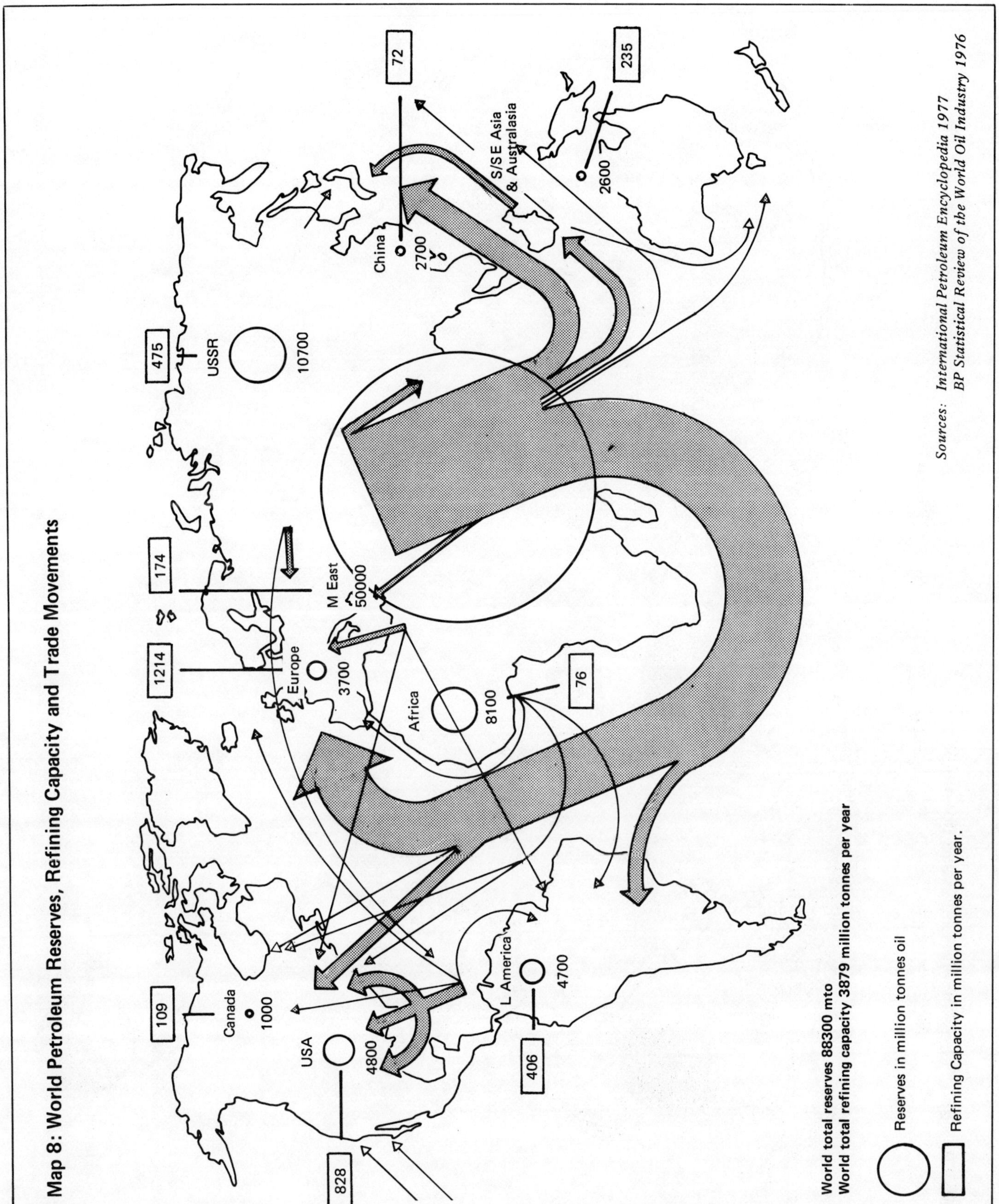

Map 8: World Petroleum Reserves, Refining Capacity and Trade Movements

Map 9: World Petroleum Production and Consumption

(million tonnes of oil)

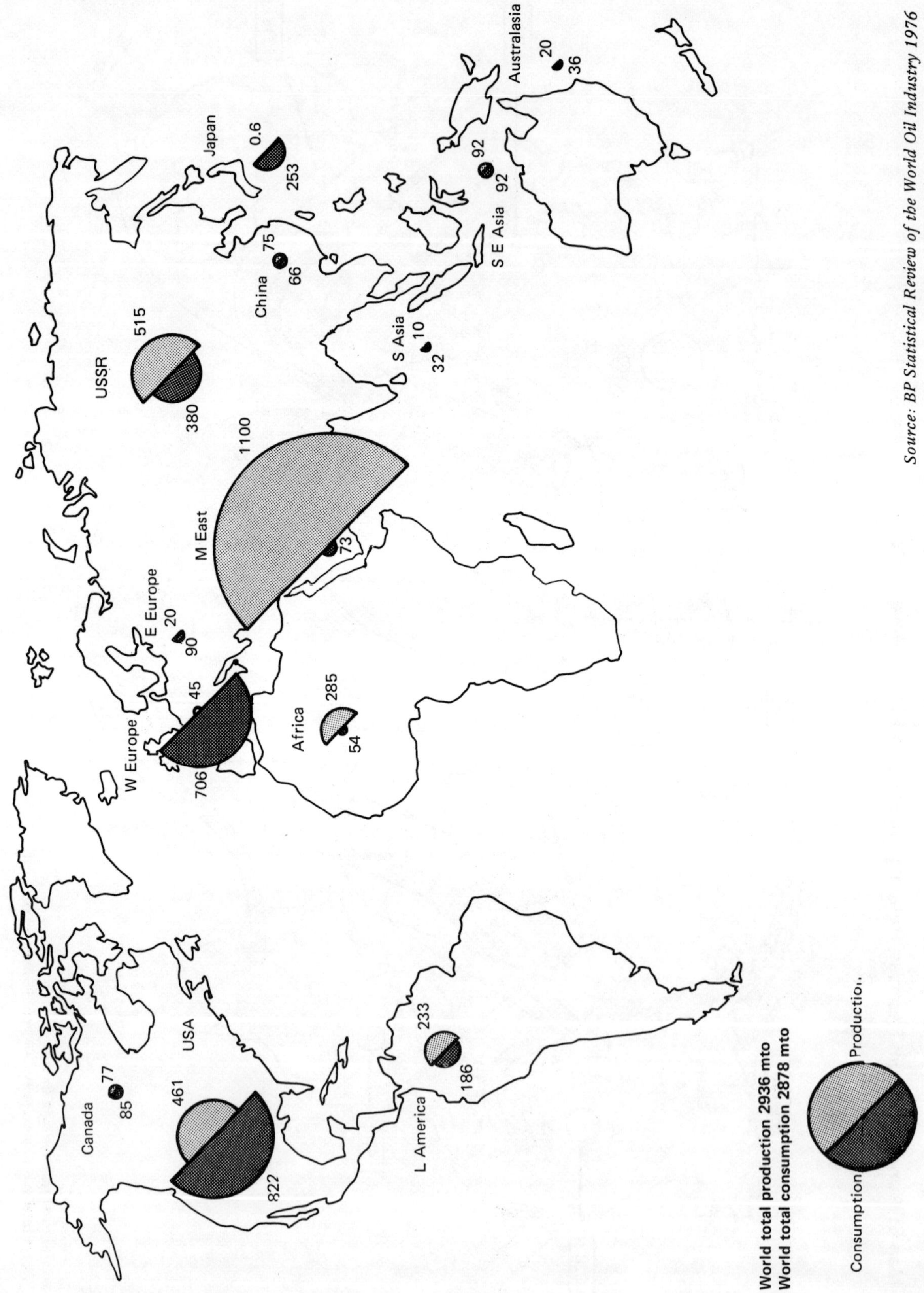

Source: BP Statistical Review of the World Oil Industry 1976

Map 10A: Oil and Gas Fields in the North Sea

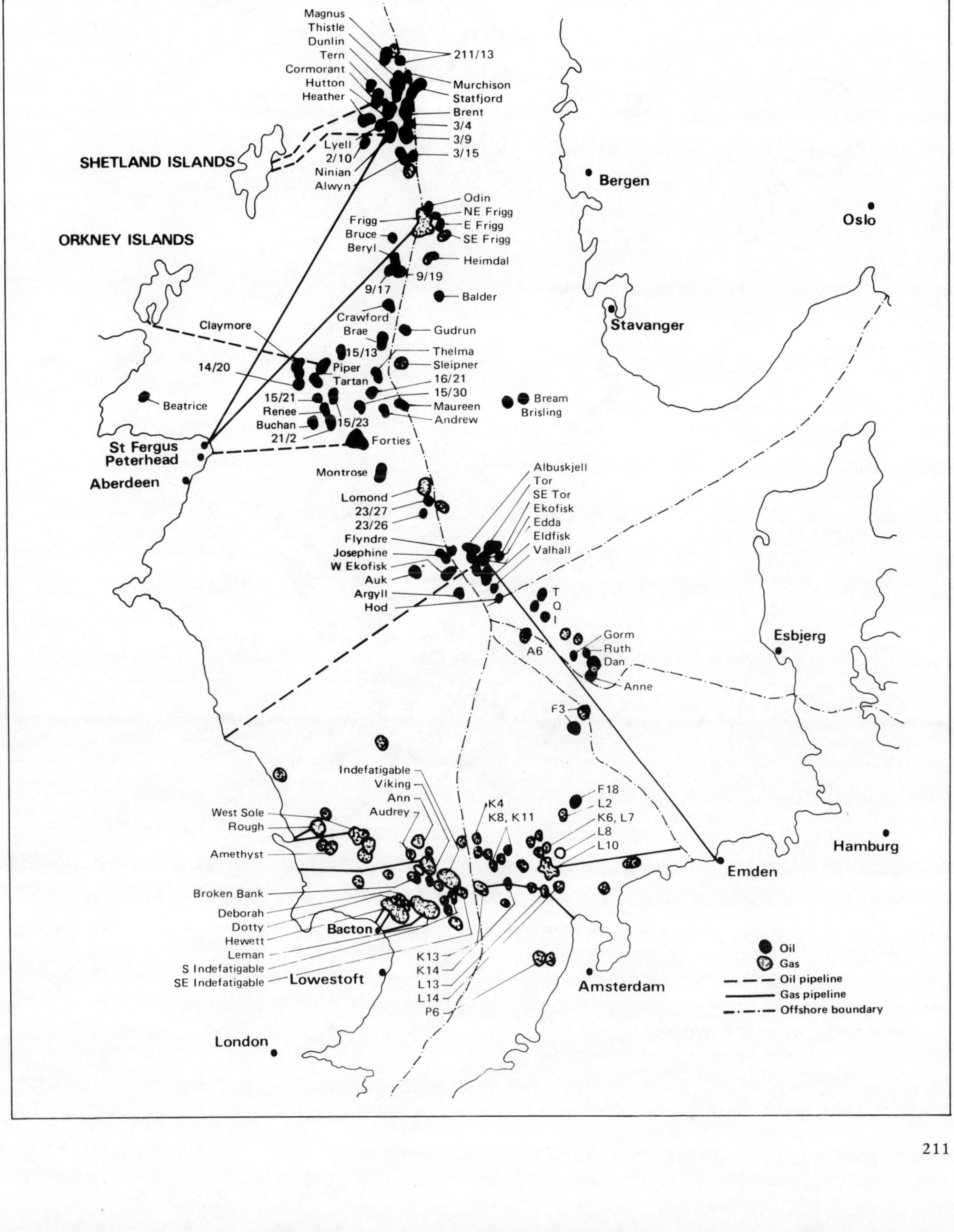

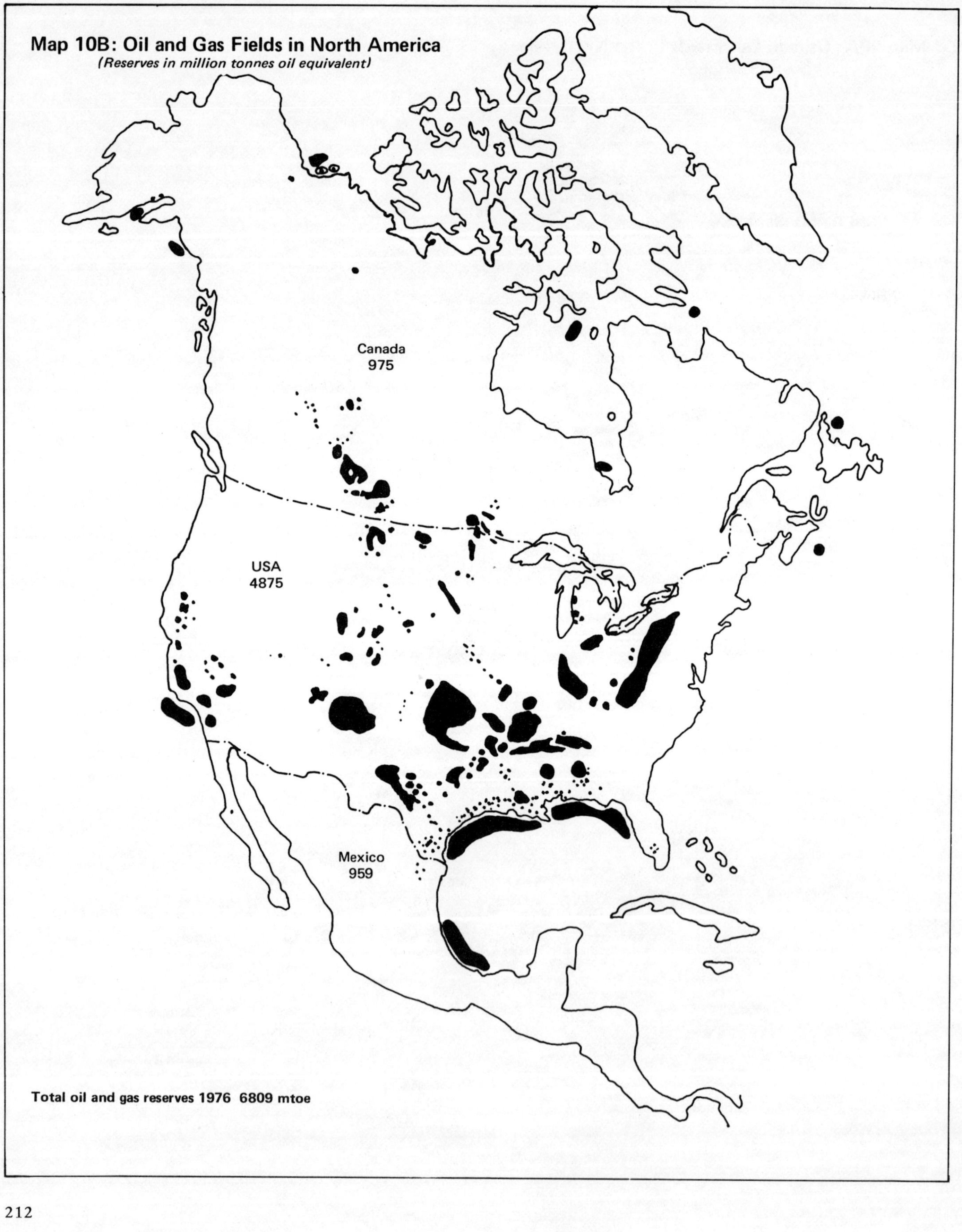
Map 10B: Oil and Gas Fields in North America
(Reserves in million tonnes oil equivalent)
Canada
975
USA
4875
Mexico
959
Total oil and gas reserves 1976 6809 mtoe

Map 10C: Oil and Gas Fields in South America
(Reserves in million tonnes oil equivalent)
Barbados
5
Trinidad
71
Venezuela
2092
Colombia
113
Ecuador
233
Peru
102
Brazil
110
Bolivia
33
Chile
25
Argentina
315
Total oil and gas reserves 1976 3099 mtoe

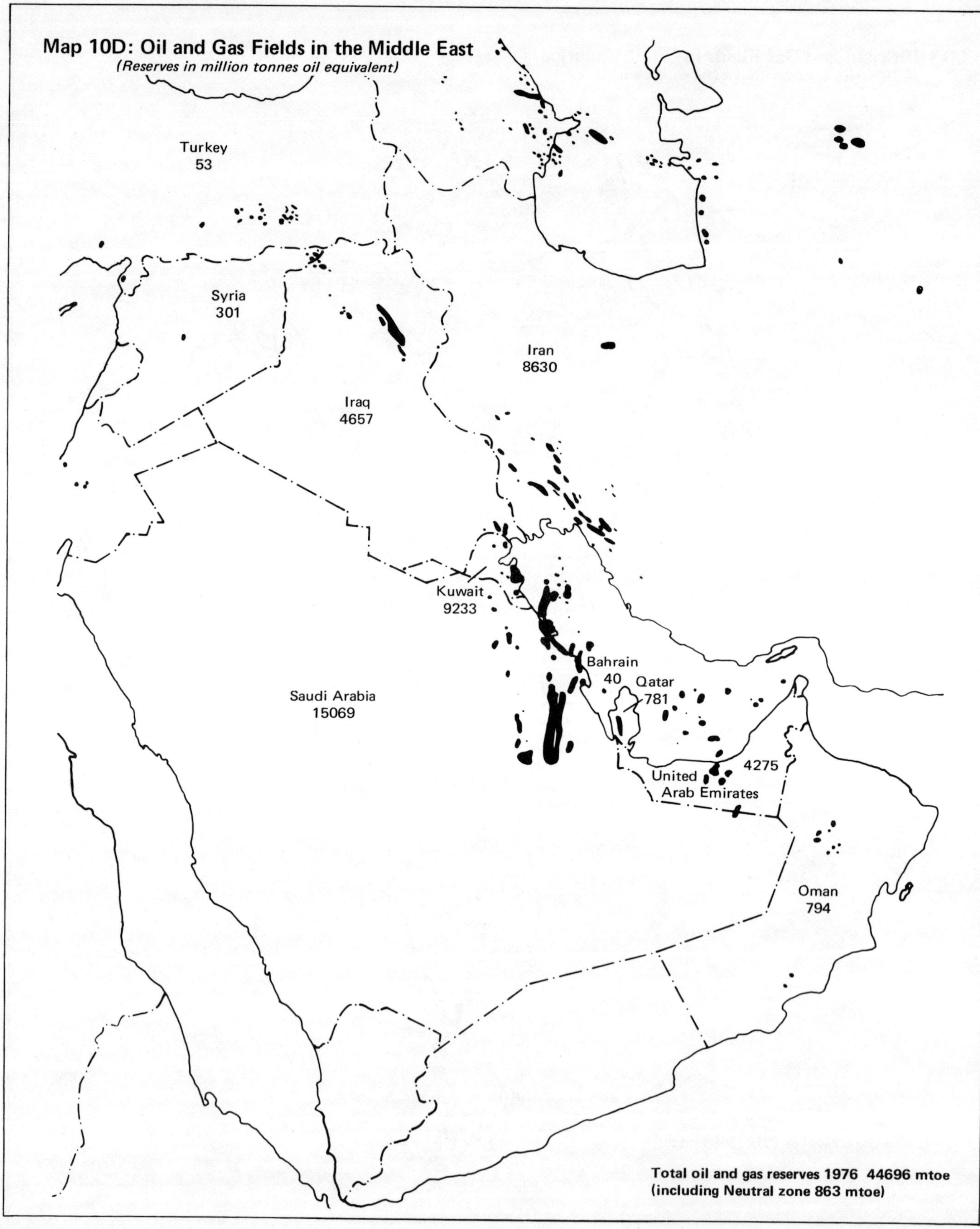
Map 10D: Oil and Gas Fields in the Middle East
(Reserves in million tonnes oil equivalent)
Turkey
53
Syria
301
Iraq
4657
Iran
8630
Kuwait
9233
Saudi Arabia
15069
Bahrain
40
Qatar
781
United
Arab Emirates
4275
Oman
794
Total oil and gas reserves 1976 44696 mtoe
(including Neutral zone 863 mtoe)

Map 10E: Oil and Gas Fields in Asia

(Reserves in million tonnes oil equivalent)

Total oil and gas reserves 1976 4065 mtoe

* Estimated Figure

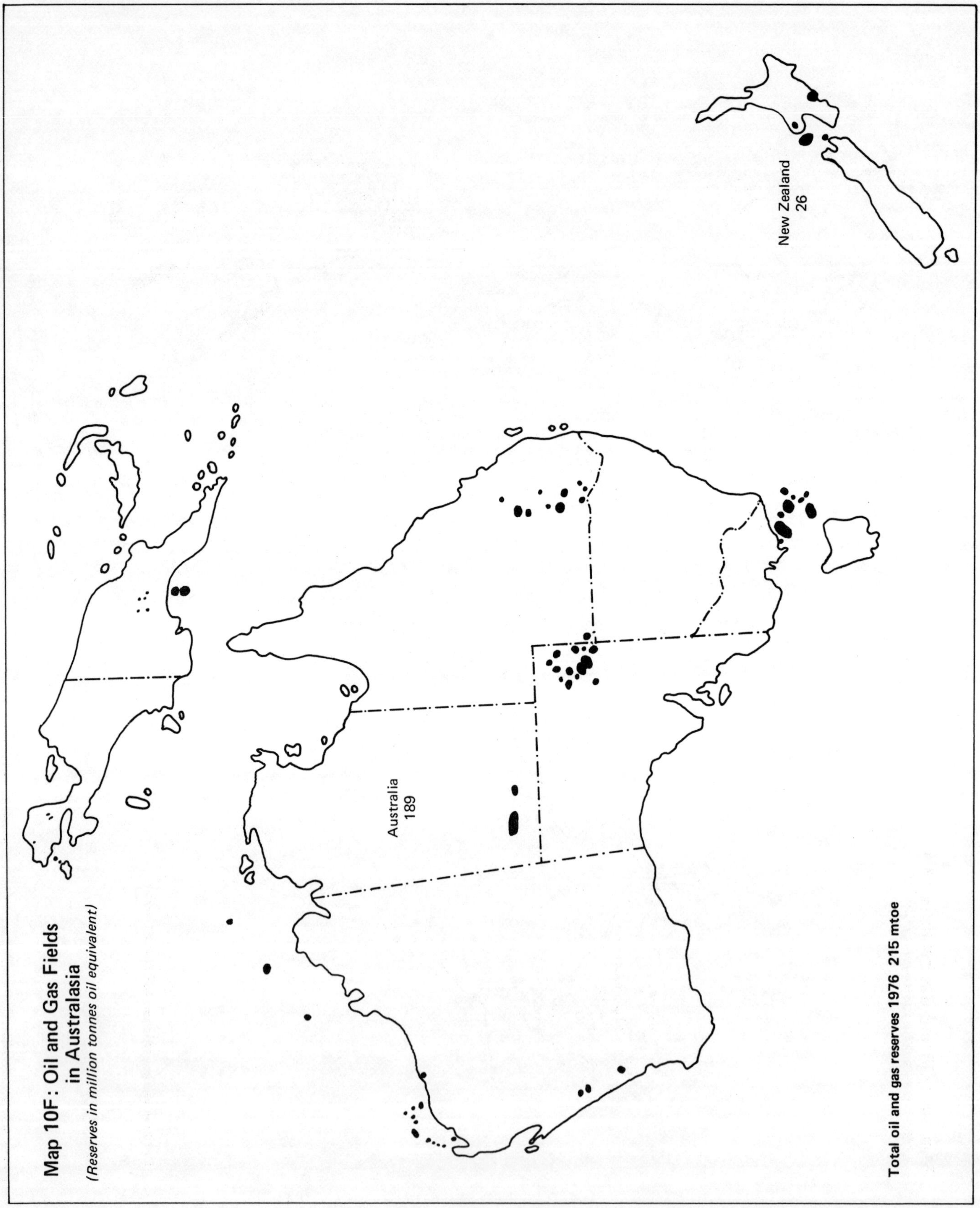

Map 10F: Oil and Gas Fields in Australasia

(Reserves in million tonnes oil equivalent)

Map 10G: Oil and Gas Fields in Africa

(Reserves in million tonnes oil equivalent)

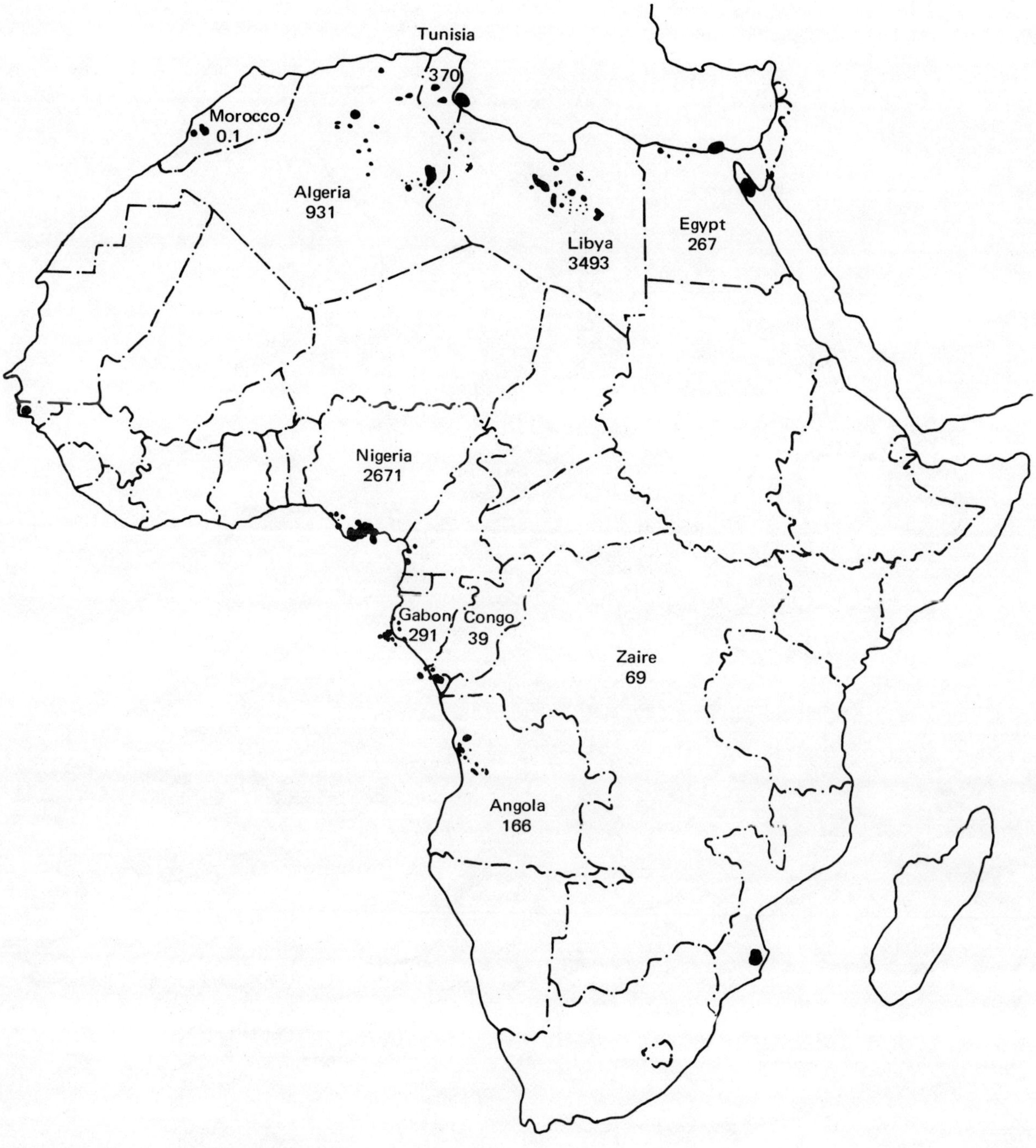

Total reserves of oil and gas 1976 8297 mtoe

Map 10H: Oil and Gas Fields in Europe

(Reserves in million tonnes oil equivalent)

Total oil and gas reserves in Western Europe 1976 2360 mtoe

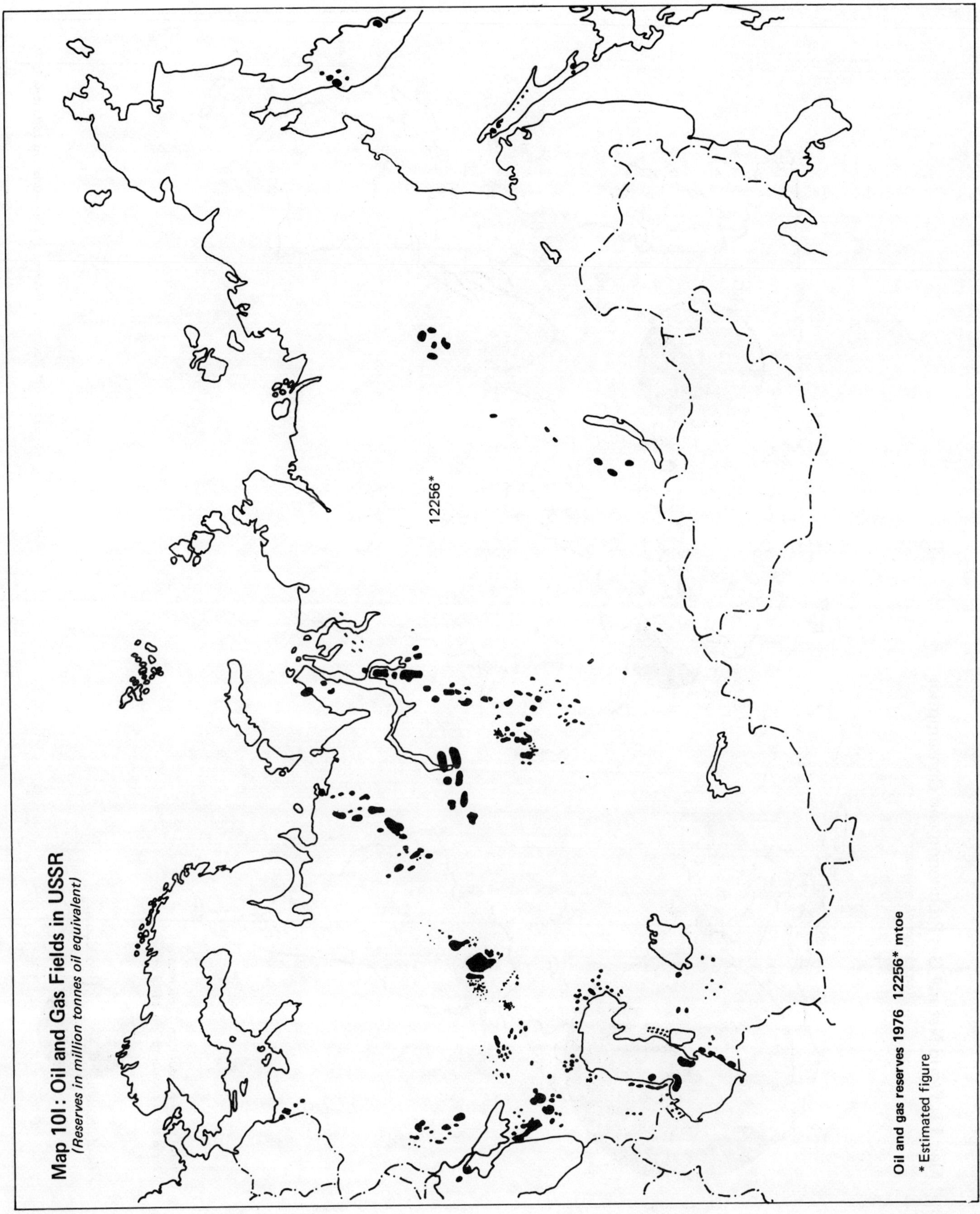

Map 10I: Oil and Gas Fields in USSR

(Reserves in million tonnes oil equivalent)

Oil and gas reserves 1976 12256* mtoe

* Estimated figure

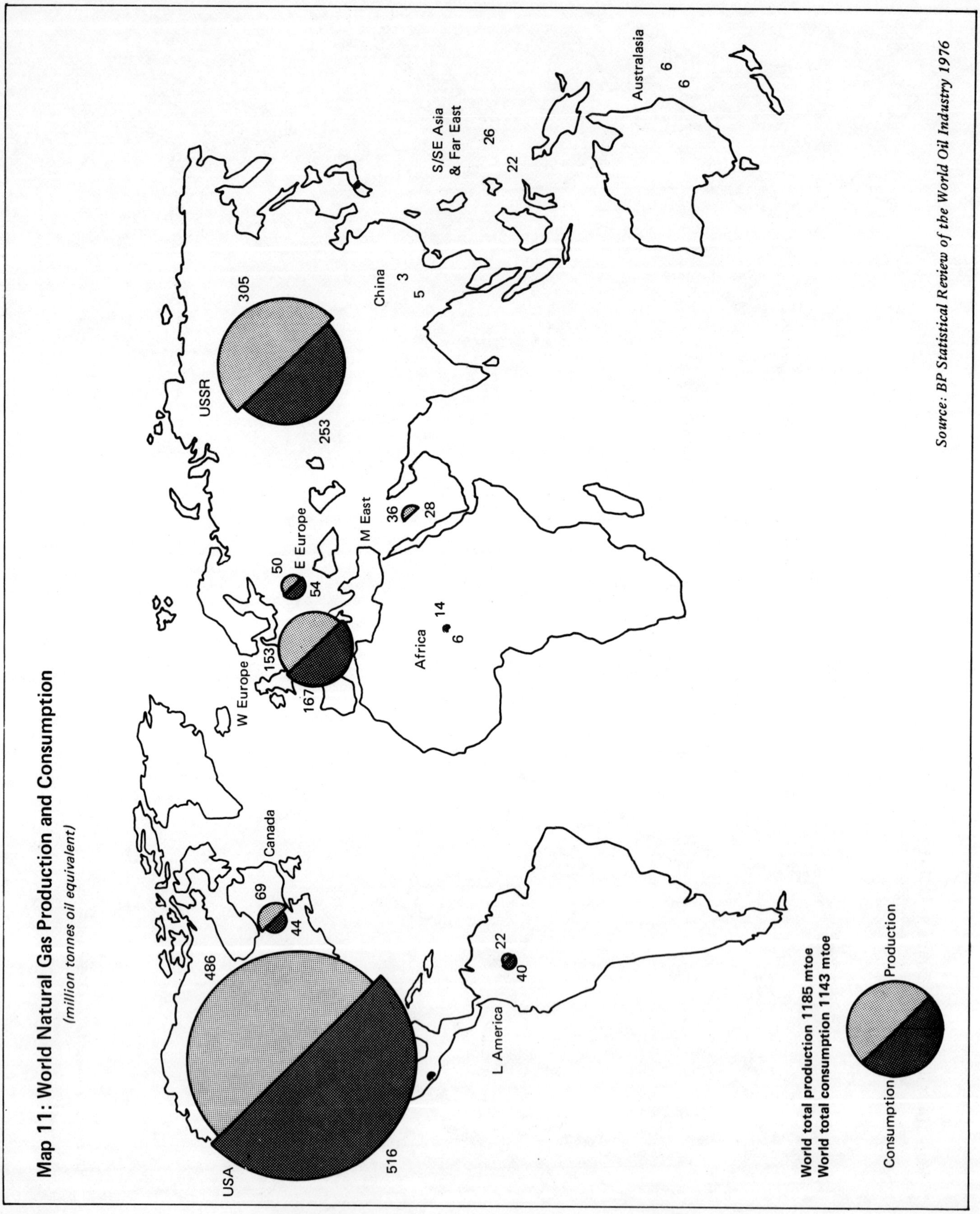

Map 11: World Natural Gas Production and Consumption

(million tonnes oil equivalent)

Source: BP Statistical Review of the World Oil Industry 1976

Map 12: World Natural Gas Reserves and Trade Movements

(million tonnes oil equivalent)

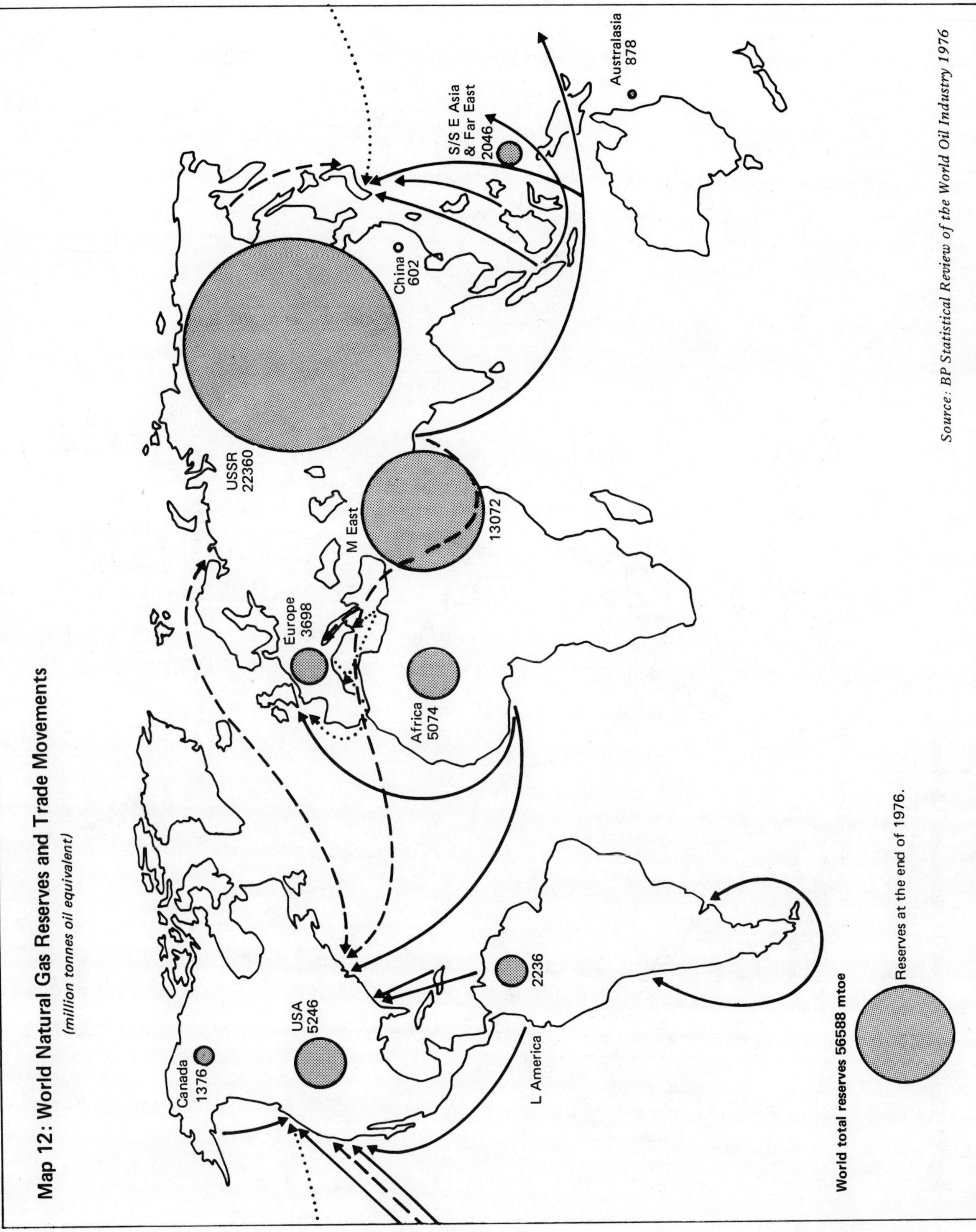

Source: BP Statistical Review of the World Oil Industry 1976

Map 13: World Deposits of Oil Shales and Tar Sands

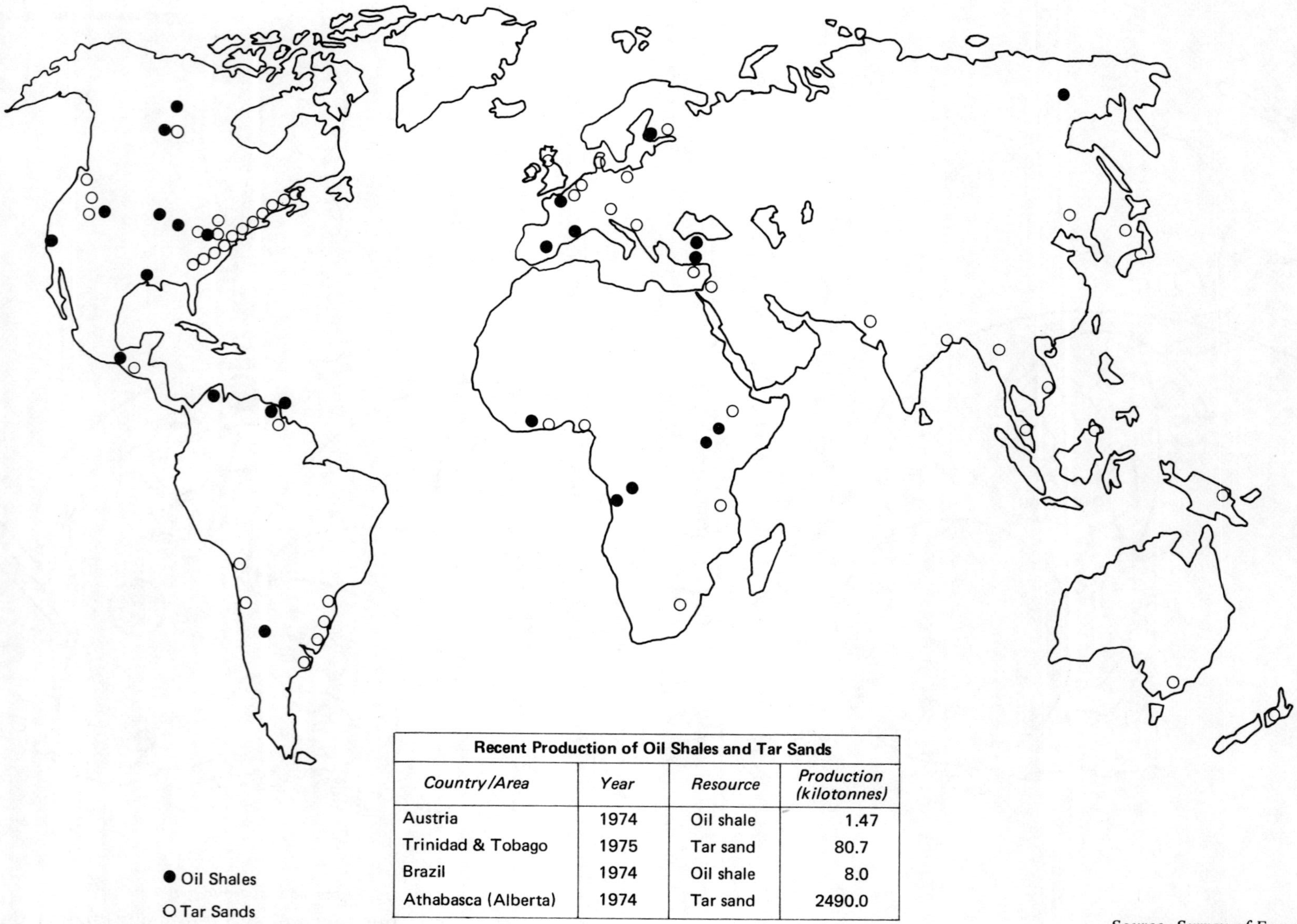

Recent Production of Oil Shales and Tar Sands

Country/Area	*Year*	*Resource*	*Production (kilotonnes)*
Austria	1974	Oil shale	1.47
Trinidad & Tobago	1975	Tar sand	80.7
Brazil	1974	Oil shale	8.0
Athabasca (Alberta)	1974	Tar sand	2490.0

Source: Survey of Energy Resources 1976

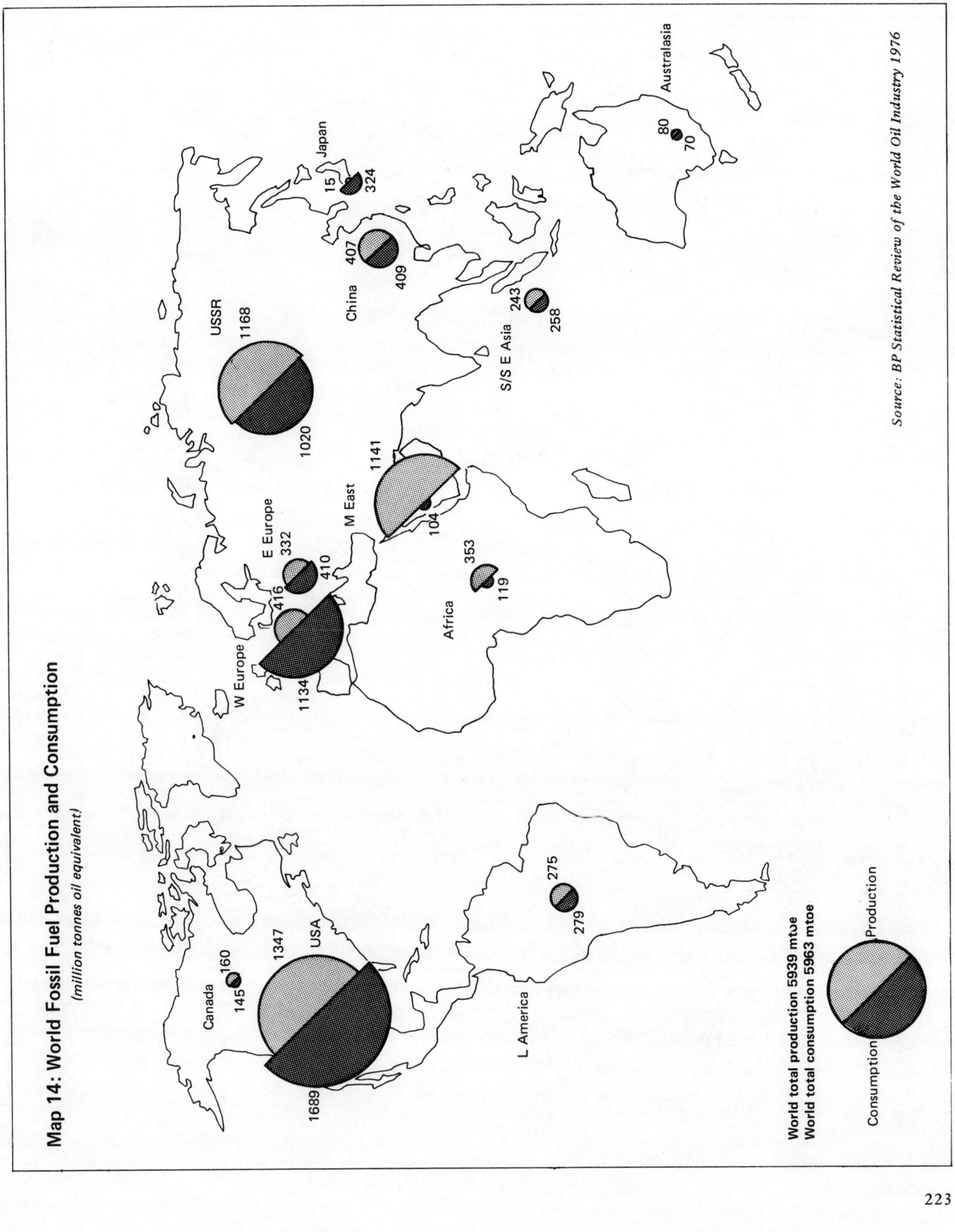

Map 14: World Fossil Fuel Production and Consumption

(million tonnes oil equivalent)

Source: BP Statistical Review of the World Oil Industry 1976

Map 15: World Nuclear Power Production and Installed Capacity

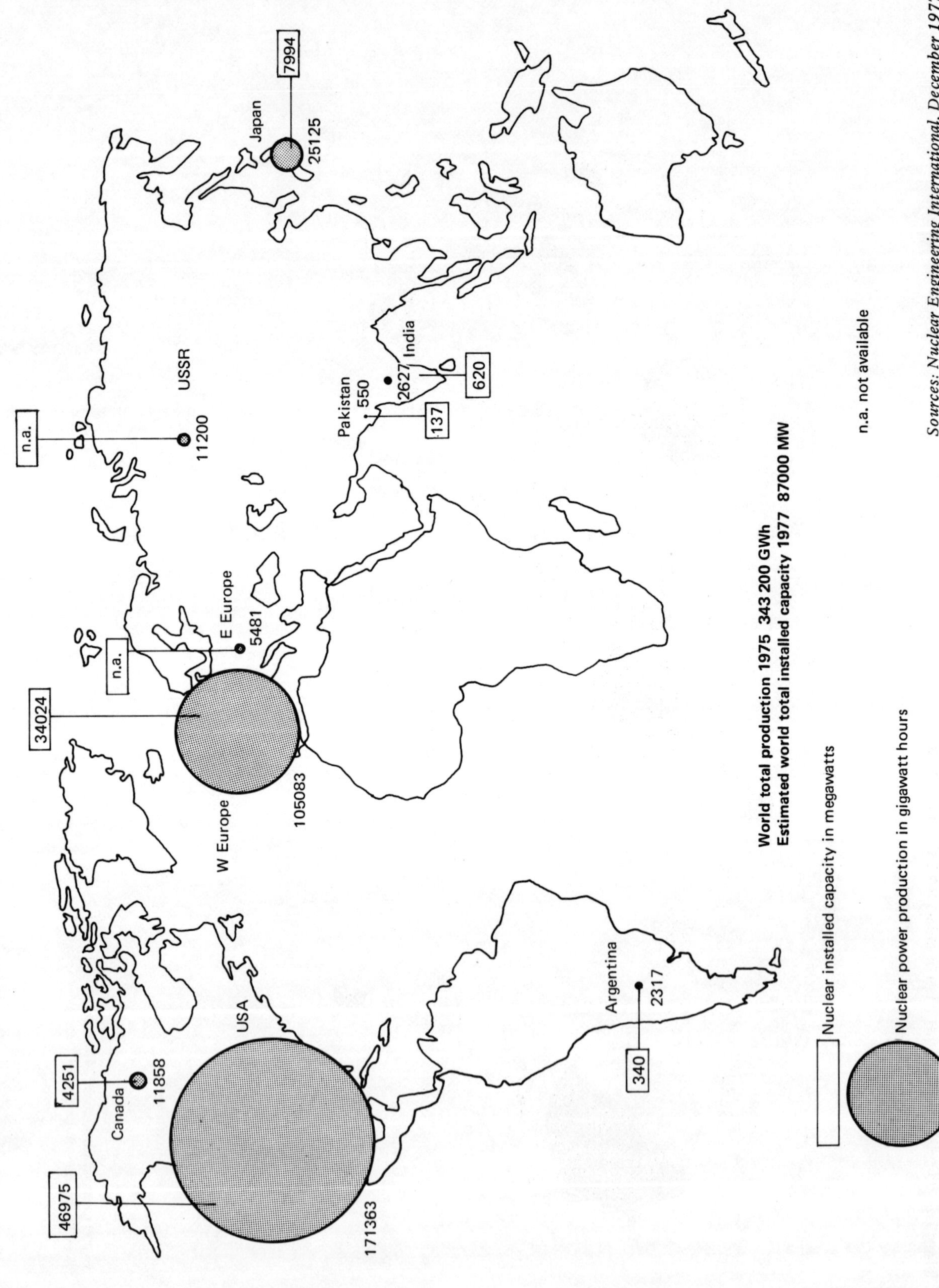

Sources: Nuclear Engineering International, December 1977
UN Statistical Yearbook 1975

Map 16: Uranium Deposits and Production

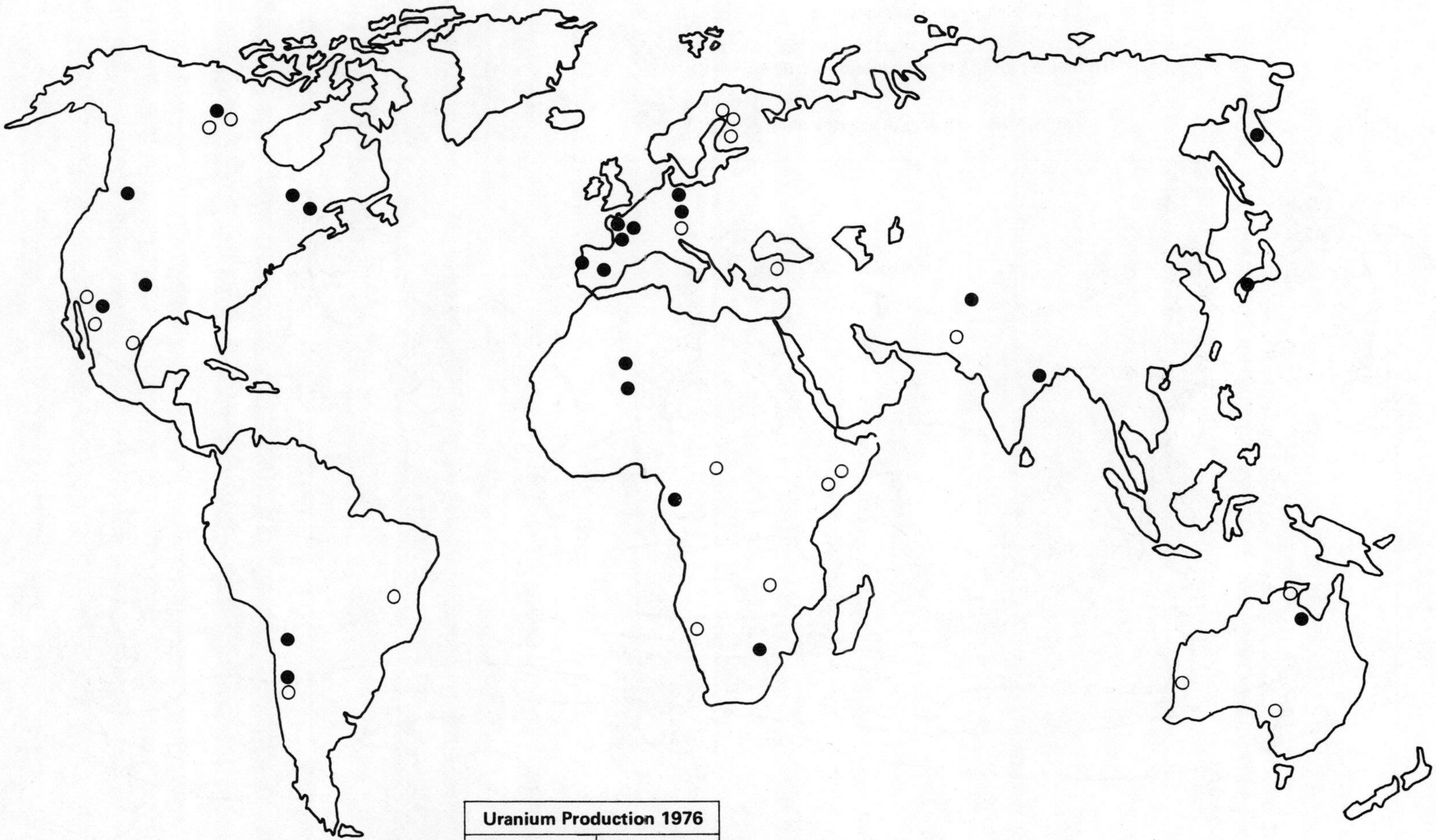

These figures exclude USSR, China and E Europe.

● Producing areas
○ Production expected by 1981

Uranium Production 1976	
Country	*Tonnes U*
USA	9800
Canada	4850
L America	2359
W Europe	4872
Africa	50
Japan	2
Australia	360
Total	22293

Source: Uranium resources, production and demand December 1977 (OECD and International Atomic Energy Agency)

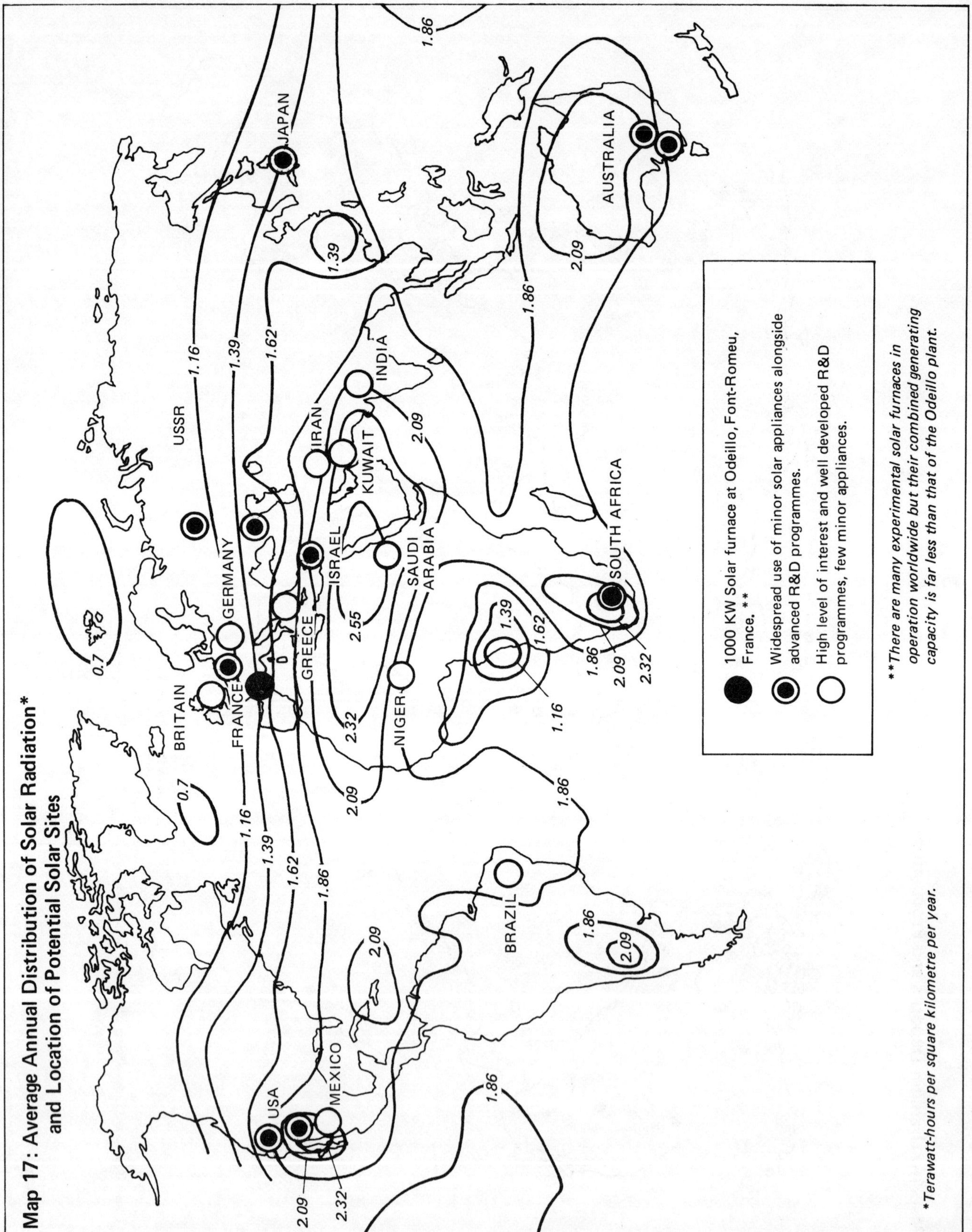

Map 17: Average Annual Distribution of Solar Radiation* and Location of Potential Solar Sites

Map 18: World Hydroelectric Production and Installed Capacity with Plant Locations

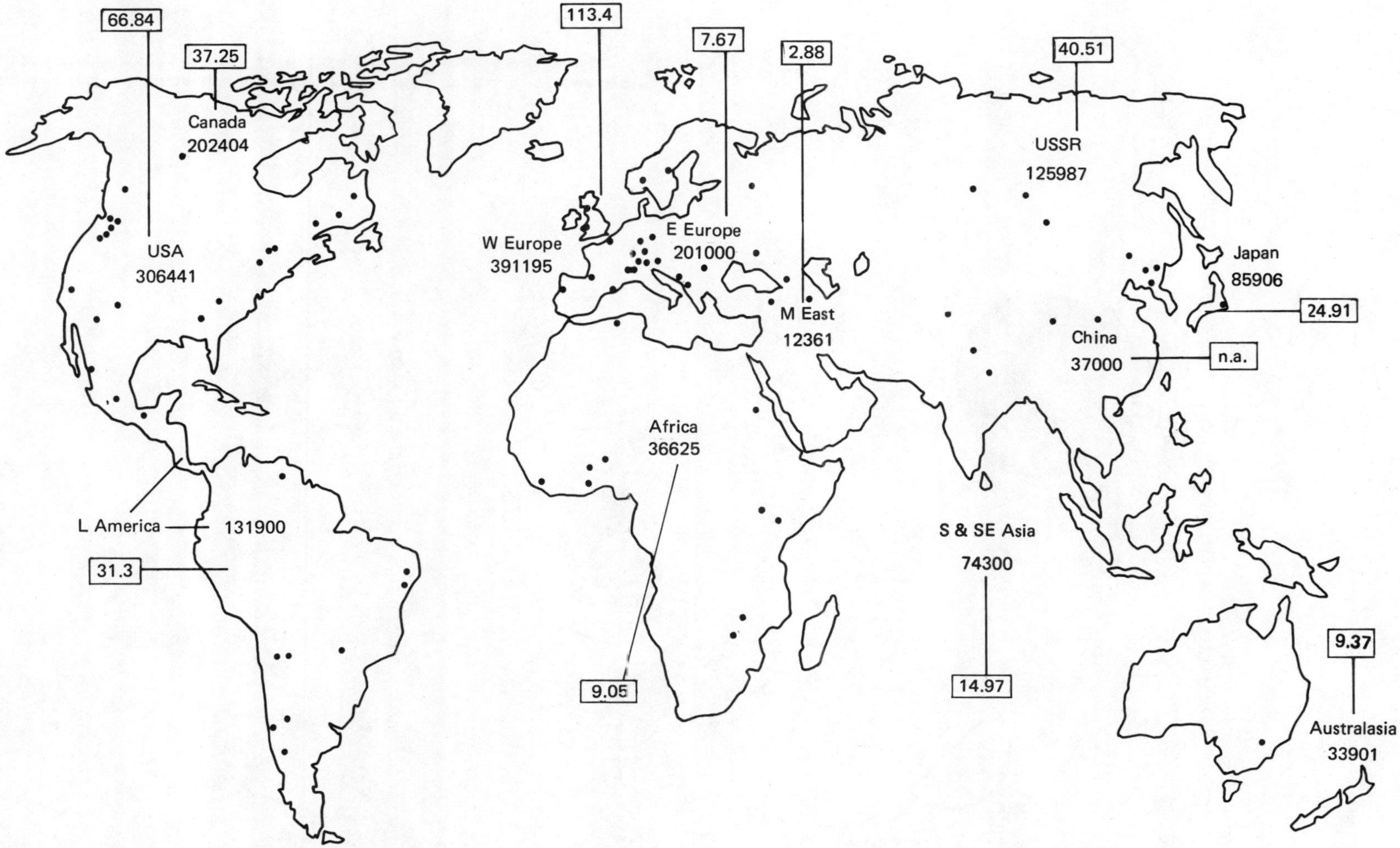

World total production 1975 1458400 GWh
World total installed generating capacity 1975 358.157 GW

- ● Plant Location
- [123] Installed generating capacity in Gigawatts
- Production figures are shown on the map in Gigawatt hours
- n.a. not available

Source: UN World Energy Supplies 1971-1975

Map 19: World Production of Wood and Consumption as Fuel

(million cubic metres of roundwood)

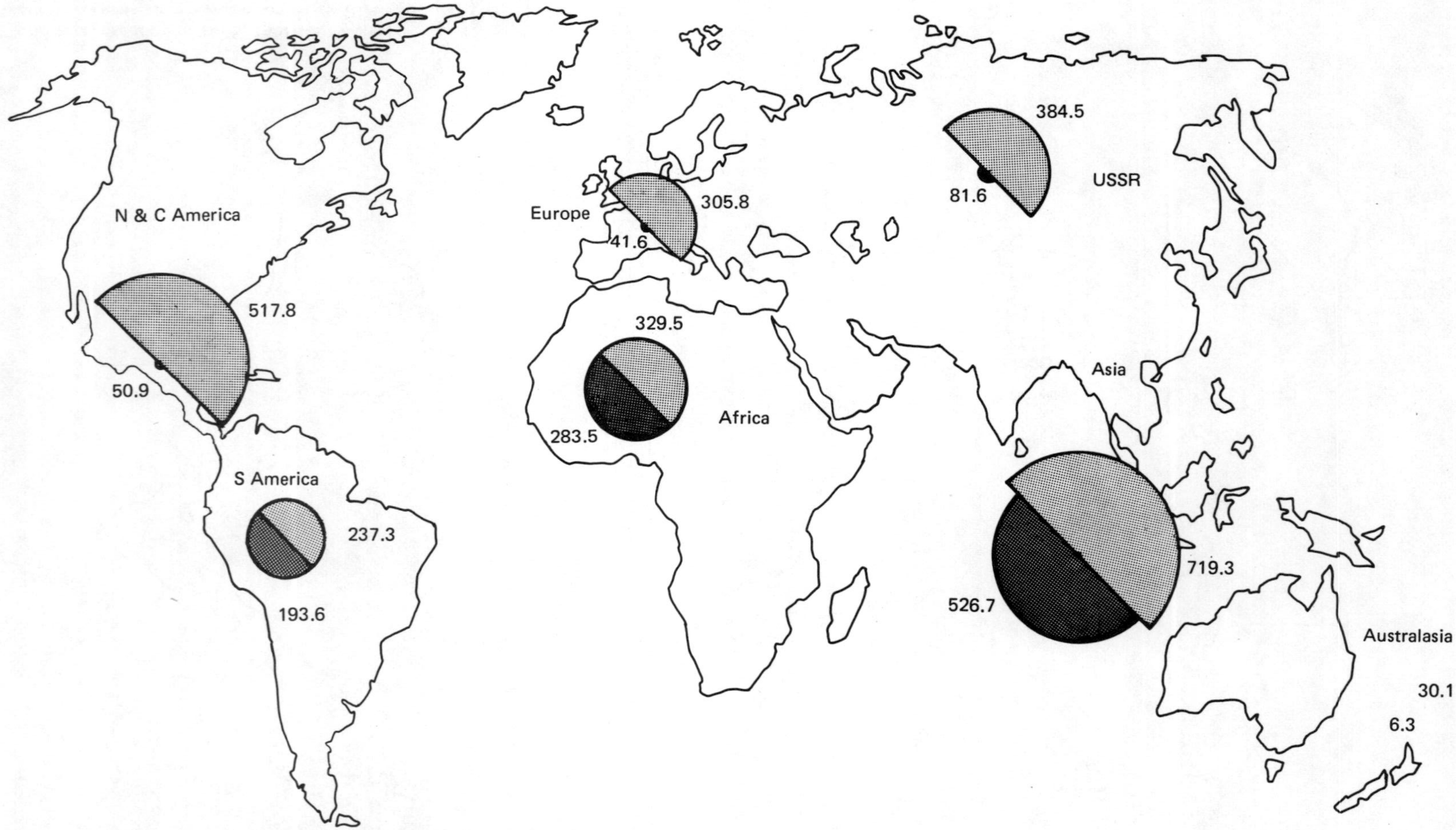

World total consumption of wood as fuel 1976 1184.2 million m^3 of roundwood
World total production of wood 1976 2524.3 million m^3 of roundwood

Source: FAO Yearbook of Forest Products

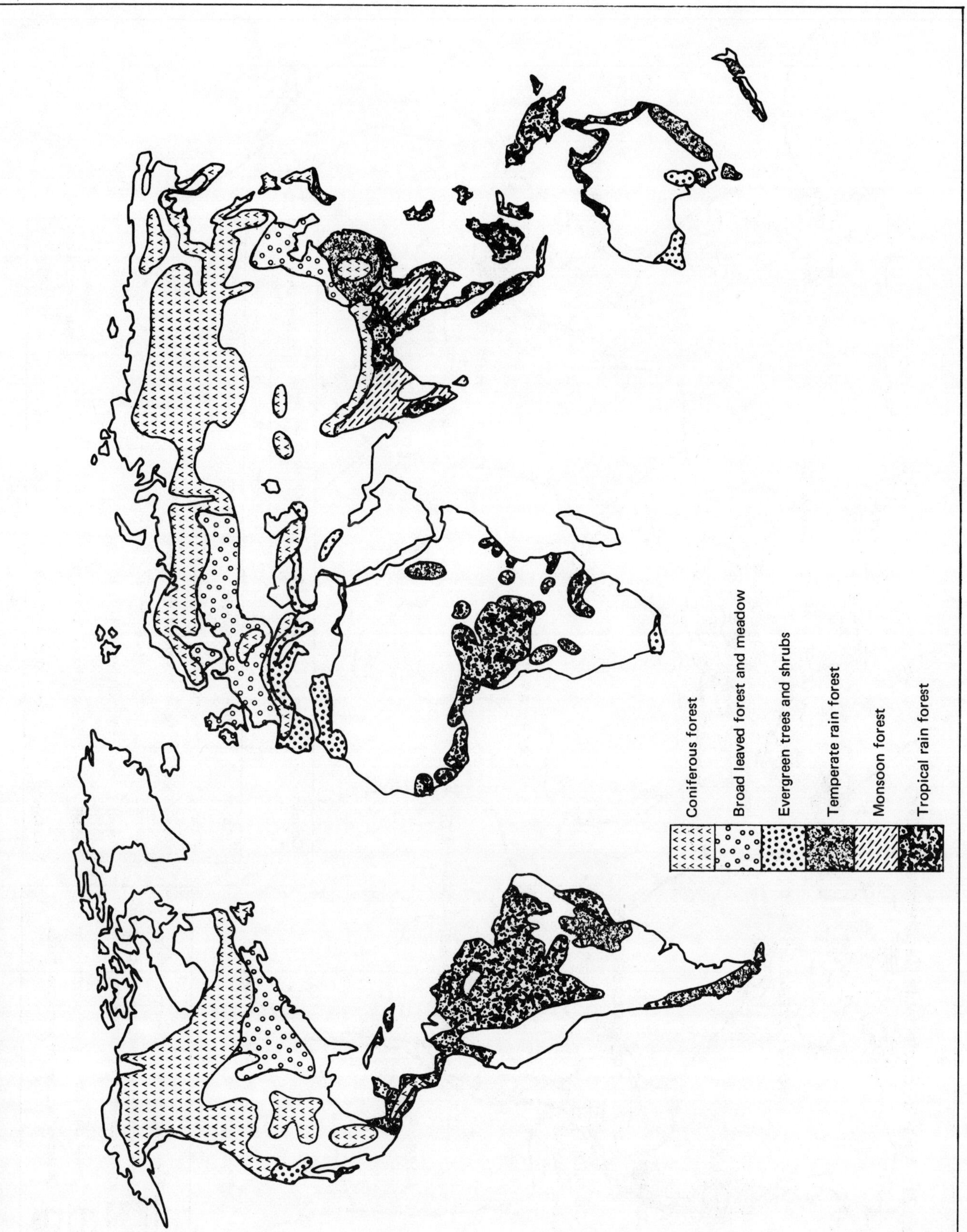

Map 20: World Distribution of Forest Areas

Map 21: Areas of Exploitable Geothermal Activity

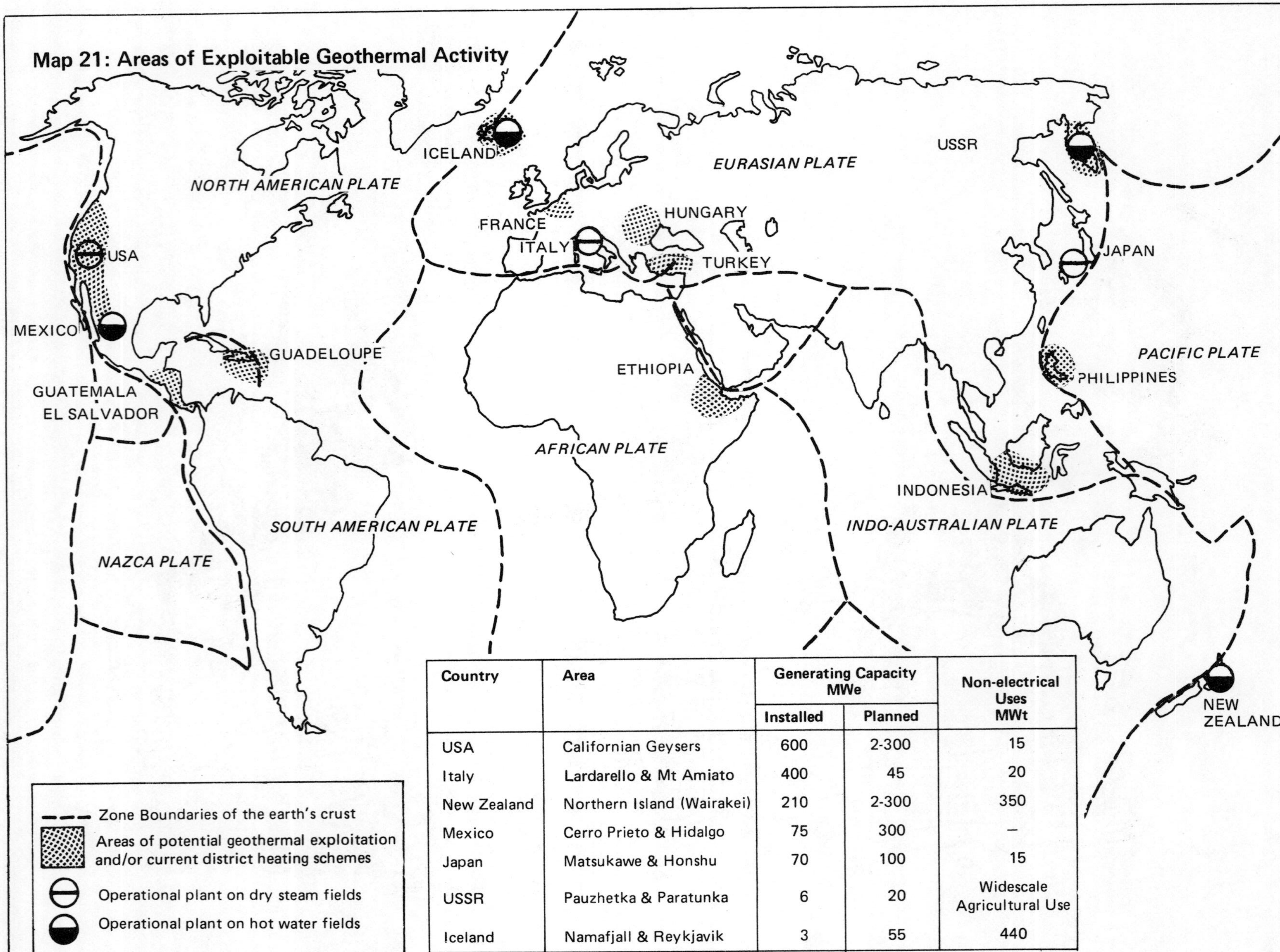

Country	Area	Generating Capacity MWe		Non-electrical Uses MWt
		Installed	Planned	
USA	Californian Geysers	600	2-300	15
Italy	Lardarello & Mt Amiato	400	45	20
New Zealand	Northern Island (Wairakei)	210	2-300	350
Mexico	Cerro Prieto & Hidalgo	75	300	–
Japan	Matsukawe & Honshu	70	100	15
USSR	Pauzhetka & Paratunka	6	20	Widescale Agricultural Use
Iceland	Namafjall & Reykjavik	3	55	440
	Total	1364*		

* *The figures listed are approximate figures; by comparison a large modern fossil-fuelled power station has a generating capacity of 1000 MWe*

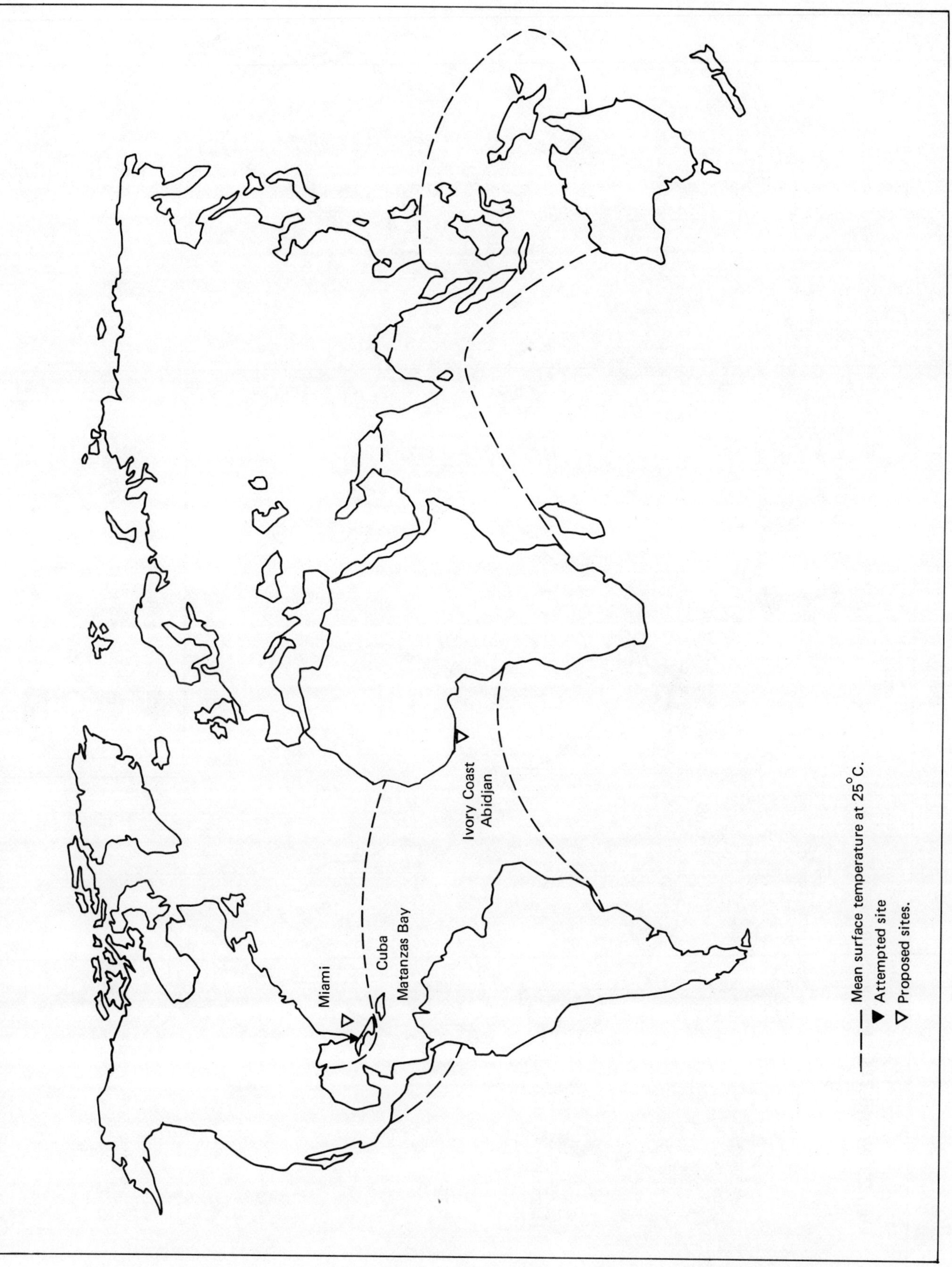

Map 22: Possible Locations for Ocean Thermal Energy

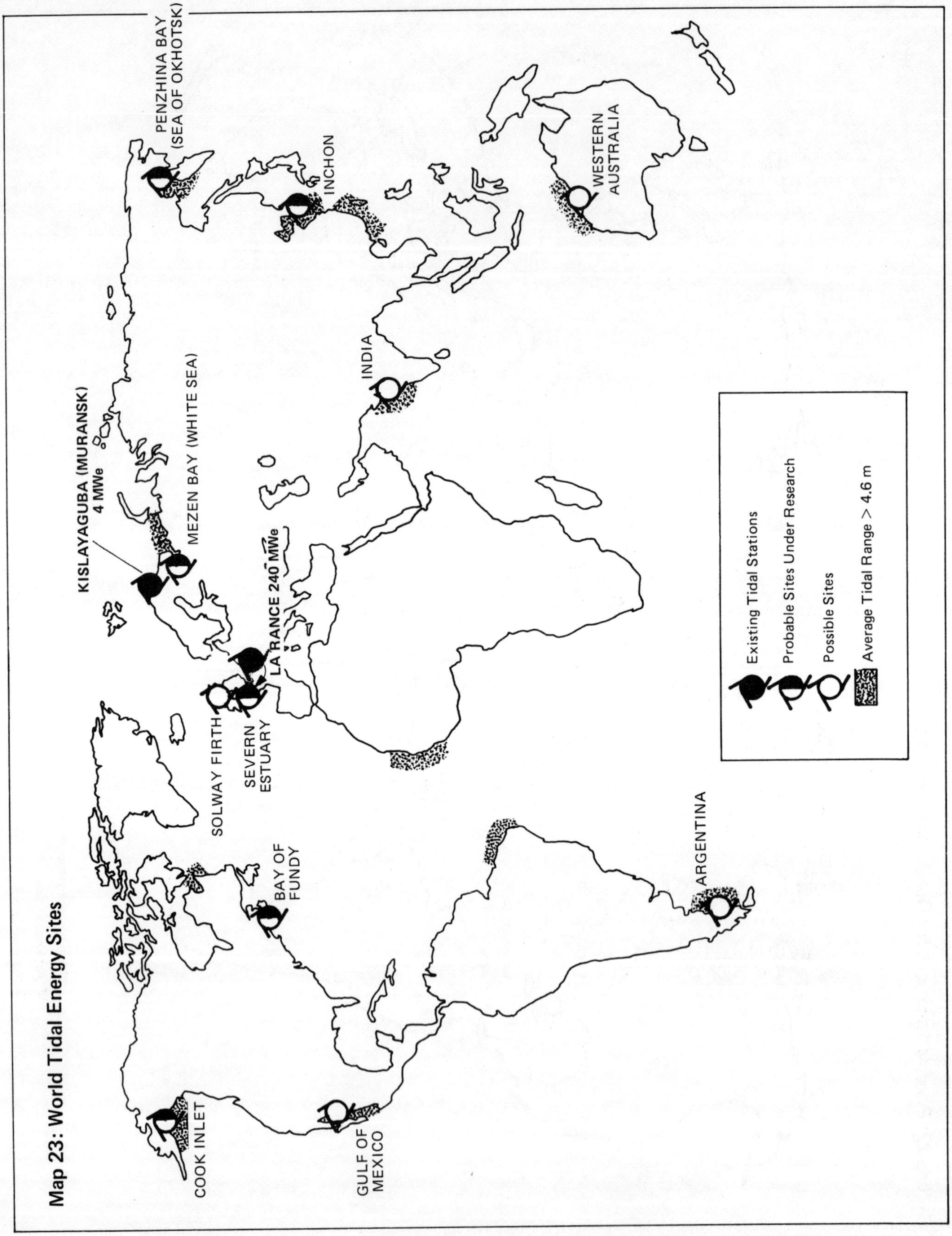

Map 23: World Tidal Energy Sites

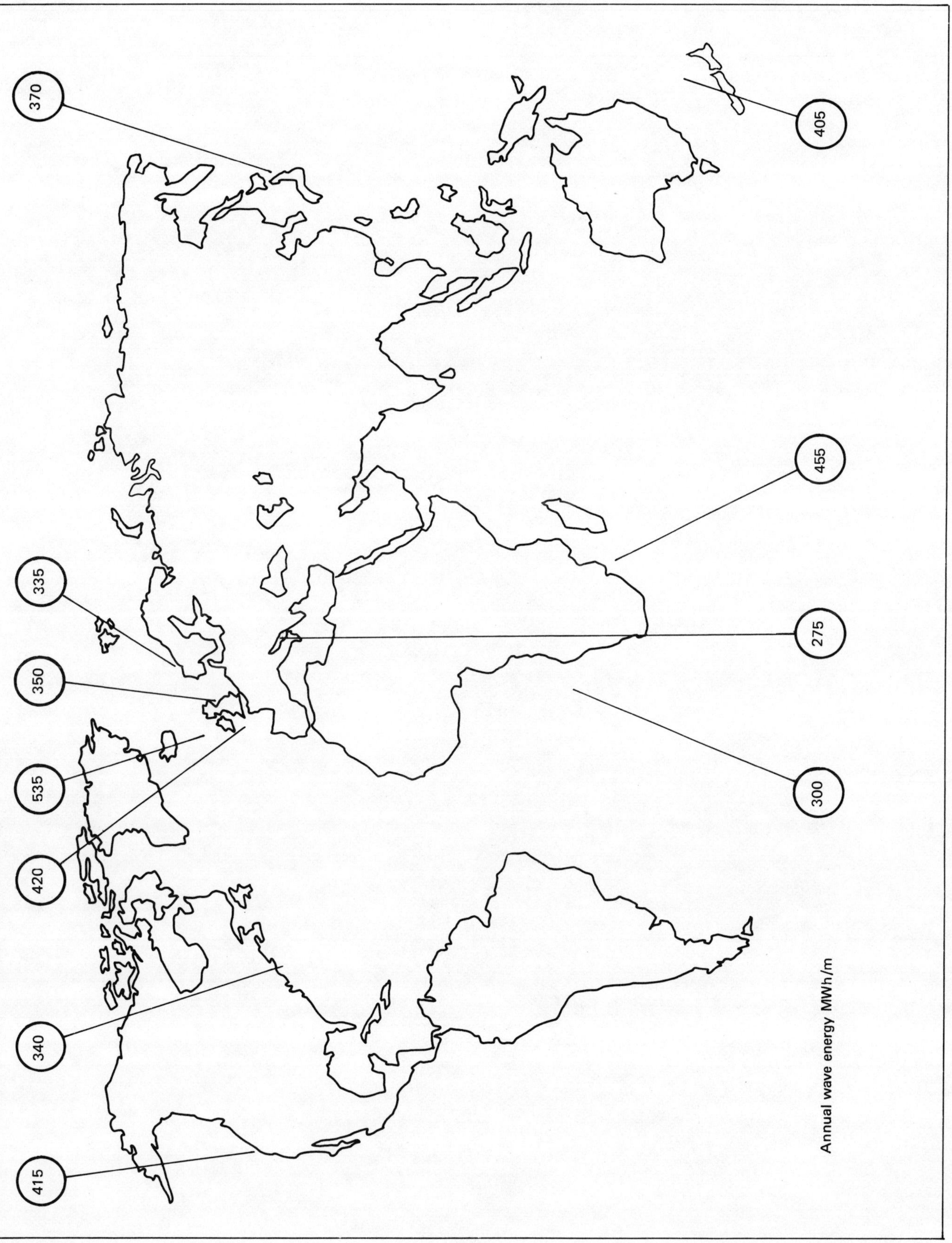

Map 24: Annual Wave Energy in Specific Sea Areas

Statistical Appendices

Appendix 1

Appendix 2

Appendix 3

Appendix 4

Note:
Some differences occur between figures where different sources have been used to describe a particular type of energy in separate tables and on different maps.

Appendix 1

Table 1: World Coal Production 1966-76

million tonnes oil equivalent

Country/Area	1966	1967	1968	1969	1970	1971	1972	1973	1974	1975	1976
USA	329.9	340.6	335.44	344.0	368.76	337.0	359.3	355.2	361.2	391.8	338.6
Canada	5.5	5.5	5.3	5.26	7.77	9.2	10.59	11.27	11.63	14.8	13.9
Total North America	**335.4**	**346.1**	**340.7**	**349.26**	**376.5**	**346.2**	**369.9**	**366.47**	**372.8**	**406.6**	**352.5**
Latin America	5.85	6.5	6.7	6.99	6.83	7.55	7.49	8.0	8.74	9.0	7.3
Total Western Hemisphere	**341.25**	**352.6**	**347.4**	**356.25**	**383.3**	**353.75**	**377.4**	**374.47**	**381.54**	**415.6**	**359.8**
Austria	0.01	0.01	–	–	–	–	–	–	–	–	–
Belgium	11.7	11.0	9.9	8.8	7.6	7.3	7.0	5.9	5.43	5.0	4.8
Eire	0.1	0.1	0.1	0.1	0.1	0.06	0.05	0.04	0.04	0.03	–
France	33.82	32.0	28.15	27.3	25.35	22.7	20.48	17.65	16.10	15.0	14.67
Italy	0.28	0.27	0.24	0.2	0.2	0.17	0.10	0.0003	–	–	–
Netherlands	6.7	5.4	4.5	3.7	2.9	2.4	1.9	1.15	0.5	0.38	–
Norway	0.27	0.29	0.217	0.26	0.31	0.29	0.32	0.28	0.29	0.26*	–
Portugal	0.28	0.3	0.27	0.28	0.18	0.17	0.17	0.15	0.15	0.14*	–
Spain	8.67	8.3	8.26	7.79	7.2	7.18	7.44	6.69	7.0 *	7.38*	7.4
Sweden	0.27	0.01	0.01	0.01	0.01	0.01	0.01	–	–	–	–
Turkey	3.27	3.09	3.2	3.14	3.06	3.10	3.11	3.1	3.3	3.35	3.0
United Kingdom	119.9	119.0	113.8	104.3	98.56	100.1	81.6	88.4	73.8	85.6	81.87
West Germany	84.6	75.23	75.15	74.89	74.67	74.4	68.8	65.39	63.8	61.9	59.8
Yugoslavia	0.89	0.61	0.57	0.46	0.43	0.47	0.39	0.39	0.4	0.4	0.4
Total Western Europe	**270.6**	**255.5**	**244.4**	**231.2**	**220.6**	**218.4**	**191.4**	**189.1**	**170.8**	**179.4**	**171.9**
Middle East	0.19	0.21	0.09	0.33	0.36	0.44	0.67	0.7	0.77	0.72	0.67
Africa	35.2	36.16	37.8	38.4	39.87	42.8	42.65	45.5	47.49	51.3	53.0
South & South East Asia	70.57	71.89	73.23	74.05	78.4	78.9	82.53	86.36	95.1	104.0	114.1
Japan	34.4	31.8	31.2	29.9	26.6	22.4	18.8	15.0	13.6	12.7	12.4
Australasia	21.63	22.4	25.74	28.8	30.29	29.7	36.8	37.46	39.03	42.7	53.0
USSR	272.4	277.4	278.9	285.28	289.9	295.75	302.2	309.0	317.16	319.46	365.8
Eastern Europe	107.54	108.1	111.47	116.52	120.82	124.81	127.5	131.4	134.9	141.24	146.4
China	219.09	152.1	201.0	217.75	256.5	262.6	268.7	286.8	297.8	309.54	321.6
Total Eastern Hemisphere	**1031.62**	**955.56**	**1003.8**	**1022.18**	**1063.3**	**1075.8**	**1071.25**	**1093.3**	**1116.65**	**1161.06**	**1238.9**
World	**1373.0**	**1308.0**	**1351.0**	**1378.5**	**1447.0**	**1429.5**	**1448.8**	**1475.8**	**1498.4**	**1576.7**	**1648.6**
World (excl USSR E Europe and China)	**774.0**	**770.4**	**759.6**	**758.9**	**779.8**	**746.3**	**750.4**	**748.6**	**748.5**	**806.5**	**814.8**

Source: 1966 and 1976 UN Statistical Yearbook
1976 International Coal Trade, US Department of the Interior. Vol 45 No 4 April 1976
**Estimated figure*

Table 2: World Solid Fuels Consumption (Coal, Peat, Lignite) 1966-76 *million tonnes oil equivalent*

Country/Area	1966	1967	1968	1969	1970	1971	1972	1973	1974	1975	1976
USA	321.6	315.8	327.9	333.1	335.0	321.3	321.6	340.0	336.9	327.9	350.5
Canada	15.2	15.0	16.3	15.8	16.9	16.1	15.2	15.6	15.9	15.5	15.3
Total North America	**336.8**	**330.8**	**344.2**	**348.9**	**351.9**	**337.4**	**336.8**	**355.6**	**352.8**	**343.4**	**365.8**
Latin America	45.7	46.4	46.7	47.4	48.1	48.4	49.2	49.1	50.5	51.5	51.9
Total Western Hemisphere	**382.5**	**377.2**	**390.9**	**396.3**	**400.0**	**385.8**	**386.0**	**404.7**	**403.3**	**394.9**	**417.7**
Austria	5.0	4.7	4.6	4.4	4.7	4.1	4.7	4.0	4.3	3.8	3.6
Belgium & Luxembourg	17.6	17.2	17.8	17.3	16.4	13.7	13.1	11.0	11.9	8.9	9.2
Denmark	3.7	3.3	3.3	2.8	2.6	1.6	1.5	2.3	2.0	2.8	2.9
Eire	2.5	2.6	3.1	2.9	2.7	2.5	1.4	1.8	1.8	1.8	1.9
Finland	5.9	5.8	6.2	6.6	6.5	6.2	6.3	6.3	6.3	5.6	6.2
France	43.9	43.6	41.7	41.5	39.7	35.9	32.4	26.3	26.7	25.2	26.5
Greece	1.4	1.4	1.5	1.8	2.1	2.8	2.5	2.8	2.9	3.7	3.9
Iceland	†	†	†	†	†	†	†	†	†	†	–
Italy	10.9	12.2	10.9	11.5	12.5	11.6	12.1	10.5	11.7	10.1	9.0
Netherlands	8.6	8.3	7.6	6.7	5.1	3.5	3.1	3.1	3.1	2.6	2.8
Norway	0.9	0.9	0.9	1.1	1.0	0.9	1.0	0.7	0.8	0.8	0.4
Portugal	0.8	0.8	0.7	0.8	0.8	0.6	1.0	1.2	0.7	0.4	0.4
Spain	10.4	10.7	10.3	10.5	10.9	11.0	11.5	11.7	10.6	11.2	11.2
Sweden	1.8	1.6	1.6	1.5	1.5	1.3	1.1	1.2	1.0	5.0	4.6
Switzerland	0.9	0.7	0.6	0.6	0.6	0.4	0.7	0.4	0.3	0.3	0.4
Turkey	13.3	12.4	12.5	12.5	12.6	11.9	12.1	12.1	12.4	14.3	16.4
United Kingdom	104.5	97.9	98.4	96.3	92.3	82.9	72.2	78.4	69.1	71.9	72.0
West Germany	95.9	89.2	91.0	94.1	94.0	88.8	81.5	87.9	88.3	81.1	76.4
Yugoslavia	13.5	12.0	12.3	12.3	12.9	14.2	13.5	12.5	12.0	11.9	12.4
Cyprus/Gibraltar/ Malta	†	†	†	†	†	†	†	†	†	†	†
Total Western Europe	**341.5**	**325.3**	**325.0**	**325.2**	**318.9**	**293.9**	**271.7**	**274.2**	**265.9**	**261.4**	**260.2**
Middle East	1.3	1.3	1.3	1.4	1.4	1.5	1.5	1.5	1.9	2.1	2.1
Africa	49.8	50.3	52.4	53.0	54.2	57.2	55.3	51.4	53.2	55.9	58.4
South Asia	49.8	50.7	52.3	55.5	52.7	53.4	55.6	56.0	57.8	60.1	62.0
South East Asia	47.6	48.7	50.7	50.1	52.7	54.6	56.2	57.1	58.7	61.2	60.6
Japan	47.2	52.8	56.0	60.4	62.6	56.0	58.0	61.8	64.8	62.1	60.0
Australasia	20.6	21.3	22.5	22.8	22.7	22.7	23.9	24.8	26.0	27.6	28.2
USSR	338.9	343.6	340.9	346.0	340.9	348.8	353.9	361.8	370.5	380.0	387.4
Eastern Europe	214.6	212.0	218.5	227.7	237.1	241.3	241.3	245.7	253.0	260.0	266.7
China	231.6	161.6	213.3	227.4	252.0	273.2	278.1	292.5	307.4	321.8	337.5
Total Eastern Hemisphere	**1342.9**	**1267.6**	**1332.9**	**1369.5**	**1395.2**	**1402.6**	**1395.5**	**1426.8**	**1459.2**	**1492.2**	**1523.1**
World	**1725.4**	**1644.8**	**1723.8**	**1765.8**	**1795.2**	**1788.4**	**1781.5**	**1831.5**	**1862.5**	**1887.1**	**1940.8**
World (excl USSR E Europe and China)	**940.3**	**927.6**	**951.1**	**964.7**	**965.2**	**925.1**	**908.2**	**931.5**	**931.6**	**925.3**	**949.2**

† = Less than 0.05 million tonnes oil equivalent.
Source: BP Statistical Review of the World Oil Industry 1976.

Table 3: Natural Gas Reserves at the end of 1976

Country/Area	Trillion* cubic feet	Trillion* cubic metres	Million tonnes oil equivalent	Share of Total
USA	216.0	6.1	5246	9.3%
Canada	58.3	1.6	1376	2.5%
Latin America	90.3	2.6	2236	3.9%
Western Europe	142.4	4.0	3440	6.1%
Middle East	535.9	15.2	13072	23.0%
Africa	209.1	5.9	5074	9.0%
USSR	918.0	26.0	22360	39.5%
Eastern Europe	10.0	0.3	258	0.4%
China	25.0	0.7	602	1.1%
Other (including South-East Asia)	120.0	3.4	2924	5.2%
World	**2325.0**	**65.8**	**56588**	**100%**

**Trillion: 10^{12}; one million million.*

Source: BP Statistical Review of the World Oil Industry 1976.

Table 4: World Natural Gas Consumption 1966-76

million tonnes oil equivalent

Country/Area	1966	1967	1968	1969	1970	1971	1972	1973	1974	1975	1976
USA	463.8	488.1	522.2	561.2	564.1	584.2	587.4	572.3	555.1	508.7	516.4
Canada	22.4	23.8	26.5	29.6	32.7	34.9	39.3	41.8	42.2	43.1	44.1
Total North America	**486.2**	**511.9**	**548.7**	**590.8**	**596.8**	**619.1**	**626.7**	**614.1**	**597.3**	**551.8**	**560.5**
Latin America	28.6	30.9	32.3	34.1	30.4	32.6	36.4	36.5	37.8	39.2	40.5
Total Western Hemisphere	**514.8**	**542.8**	**581.0**	**624.9**	**627.2**	**651.7**	**663.1**	**650.6**	**635.1**	**591.0**	**601.0**
Austria	1.7	1.7	1.8	2.2	2.6	3.0	3.3	3.4	3.8	3.6	4.0
Belgium & Luxembourg	0.1	0.5	1.3	2.6	4.2	5.8	6.7	8.2	9.8	9.6	10.2
Denmark	–	–	–	–	–	–	–	–	–	–	–
Eire	–	–	–	–	–	–	–	–	–	–	–
Finland	–	–	–	–	–	–	–	–	0.4	0.7	0.8
France	5.3	5.9	7.1	8.5	9.3	11.1	13.2	15.7	17.2	17.0	18.5
Greece	–	–	–	–	–	–	–	–	–	–	–
Iceland	–	–	–	–	–	–	–	–	–	–	–
Italy	8.2	8.7	9.7	11.2	12.3	12.5	12.3	14.4	15.8	18.0	25.1
Netherlands	3.0	5.5	9.1	13.4	19.1	24.6	29.0	32.2	32.1	33.2	34.6
Norway	–	–	–	–	–	–	–	–	–	–	–
Portugal	–	–	–	–	–	–	–	–	–	–	–
Spain	–	–	–	0.1	0.1	0.4	1.0	1.0	1.3	1.3	1.5
Sweden	–	–	–	–	–	–	–	–	–	–	–
Switzerland	–	–	–	–	†	†	0.1	0.2	0.4	0.6	0.5
Turkey	–	–	–	–	–	–	–	–	–	–	–
United Kingdom	0.8	1.3	3.0	5.9	11.2	18.1	25.2	26.1	31.8	32.9	34.4
West Germany	3.6	4.7	8.0	11.1	15.6	20.7	21.5	27.0	32.5	33.3	36.2
Yugoslavia	0.4	0.4	0.5	0.7	0.9	1.1	1.4	1.7	2.3	2.2	1.5
Cyprus/Gibraltar/Malta	–	–	–	–	–	–	–	–	–	–	–
Total Western Europe	**23.1**	**28.7**	**40.5**	**55.7**	**75.3**	**97.3**	**113.7**	**129.9**	**147.4**	**152.4**	**167.3**
Middle East	4.7	5.7	7.6	11.5	19.0	18.5	20.0	22.5	26.1	26.7	28.8
Africa	1.2	1.1	1.2	1.4	1.5	1.7	2.4	3.2	5.8	5.8	6.5
South Asia	1.9	2.5	2.8	3.5	3.7	3.9	4.2	8.0	8.5	6.0	7.0
South East Asia	1.7	1.7	2.0	2.0	2.3	2.3	2.4	3.6	4.3	4.1	4.2
Japan	1.9	2.0	2.2	2.4	3.6	3.8	3.7	5.3	7.0	8.5	10.9
Australasia	†	†	†	0.2	1.5	2.2	3.2	3.9	4.6	4.9	6.0
USSR	132.5	145.8	157.5	168.3	184.8	201.4	188.5	200.4	215.7	235.0	252.6
Eastern Europe	20.4	23.4	26.3	31.4	34.5	37.5	35.3	39.4	43.7	50.0	53.7
China	0.9	0.9	1.1	1.3	1.6	2.1	3.7	4.0	4.3	4.7	5.2
Total Eastern Hemisphere	**188.3**	**211.8**	**241.2**	**277.7**	**327.8**	**370.7**	**377.1**	**420.2**	**467.4**	**498.1**	**542.2**
World	**703.1**	**754.6**	**822.2**	**902.6**	**955.0**	**1022.4**	**1040.2**	**1070.8**	**1102.5**	**1089.1**	**1143.2**
World (excl USSR E Europe & China)	**549.3**	**584.5**	**637.3**	**701.6**	**734.1**	**781.4**	**812.7**	**827.0**	**838.8**	**799.4**	**831.7**

† = *Less than 0.05 million tonnes oil equivalent.*

Source: BP Statistical Review of the World Oil Industry 1976.

Table 5: World Oil Production 1966-76

million tonnes

Country/Area	1966	1967	1968	1969	1970	1971	1972	1973	1974	1975	1976
USA – Crude Oil	411.9	437.5	452.9	458.7	478.6	469.9	470.1	457.3	436.8	415.9	404.3
–Natural Gas Liquids	46.1	50.4	53.8	56.4	58.9	60.1	62.1	61.7	59.9	58.0	57.1
Total	458.0	487.9	506.7	515.1	537.5	530.0	532.2	519.0	496.7	473.9	461.4
Canada	49.2	53.8	58.2	62.2	71.5	76.6	88.8	102.3	96.5	83.5	77.2
Total North America	**507.2**	**541.7**	**564.9**	**577.3**	**609.0**	**606.6**	**621.0**	**621.3**	**593.2**	**557.4**	**538.6**
Latin America											
Argentina	15.0	16.4	17.9	18.6	20.4	22.1	22.6	22.0	21.6	20.3	20.4
Brazil	5.6	7.1	8.0	8.5	8.0	8.3	8.1	8.1	8.5	8.4	8.3
Colombia	10.0	9.6	8.8	10.7	11.2	11.0	10.0	9.4	8.7	8.1	7.4
Ecuador	0.3	0.3	0.2	0.2	0.2	0.2	3.5	10.2	8.7	7.9	9.1
Mexico	18.3	20.2	21.7	22.8	23.9	23.8	24.8	26.9	31.6	39.8	46.0
Trinidad	7.9	9.3	9.5	8.2	7.3	6.7	7.3	8.6	9.4	11.2	11.7
Venezuela	177.0	186.1	189.9	188.7	195.2	187.7	171.5	179.0	158.5	125.3	122.9
Other Latin America	5.5	7.0	7.8	7.2	6.5	6.4	7.5	7.9	7.7	7.1	7.2
Total Latin America	**239.6**	**256.0**	**263.8**	**264.9**	**272.7**	**266.2**	**255.3**	**272.1**	**254.7**	**228.1**	**233.0**
Total Western Hemisphere	**746.8**	**797.7**	**828.7**	**842.2**	**881.7**	**872.8**	**876.3**	**893.4**	**847.9**	**785.5**	**771.6**
Western Europe											
Austria	2.8	2.7	2.7	2.8	2.8	2.5	2.5	2.6	2.2	2.0	1.9
France	2.9	2.8	2.7	2.5	2.3	1.9	1.5	1.3	1.1	1.0	1.1
Italy	1.9	1.7	1.6	1.8	1.6	1.4	1.2	1.0	1.0	1.0	1.1
Norway	–	–	–	–	–	0.3	1.6	1.8	1.7	9.3	13.8
Turkey	2.0	2.8	3.1	3.6	3.5	3.5	3.4	3.5	3.3	3.1	2.6
United Kingdom	0.1	0.1	0.1	0.1	0.1	0.1	0.1	0.1	0.1	1.2	11.5
West Germany	7.9	7.9	8.0	7.9	7.5	7.4	7.1	6.6	6.2	5.7	5.5
Yugoslavia	2.2	2.4	2.5	2.7	2.9	3.0	3.2	3.3	3.5	3.9	3.9
Other Western Europe	2.4	2.3	2.3	2.2	2.1	1.8	1.7	2.4	3.5	3.4	3.6
Total Western Europe	**22.2**	**22.7**	**23.0**	**23.6**	**22.8**	**21.9**	**22.3**	**22.6**	**22.6**	**30.6**	**45.0**
Middle East											
Abu Dhabi	17.3	18.3	23.9	28.9	33.4	44.9	50.6	62.6	67.7	67.3	76.8
Dubai	–	–	–	0.5	4.3	6.2	7.6	10.8	12.0	12.6	15.6
Iran	105.2	129.6	141.8	168.1	191.3	227.0	251.9	293.2	301.2	267.7	295.0
Iraq	68.1	60.2	73.9	74.9	76.9	83.5	72.1	99.0	96.9	110.9	112.2
Kuwait	114.4	115.2	122.1	129.5	137.5	147.1	151.2	138.4	114.4	92.4	98.4
Neutral Zone	21.7	21.5	21.0	21.7	26.0	28.3	29.3	27.6	28.0	25.8	24.4
Oman	–	2.9	12.1	16.4	16.6	14.4	14.2	14.7	14.5	17.1	18.4
Qatar	13.8	15.5	16.3	17.0	17.7	20.5	23.2	27.3	24.9	21.0	23.4
Saudi Arabia	118.8	129.0	140.9	148.6	176.2	223.4	285.4	364.7	412.4	343.9	421.6
Sharjah	–	–	–	–	–	–	–	–	1.4	1.9	1.8
Other Middle East	3.3	3.6	5.3	9.0	8.2	10.0	9.9	8.8	9.8	12.1	12.5
Total Middle East	**462.6**	**495.8**	**557.3**	**614.6**	**688.1**	**805.3**	**895.4**	**1047.1**	**1083.2**	**972.7**	**1100.1**
Africa											
Algeria	34.2	39.1	42.9	44.5	48.5	36.5	50.1	51.2	48.5	45.8	50.1
Egypt	6.3	6.2	11.2	17.1	23.5	21.0	17.6	13.0	11.5	14.8	16.4
Gabon	1.4	3.5	4.6	5.0	5.4	5.8	6.3	8.1	10.0	10.3	10.8
Libya	72.4	84.1	125.7	149.9	159.8	133.1	108.2	104.9	73.5	71.3	93.3
Nigeria	20.4	15.6	7.2	26.4	52.9	74.7	88.9	100.1	112.2	88.8	102.9
Other North Africa	0.7	2.3	3.3	3.8	4.3	4.2	4.0	3.8	4.0	4.5	3.5
Other West Africa	0.8	0.6	1.2	2.5	5.5	5.8	7.3	9.5	11.3	10.4	8.4
Total Africa	**136.2**	**151.4**	**196.1**	**249.2**	**299.9**	**281.1**	**282.4**	**290.6**	**271.0**	**245.9**	**285.4**
South Asia	5.8	6.8	6.9	7.5	8.2	8.6	9.2	9.0	8.9	9.6	10.0
South East Asia											
Brunei	4.6	5.3	6.0	6.3	6.9	7.5	8.6	11.6	10.0	9.4	10.2
Indonesia	23.5	25.2	29.7	37.1	42.2	44.1	53.4	66.0	69.0	65.0	74.6
Other South East Asia	0.1	†	0.2	0.5	0.9	3.5	3.7	4.4	4.0	4.5	7.7
Total South East Asia	**28.2**	**30.5**	**35.9**	**43.9**	**50.0**	**55.1**	**65.7**	**82.0**	**83.0**	**78.9**	**92.5**
Japan	0.7	0.7	0.7	0.7	0.8	0.7	0.7	0.7	0.7	0.6	0.6
Australasia	0.5	1.0	1.8	2.0	8.6	15.0	16.8	18.5	18.4	19.9	20.5
USSR	265.1	288.1	309.2	328.3	353.0	372.0	394.0	421.0	452.0	485.0	515.0
Eastern Europe	16.4	17.1	17.4	17.3	17.6	18.2	19.3	19.2	19.7	20.0	20.0
China	12.5	11.0	13.0	14.5	20.0	25.5	29.5	40.0	54.0	65.0	75.0
Total Eastern Hemisphere	**950.2**	**1025.1**	**1161.3**	**1301.6**	**1469.0**	**1603.4**	**1735.3**	**1950.7**	**2013.5**	**1928.2**	**2164.1**
World	**1697.0**	**1822.8**	**1990.0**	**2143.8**	**2350.7**	**2476.2**	**2611.6**	**2844.1**	**2861.4**	**2713.7**	**2935.7**
World (excl USSR E Europe & China)	**1403.0**	**1506.6**	**1650.4**	**1783.7**	**1960.1**	**2060.5**	**2168.8**	**2363.9**	**2335.7**	**2143.7**	**2325.7**

Note:— Egypt (UAR) includes onshore Gulf of Suez and Sinai production. † = Less than 0.05 million tonnes.

Source: BP Statistical Review of the World Oil Industry 1976

Table 6: World Oil Consumption 1966-76

million tonnes

Country/Area	1966	1967	1968	1969	1970	1971	1972	1973	1974	1975	1976
USA	575.7	595.8	635.5	667.8	694.6	719.3	775.8	818.0	782.6	765.9	822.4
Canada	58.0	61.7	66.1	69.1	73.0	75.8	79.3	83.7	84.8	83.1	85.9
Total North America	**633.7**	**657.5**	**701.6**	**736.9**	**767.6**	**795.1**	**855.1**	**901.7**	**867.4**	**849.0**	**908.3**
Latin America	106.8	110.6	118.9	126.8	137.2	147.6	151.5	163.7	171.3	174.0	186.4
Total Western Hemisphere	**740.5**	**768.1**	**820.5**	**863.7**	**904.8**	**942.7**	**1006.6**	**1065.4**	**1038.7**	**1023.0**	**1094.7**
Austria	6.1	6.5	7.5	8.3	9.3	10.3	10.8	11.5	10.7	10.6	11.5
Belgium & Luxembourg	17.5	19.5	21.9	25.1	27.9	28.4	31.1	31.5	28.1	26.5	27.6
Denmark	11.8	12.3	13.8	16.5	18.6	18.6	19.5	17.9	16.0	15.7	16.8
Eire	2.7	3.0	3.4	3.7	4.1	4.5	5.0	5.4	5.4	5.2	5.3
Finland	6.9	7.2	8.2	9.6	10.8	11.1	11.9	13.3	11.6	11.9	12.8
France	57.7	66.2	71.8	83.0	94.3	102.8	114.1	127.3	121.0	110.4	117.3
Greece	4.8	5.5	5.9	6.1	6.7	7.4	8.6	10.0	9.4	9.9	10.7
Iceland	0.5	0.5	0.5	0.5	0.6	0.5	0.6	0.7	0.6	0.6	0.6
Italy	57.7	63.7	70.3	77.3	87.3	93.8	98.2	103.6	100.8	94.5	98.5
Netherlands	27.4	28.0	30.1	32.9	36.5	36.0	40.1	41.3	35.4	34.8	39.4
Norway	5.9	6.1	6.7	7.4	8.3	8.2	8.5	8.6	7.7	8.0	8.7
Portugal	2.9	3.2	3.4	3.9	4.6	5.4	5.8	6.5	6.5	6.8	7.2
Spain	16.8	20.3	21.6	24.6	28.1	30.9	32.5	39.1	41.1	42.7	47.2
Sweden	21.5	21.0	23.8	26.8	29.9	28.2	28.6	29.4	27.1	26.6	29.4
Switzerland	8.6	9.3	10.3	11.2	12.5	13.3	13.6	14.7	13.0	12.5	12.9
Turkey	4.5	5.6	6.4	7.1	7.7	9.0	10.0	12.4	12.5	13.4	15.2
United Kingdom	79.0	83.2	89.8	95.7	101.8	103.1	109.7	113.4	105.9	92.1	91.6
West Germany	89.2	92.1	103.8	117.1	128.6	133.5	140.9	149.7	134.3	128.9	139.2
Yugoslavia	3.4	4.1	4.8	6.0	7.0	9.0	10.1	11.3	11.5	12.2	13.2
Cyprus/Gibraltar/ Malta	0.7	0.8	0.9	1.0	1.0	1.2	1.3	1.4	1.3	1.1	1.3
Total Western Europe	**425.6**	**458.1**	**504.9**	**563.8**	**625.6**	**655.2**	**700.9**	**749.0**	**699.9**	**664.4**	**706.4**
Middle East	39.3	38.9	41.1	47.2	49.5	54.9	59.6	64.6	68.6	66.8	73.3
Africa	31.1	34.9	36.4	39.0	41.6	44.2	44.5	47.5	50.3	51.3	54.2
South Asia	18.7	19.8	23.0	27.1	27.1	28.0	30.8	33.2	27.7	31.0	32.4
South East Asia	29.2	42.4	48.5	53.4	59.6	63.9	71.2	77.6	79.2	81.2	91.7
Japan	100.0	122.9	142.7	169.0	199.1	219.7	234.4	269.1	258.9	244.0	253.8
Australasia	21.7	23.5	25.8	27.7	30.1	31.5	32.1	34.8	35.9	35.3	36.3
USSR	193.2	210.7	225.7	241.4	262.1	273.8	296.1	317.7	341.5	362.0	380.0
Eastern Europe	35.1	39.0	44.1	50.3	55.3	60.7	67.0	76.7	81.0	86.0	90.0
China	12.5	11.0	13.0	14.5	20.0	24.8	28.6	37.6	48.5	55.6	66.0
Total Eastern Hemisphere	**906.4**	**1001.2**	**1105.2**	**1233.4**	**1370.0**	**1456.7**	**1565.2**	**1707.8**	**1691.6**	**1677.6**	**1784.1**
World	**1646.9**	**1769.3**	**1925.7**	**2097.1**	**2274.8**	**2399.4**	**2571.8**	**2773.2**	**2730.2**	**2700.6**	**2878.8**
World (excl USSR E Europe & China)	**1406.1**	**1508.6**	**1642.9**	**1790.9**	**1937.4**	**2040.1**	**2180.1**	**2341.2**	**2259.2**	**2197.0**	**2342.8**

Source: BP Statistical Review of the World Oil Industry 1976

Table 7: Per Capita Consumption of Petroleum Products, 1976

Country	Tonnes per capita
USA	3.5
Canada	3.4
Sweden	3.2
West Germany	2.2
Japan	2.0
France	2.0
Netherlands	2.0
Australia	1.9
UK	1.4
Argentina	0.8
Venezuela	0.8
India	0.1

Source: Shell information handbook 1977-1978

Table 8: World Primary Energy Consumption 1966-76

million tonnes oil equivalent

Country/Area	1966	1967	1968	1969	1970	1971	1972	1973	1974	1975	1976
USA	1414.5	1460.8	1548.3	1632.5	1665.3	1705.6	1772.8	1827.7	1774.1	1727.1	1814.6
Canada	128.8	134.8	144.1	152.7	162.7	168.3	179.7	190.9	197.6	195.0	201.7
Total North America	**1543.3**	**1595.6**	**1692.4**	**1785.2**	**1828.0**	**1873.9**	**1952.5**	**2018.6**	**1971.1**	**1922.1**	**2016.3**
Latin America	196.0	203.8	214.8	226.6	236.8	250.8	261.0	277.4	291.8	299.0	315.4
Total Western Hemisphere	**1739.3**	**1799.4**	**1907.2**	**2011.8**	**2064.8**	**2124.7**	**2213.5**	**2296.0**	**2263.5**	**2221.1**	**2331.7**
Austria	17.3	17.5	18.7	19.3	22.2	21.8	22.9	23.9	24.8	24.2	24.1
Belgium & Luxembourg	35.5	37.5	41.3	45.3	48.8	48.2	51.3	50.8	49.9	46.7	49.3
Denmark	15.5	15.6	17.1	19.3	21.2	20.2	21.0	20.2	18.0	18.5	19.7
Eire	5.4	5.8	6.7	6.8	7.0	7.2	6.5	7.4	7.4	7.2	7.4
Finland	15.5	16.1	17.1	18.5	19.8	20.1	21.0	22.2	21.4	21.3	22.1
France	120.9	128.2	134.6	148.1	159.6	164.9	174.3	182.9	180.5	169.8	178.8
Greece	6.6	7.3	7.8	8.4	9.5	10.9	11.8	13.4	12.9	14.2	15.1
Iceland	1.0	1.0	1.1	1.1	1.6	1.5	1.8	2.1	1.8	2.0	2.1
Italy	90.1	97.4	103.7	112.2	124.5	130.0	135.7	139.4	138.8	133.2	143.6
Netherlands	39.0	41.8	46.8	53.1	60.8	64.2	72.3	76.9	71.4	71.4	77.7
Norway	19.4	20.8	23.1	23.3	24.2	25.6	27.1	28.3	28.5	29.0	30.5
Portugal	5.1	5.4	5.5	6.4	6.9	7.6	8.8	9.8	9.2	8.8	8.9
Spain	34.4	37.0	38.3	43.5	46.6	51.6	55.4	61.0	63.1	63.9	67.6
Sweden	35.3	35.5	38.2	39.3	42.3	43.2	44.3	46.6	43.4	49.2	51.9
Switzerland	16.7	17.8	18.6	19.3	21.4	22.1	21.8	24.1	22.8	23.8	22.6
Turkey	18.4	18.6	19.7	20.5	21.1	21.6	23.0	25.5	25.2	29.1	33.6
United Kingdom	191.0	190.3	199.5	206.6	213.5	212.3	214.8	225.0	215.1	204.4	206.7
West Germany	193.3	190.6	207.7	227.4	244.5	248.2	249.9	271.6	262.7	252.0	260.6
Yugoslavia	19.9	19.3	20.7	22.9	24.7	28.4	29.8	30.0	31.1	31.3	32.4
Cyprus/Gibraltar/ Malta	0.7	0.8	0.9	1.0	1.0	1.2	1.3	1.4	1.3	1.1	1.3
Total Western Europe	**881.0**	**904.3**	**967.1**	**1042.3**	**1121.2**	**1150.8**	**1194.8**	**1262.5**	**1229.3**	**1201.1**	**1256.0**
Middle East	45.6	46.3	50.4	60.7	70.6	75.8	81.9	89.4	98.2	97.2	105.6
Africa	85.8	90.5	95.0	99.3	103.8	110.1	108.6	109.0	116.6	120.3	128.0
South Asia	75.2	78.4	84.3	93.4	91.5	93.9	99.5	106.2	103.7	105.5	111.0
South East Asia	83.8	98.2	106.8	111.1	120.7	127.1	136.0	144.7	149.6	153.8	163.9
Japan	170.7	196.7	220.0	253.0	287.5	303.5	318.3	355.8	354.0	339.9	356.0
Australasia	47.1	49.9	53.5	56.1	60.3	63.5	66.1	68.5	71.6	73.2	76.0
USSR	698.1	723.8	752.1	786.7	821.3	858.2	875.6	919.8	970.4	1022.0	1070.3
Eastern Europe	272.6	277.0	291.3	311.9	330.2	343.0	348.7	367.2	383.9	401.0	418.4
China	253.6	180.8	235.0	250.0	280.7	308.0	315.7	339.7	366.0	388.4	415.5
Total Eastern Hemisphere	**2604.5**	**2645.9**	**2855.5**	**3064.5**	**3287.8**	**3433.9**	**3545.2**	**3762.8**	**3843.3**	**3902.4**	**4100.7**
World	**4343.8**	**4445.3**	**4762.7**	**5076.3**	**5352.6**	**5558.6**	**5758.7**	**6058.8**	**6106.8**	**6123.5**	**6432.4**
World (excl USSR E Europe & China)	**3128.5**	**3263.7**	**3484.3**	**3727.7**	**3920.4**	**4049.4**	**4218.7**	**4432.1**	**4386.5**	**4312.1**	**4528.2**

Source: BP Statistical Review of the World Oil Industry 1976.

Table 9: World Energy Production by Region 1950 and 1975

million tonnes oil equivalent

Country/Area	Solid fuels	Liquid fuels	Natural gas	Hydro/ nuclear electricity	Total 1950	Solid fuels	Liquid fuels	Natural gas	Hydro/ nuclear electricity	Total 1975	% changes
USA	339.2	282.3	150.7	8.3	780.6	384.7	466.9	473.6	39.4	1364.6	+ 74.8%
Canada	10.7	3.89	4.19	4.35	20.68	15.3	80.6	66.2	17.6	179.7	+ 768%
Latin America	3.96	100.4	7.9	1.14	108	8.7	226.9	34.5	11.05	281.1	+ 160%
Western Europe	328.01	3.97	0.87	3.1	342	227.4	32.0	144.3	41.9	445.6	+ 30%
Middle East	0.13	84.53	negl	–	84.6	0.8	966.4	34.3	0.53	1002.0	+1095%
Africa	20.13	2.6	negl	0.115	22.9	50.5	240.0	12.26	3.015	305.8	+1235%
South & South East Asia	25	11.2	0.74	0.4	37.3	110.3	86.0	20.6	6.34	223.3	+ 502%
Japan	26	0.29	0.06	3.15	29.56	12.7	0.6	2.5	9.18	25	– 13.7%
Australia	14.6	negl	–	–	14.6	48.8	21.4	4.6	2.8	77.6	+ 431%
USSR	147.7	37.5	5.1	0.7	191.2	339.9	483.6	271.0	11.3	1105.8	+ 478%
Eastern Europe	114.6	5.7	3.4	0.1	124	254.7	20.2	43.8	2.2	320	+ 158%
China	28.7	0.201	n.a.	0.07	29	314.9	78.8	3.123	3.05	399.9	+1279%
World	1058.6	532.8	165.15	28.3	1785	1768.7	2703	1110.7	148.4	5730	+ 221%

Source: UN World Energy Supplies 1950-1974, 1971-1975

Table 10: World Energy Production by Region 1955, 1965 and 1975

million tonnes oil equivalent

Country/Area	1955 Solid fuels	1955 Liquid fuels	1955 Natural gas	1955 Hydro/ nuclear electricity	Total 1955	1965 Solid fuels	1965 Liquid fuels	1965 Natural gas	1965 Hydro/ nuclear electricity	Total 1965	% Change 1965 over 1955	1975 Solid fuels	1975 Liquid fuels	1975 Natural gas	1975 Hydro/ nuclear electricity	Total 1975	% Change 1975 over 1965
USA	297.0	360.4	225.2	9.58	892.2	319.0	425.0	386.3	16.5	1146.8	+ 28.5%	384.7	466.9	473.6	39.4	1364.6	+ 19%
Canada	8.07	17.3	3.7	6.3	35.4	6.12	44.1	33.4	9.6	93.2	+163%	15.3	80.6	66.2	17.6	179.7	+ 92.8%
Latin America	4.69	141.17	5.7	1.8	153.4	6.1	239.7	19.8	4.2	269	+ 75.3%	8.7	226.9	34.5	11.05	281.1	+ 4.5%
Western Europe	358.6	9.6	5.0	13.2	386.4	326.2	21.8	18.2	25.9	392.1	+ 1.5%	227.4	32.0	144.3	41.9	445.6	+ 13.6%
Middle East	2.7	157.18	0.54	0.02	160.5	3.69	410.4	4.12	0.25	418.5	+160.1%	0.8	966.4	34.3	0.53	1002.0	+139%
Africa	25.2	2.0	0.01	0.26	27.5	35.9	105.5	1.7	1.1	144.2	+424%	50.5	240.0	12.26	3.015	305.8	+112%
South & South-East Asia	32.0	18.2	1.92	0.84	53.0	71.8	31.6	5.0	3.1	111.5	+110%	110.3	86.0	20.6	6.34	223.3	+100%
Japan	28.7	0.3	0.15	4.02	33.2	33.3	0.7	1.7	5.8	41.6	+ 25.3%	12.7	0.6	2.5	9.18	25	– 40%
Australasia	17.0	–	–	–	17.0	27.4	0.5	–	–	27.9	+ 64%	48.8	21.4	4.6	2.8	77.6	+178%
USSR	222.6	69.7	8.0	1.88	302.2	251.6	96.88	16.6	3.2	368.2	+ 21.8%	339.9	483.6	271.0	11.3	1105.8	+200%
Eastern Europe	150.8	12.86	6.06	0.4	170.0	240.1	158.6	115.4	4.3	518.4	+204.9%	254.7	20.2	43.8	2.2	320	– 38.7%
China	65.9	0.95	0.098	0.19	67.1	200.3	9.85	0.76	2.06	213.0	+217.4%	314.9	78.8	3.123	3.05	399.9	+ 87.7%
World	1213.3	789.7	256.4	38.5	2298	1521	1544	603	76.0	3744	+ 63%	1768.7	2703	1110.7	148.4	5730	+ 53%

Source: UN World Energy Supplies 1950-1974, 1971-1975

Table 11: World Energy Consumption 1956, 1966 and 1976

million tonnes oil equivalent

Country/Area	1956 Solid fuels	1956 Liquid fuels	1956 Natural gas	1956 Hydro/nuclear electricity	Total 1956	1966 Solid fuels	1966 Liquid fuels	1966 Natural gas	1966 Hydro/nuclear electricity	Total 1966	% Change 1966 over 1956	1976 Solid fuels	1976 Liquid fuels	1976 Natural gas	1976 Hydro/nuclear electricity	Total 1976	% Change 1976 over 1966
USA	275.8	389.7	237.7	10.7	913.9	300.56	530.6	425.9	16.88	1273.9	+ 39%	350.5	822.4	516.4	125.3	1814.6	+ 42.4%
Canada	21.7	31.7	4.1	6.4	63.8	14.6	52.5	20.3	10.6	98.0	+ 54%	15.3	85.9	44.1	56.4	201.7	+106%
Latin America	6.619	51.69	6.4	2.1	66.86	8.0	87.64	19.7	4.7	120.0	+ 79%	51.9	186.4	40.5	36.6	315.4	+162.8%
Western Europe	396.4	104.8	6.02	13.8	521.0	329.05	359.0	21.8	28.46	738.3	+ 41.7%	260.2	706.4	167.3	122.1	1256.0	+ 70%
Middle East	0.16	7.3	0.71	0.01	8.17	0.2	17.7	5.2	0.16	23.3	+185.2%	2.1	73.3	28.8	1.4	105.6	+353%
Africa	26.6	12.5	0.01	0.3	39.4	35.0	24.1	1.07	1.1	61.3	+ 55.6%	58.5	54.2	6.5	8.9	128.0	+108.8%
South & South-East Asia	34.14	18.25	2.2	1.04	55.69	73.6	51.0	5.4	4.0	134.0	+140.6%	122.6	124.1	11.2	17.0	274.9	+105%
Japan	34	10.6	0.21	4.4	49.2	44	67.5	1.7	5.8	119.0	+141%	60.0	253.8	10.9	31.3	356.0	+199%
Australasia	16.5	9.5	–	0.63	26.6	21.5	19.17	0.003	1.52	42.2	+ 58.6%	28.2	36.3	6.0	5.5	76.0	+ 80%
USSR	230.6	70.505	10.64	2.39	314.1	286.06	181.93	126.8	7.6	602.4	+ 91.8%	387.4	380.0	252.6	50.3	1070.3	+ 77.7%
Eastern Europe	205.5	29.5	20.27	1.0	256.3	205.5	29.5	20.27	1.0	256.3	none	266.7	90.0	53.7	8.0	418.4	+ 63.2%
China	73.2	3.0	0.205	0.285	76.7	218.0	12.49	0.816	2.653	233.9	+205%	337.5	66.0	5.2	6.8	415.5	+ 77.6%
World	1321	739.0	288.5	43.0	2391	1536	1433	648.9	84.5	3702	+ 54.8%	1940.8	2878.8	1143.2	469.6	6432.4	+ 73.7%

1956/66: UN World Energy Supplies 1950-1974. 1976: BP Statistical Review of World Oil Industry 1976

Table 12: Nuclear Installed Capacity 1960–77

(megawatts electric)

Country/Area	1960	1961	1962	1963	1964	1965	1966	1967	1968	1969	1970	1971	1972	1973	1974	1975	1976	1977
USA	297	442	672	749	906	926	1942	2887	2817	3980	6493	8687	15301	21118	31662	38943	38323	46975
Canada	–	–	20	20	20	20	20	240	240	240	240	1570	2126	2400	2666	2666	2403	4251
Argentina	–	–	–	–	–	–	–	–	–	–	–	–	–	–	240	240	340	340
Puerto Rico	–	–	–	–	–	–	16	16	16	–	–	–	–	–	–	–	–	–
Panama CZ	–	–	–	–	–	–	–	–	–	10	10	10	10	10	10	10	–	–
Japan	–	–	–	–	–	13	138	179	179	510	1336	1336	1836	2296	3906	6615	6602	7994
India	–	–	–	–	–	–	–	–	–	420	420	420	620	640	640	640	620	620
Pakistan	–	–	–	–	–	–	–	–	–	–	–	137	137	137	137	137	137	137
France	97	97	97	146	166	416	416	1025	1271	1771	1771	2301	2895	3076	3076	3098	3061	5868
West Germany	–	15	15	15	15	15	322	338	888	933	958	962	2307	2414	3504	3504	3472	9487
Belgium	–	–	11	11	11	11	11	11	11	11	11	11	11	11	2	1663	1750	1750
Netherlands	–	–	–	–	–	–	–	–	54	54	54	55	55	524	524	524	532	532
Italy	–	–	–	210	556	642	642	642	642	642	642	670	670	670	670	670	577	1439
UK	360	360	1089	1138	1503	3387	3881	4168	4148	4647	4813	5607	5614	5814	5814	5734	5971	8870
Sweden	–	–	–	–	10	10	10	10	10	10	10	12	472	472	1062	2522	4366	3892
Finland	–	–	–	–	–	–	–	–	–	–	–	–	–	–	–	–	–	440
Switzerland	–	–	–	–	–	–	–	–	–	350	350	700	1006	1006	1006	2422	1054	1054
Spain	–	–	–	–	–	–	–	–	153	153	153	613	1120	1120	1120	1120	1132	1132
Bulgaria	–	–	–	–	–	–	–	–	–	–	–	–	–	–	440	880	n.a.	n.a.
Czechoslovakia	–	–	–	–	–	–	–	–	–	–	–	–	100	100	100	150	n.a.	n.a.
East Germany	–	–	–	–	–	–	70	75	75	75	75	75	75	75	515	950	n.a.	n.a.
USSR	105	105	105	966	966	1016	1016	1216	1226	1591	1591	2031	2621	3509	4500	5600	n.a.	n.a.
World	859	1019	2009	3255	4153	6456	8484	10807	12230	15397	18927	25197	36976	45392	61714	78188		

1960 – 1975: UN World Energy Supplies 1950-1974 and 1971-1975; 1976-1978: Nuclear Engineering International. Dec issues for Sept '76, '77.

n.a. – figures not available.

Table 13: Nuclear Power Growth Estimate 1978–2000

megawatts electric

OECD countries	1978	1979	1980	1981	1982	1983	1984	end of year 1985	1986	1987	1988	1989	1990	1995	2000
Australia	700	700	700	700	700	700	700	700	1000	2000	2000	2000	2000		
Belgium	1700	1700	1700	2600	2600	3500	3500	3500	5000	5000	7000	8000	8000		
Canada	4000	5000	6000	6000	7000	8000	9000	10000	12000	14000	16000	18000	20000		
Denmark											1000	1000	2000		
Finland	1500	1500	2200	2200	2200	2200	2200	2200	2200	2700	2700	3500	3500		
France	6500	12000	15000	19000	23000	27000	31000	34000	38000	42000	46000	50000	53000		
West Germany	9000	10300	12000	14000	16000	18000	22000	25000	29000	34000	38000	43000	47000		
Greece												1000	1000		
Italy	1400	1400	1400	1400	1400	1400	2400	5400	7000	12000	15000	20000	25000		
Japan	12000	13000	15000	17000	19000	22000	24000	27000	31000	35000	39000	44000	50000		
Luxembourg										1000	1000	1000	1000		
Netherlands	500	500	500	500	500	500	500	500	1000	2000	2000	2000	3000		
Portugal									900	900	900	1800	1800		
Spain	2100	4100	8000	8000	11000	13000	14000	15000	16000	17000	18000	19000	20000		
Sweden	4700	5600	6500	6500	7400	7400	7400	7400	7000	7000	7000	7000	8000		
Switzerland	1000	1000	1900	1900	1900	1900	2800	2800	3000	3000	3000	3000	3000		
Turkey											1000	1000	1000		
United Kingdom	6600	10300	10300	10300	10300	10300	10300	10300	11600	12800	14100	15300	15300		
United States	50000	55000	60000	68000	77000	88000	101000	115000	130000	146000	162000	178000	194000		
OECD															
Present trend estimate	102000	122000	141000	158000	180000	204000	231000	259000	295000	336000	376000	419000	459000	660000	850000
Accelerated nuclear estimate	102000	122000	141000	167000	199000	242000	287000	343000	394000	449000	510000	570000	640000	1090000	1640000
NON OECD															
Present trend	3000	4000	5000	7000	9000	12000	15000	19000	23000	28000	33000	39000	45000	90000	150000
Accelerated	3000	4000	5000	9000	12000	16000	21000	25000	31000	38000	45000	52000	60000	130000	250000
TOTAL															
Present trend	105000	126000	146000	165000	189000	216000	246000	278000	318000	364000	409000	458000	504000	750000	1000000
Accelerated	105000	126000	146000	175000	211000	258000	308000	368000	425000	487000	555000	622000	700000	1220000	1890000

Source: OECD and International Atomic Energy Agency report on uranium resources, production and demand December 1977.

Table 14: Uranium Production (World; excluding USSR, E Europe and China)

(tonnes U)

Country	Cumulative pre-1972	1972	1973	1974	1975	1976	1977 (estimated)
Argentina	188	25	24	30	23	50	130
Australia	7080	0	0	0	0	360	400
Canada	92540	4000	3710	3420	3510	4850	6100
France	16600	1545	1616	1673	1742	2063	2200
Gabon	4400	210	402	436	800	n.a.	n.a.
Japan	8	8	10	7	3	2	2
Mexico	42	0	0	0	0	0	0
Niger	410	867	948	1117	1306	1460	1609
Portugal	1483	73	73	92	115	88	85
South Africa	61433	3197	2735	2711	2488	3412	6700
Spain	166	55	55	60	136	170	191
United States	162000	9900	10200	8900	8900	9800	11200
West Germany	150	0	0	26	57	38	n.a.
Zaire*	25600	0	0	0	0	0	0
Total	372100	19880	19773	18472	19080	22293	28617

n.a. – *not available.*

* *Estimated by the Steering Group of the Joint NEA/IAEA Working Party on Uranium Resources.*

Sweden: total cumulative production to 1976 was 200 tonnes U.

Source: Uranium resources, production and demand, December 1977; Report by OECD and International Atomic Energy Agency.

Appendix 2

Table 1: Conversion Tables

Mass

	Metric tonnes	Long tons	Short tons	Pounds
Metric tonne*	–	0.984	1.102	2205
Long ton	1.016	–	1.120	2240
Short ton	0.907	0.893	–	2000
Pound	0.000453	0.000446	0.0005	–

* *In converting crude oil to kilolitres or cubic metres, one metric tonne is approximately equal to 1.16 kilolitres, or 1.16 cubic metres.*

Volume

	Barrels	Imperial gallons	US gallons	Kilolitres
Barrel	–	34.973	42	0.159
Imperial gallon	0.0286	–	1.201	0.00455
US gallon	0.0238	0.833	–	0.00379
Kilolitre*	6.2898	219.97	264.17	–

* *In converting crude oil to tonnes one kilolitre is approximately equal to 0.863 tonnes.*

	Cubic feet	Cubic metres
Cubic foot	–	0.0283
Cubic metre	35.315	–

Length

	Fathoms	Metres	Feet
Fathom	–	1.8288	6
Metre	0.549	–	3.2808
Foot	0.166	0.3048	–

Table 2: Calorific Equivalents – Conversion Factors (multiply by the figure shown)

To \ From	Million tonnes oil equivalent	Million tonnes coal equivalent	Million therms	Terawatt hours (thermal)	Terawatt hours (electrical)	Terajoules
Million tonnes oil equivalent (mtoe)	1	0.60	0.0024	0.0800*	0.2667**	2×10^{-5}
Million tonnes coal equivalent (mtce)	1.67	1	0.0039	0.1335*	0.4450**	4×10^{-5}
Million therms (m th)	425	255	1	34.13	114	0.0095
Terawatt hours (thermal) (TWh[t])	12.50	7.49	0.0293	1	3.33	0.0003
Terawatt hours (electrical) (TWh[e])	3.75	2.25	0.0088	0.30	1	0.0001
Terajoules (TJ)	44,800	26,900	105.5	3600	12000	1

* *The quantity of fuel equivalent to 1 TWh of energy*
** *The quantity of fuel required to generate 1 TWh of electricity*

Table 3: Approximate Calorific Equivalents for Conversion into Million Tonnes of Oil

One million tonnes of oil equals approximately	Heat units and other fuels expressed in terms of million tonnes of oil	
Heat Units		**million tonnes of oil**
43 million million Btu	10 million million Btu approximates to	0.24
425 million therms	100 million therms approximates to	0.24
44800 Terajoules	10,000 Terajoules approximates to	0.22
Solid Fuels		
1.67 million tonnes of coal	1 million tonnes of coal approximates to	0.60
3.04 million tonnes of lignite	1 million tonnes of lignite approximates to	0.33
5.67 million tonnes of peat	1 million tonnes of peat approximates to	0.18
Natural Gas (1 ft^3 = 1000 Btu) (1 m^3 = 37.1 MJ)		
1.17 thousand million m^3	1 thousand million m^3 approximates to	0.85
41.4 thousand million ft^3	10 thousand million ft^3 approximates to	0.24
113 million ft^3/day for a year	100 million ft^3/day for a year approximates to	0.88

Table 4: Approximate Calorific Values for Different Petroleum Products

	Therms per tonne	GJ/tonne
Crude oil (average)	425	45
Petroleum gases	500	53
Liquified petroleum gas	470	50
Naphtha	455	48
Motor spirit	445	47
Burning oil	440	46
Gas/Diesel oil	430	45
Fuel oil	405	43

1 GJ is approximately equal to 9.5 therms

Table 5: Approximate Calorific Values for Solid Fuels

	Therms per tonne	GJ/tonne
Anthracite	315	33
Bituminous coal (average)	275	29
Sub-bituminous coal (average)	235	25
Brown coal and lignite	140	15
Peat	75	8
Coke	265	28

In the UK, power stations burn the lowest grade of hard coal (220 therms per tonne), industrial coal has an average calorific value of 260 therms per tonne, and domestic coal has a calorific value of 285 therms per tonne.

Appendix 3

Table 1: Geological Timescale

ERA	Period	Duration millions of years	Time to commencement millions of years ago
Cenozoic	Pleistocene	to present	1.0
	Quaternary	1.5	2.5
	Tertiary	62.5	65
Mesozoic	Cretaceous	71	136
	Jurassic	54	190
	Triassic	35	225
Paleozoic	Permian	55	280
	Upper Carboniferous	45	325
	Lower Carboniferous	20	345
	Devonian	50	395
	Silurian	35	430
	Ordovician	70	500
	Cambrian	70	570
Precambrian			

Source: *The Energy Question, G Foley with C Nassim (Pelican Books 1976) p114, © Gerald Foley 1976. Reprinted by permission of Penguin Books Ltd*

Table 2: The Age of Coal Resources of the World

Era	System		Age in million years	Localities
CENOZOIC	Pleistocene			Peat deposits in most areas
			1	
	TERTIARY	Pliocene		Lignites in: Hungary (Zala), Indonesia (Sumatra-Lematang, E.Borneo), Italy (Tuscany), Japan, Rumania (Comanesti), USA (Alaska)
			11	
		Miocene		Lignites in: Argentina (Patagonia), Austria (Upper Austria, Styria), Canada (Columbia), Czechoslovakia (Eger Valley, Teplice), Denmark (Jutland), Germany (Cologne Cottbus), Greenland, Holland (S.Limburg), Hungary (Paszlo), Japan, New Zealand (Otago), Yugoslavia (Bosnia)
			25	
		Oligocene		Lignites in: Australia (Victoria), Canada (Columbia), Chile (Arauca), Great Britain (Devon), New Zealand (N. & S. Auckland, Southland), Rumania (Cluj, Petrosam), Spitzbergen, Turkey (Kulahya), Yugoslavia (Ljubliana)
			40	
		Eocene		Mainly lignites in: Canada (Saskatchewan, Vancouver), Germany (Halle, Leipzig, Magdeburg, Saxony, Thuringia), India (Assam), Indonesia (Sumatra-Umbilm), E.Java, New Guinea (Vogelkop), Japan, New Zealand (Nelson, Westland), Pakistan (W.Punjab), Spitzbergen (King's Day), USA (Alaska, Dakota, Montana), Yugoslavia (Bosnia, Dalmatia, Istria)
			70	
MESOZOIC	Cretaceous			Lignites and bituminous coals in: Australia (Queensland), Bulgaria (Balkans), Canada (Alberta, Columbia, Saskatchewan), Columbia (Bogota), France (Basse-Provence), Germany (Hanover), Greenland (Disko), Japan, Mexico (Barroteran, Eagle Pass, Sabinas), New Zealand (Nelson, Southland), Nigeria (Enugu), Peru (Andean Provinces), USA (Rocky Mt. states), USSR (Burega, Chita, Sakhalm, Vilui), Yugoslavia (Serbia)
			135	
	Jurassic			Coals in: Australia (New South Wales, Queensland), China (Szechwan), Egypt (N.Sinai), Hungary (Pecs), Iran (Demavend, Kerman), Sweden (Skane), USSR (Bergana, Georgia, Irkutsk, Kansk)
			180	
	Triassic			Coals in: Australia (Tasmania), Mexico (Santa Clara), Poland (Katowice, Kielce), USA (N.Carolina, Virginia), USSR (Urals)
			225	
PALEOZOIC	Permian			Coals in: Antarctica, Australia (New South Wales, Queensland, Tasmania), China (Honan, Hopeh, Shansi), France (Central Plateau), Germany (Saxony), India (Bengal, Bihar, Orissa), Korea (Samchok, Yongwol), Rhodesia (Wankie), USA (Maryland, Ohio, Pennsylvania), USSR (Kuzbass, Pechora), S. Africa (Natal, Transvaal)
			270	
	Carboniferous	Upper		Coals in: Algeria (Abadia, Colomb-Bechat), Belgium (Kempen, Sambre-Meuse), Brazil (Rio Bonito, Rio Grande do Sul), Canada (New Brunswick, Nova Scotia), Czechoslovakia (L.Bohemia), France (Nord, Pas de Calais), Germany (Aachen, Ruhr, Saar), Great Britain, Holland (S.Limburg), Morocco (Djerada), Poland (Silesia), Spain (Granada, Oviedo, Santanander), Turkey, USA (Appalachian, Gulf and Interior States), USSR (Donbuss, Karaganda)
			320	
		Lower		Coals in: Canada (Arctic Isles), Great Britain (Northumberland), Spitzbergen, USA (Pennsylvania), USSR (Moscow and Ural Basins)
			350	
	Devonian			Coals in: Bear Island, Canada (Arctic Isles)

Source: Coal Mining Geology, Williamson (OUP 1967)

Appendix 4

Table 1: Classification of Crude Petroleum and its Components

Boiling Point Range °C	-150 · 0 · 50 · 100 · 150 · 200 · 250 · 350 · 500 · 1000+
General Classification	← Petroleum Gases → · ← Light Distillate → · ← Middle Distillate → · ← Heavy Distillate → · ← Residue →
Main Components	← Petroleum Gases → · ← Gasolines → (*light* … *heavy*) · ← Gas Oils → (*light* … *heavy*) · ← Lubricating Oils → ← Kerosines → (*light* … *heavy*) · ← Fuel Oils → (*light* … *heavy*) · ← Asphaltenes → ← Naphthas →
Hydrocarbon Range	← C_4 and lower → · ← Pentane Plus → · ← Liquid → · ← Solid → C_1 · C_4 · C_5 · C_8 · C_{14} · C_{16} · C_{60}
US Bureau of Mines Correlation Index	Paraffinic-Paraffinic · Paraffinic-Naphthenic · Naphthenic-Paraffinic · Naphthenic-Naphthenic
Base Classification	Paraffinic (Light) · Mixed (Aromatic) · Naphthenic (Heavy) · Asphaltic
Typical API Gravity Range	38°–47° · 37°–30° · 25°–15°
Specific Gravity	0.835–0.800 · 0.840–0.876 · 0.900–0.970

Note: The classifications shown in this table are intended to be representative, and no precise demarcations are implied.

Table 2: Natural and Manufactured Hydrocarbons derived from Petroleum

Structure	Class	Sub-class	Hydrocarbon	No of atoms C	No of atoms H	State at STP	Molecular arrangement	Boiling point °C	Melting point °C	Specific gravity	Refinery process
acyclic	aliphatic	alkane (naturally occurring paraffins)	methane	1	4	dry gas	straight chain	−161.5	−184.0	0.424	
			ethane	2	6	dry/wet gas		− 89.0	−172.0	0.546	
			propane	3	8	wet/dry gas	single bond straight chain	− 42.0	−187.0	0.582	distillation, fractionation, molecular sieve
			butane	4	10	wet gas		+ 0.5	−138.0	0.579	
			isobutane	4	10	wet gas	(isobutane, isopentane branched chain)	− 11.7	−160.0	0.557	
			pentane	5	12	liquid		+ 36.0	−130.0	0.626	
			isopentane	5	12	liquid		+ 28.0	−160.0	0.620	
		higher alkane (naturally occurring paraffins)	hexane	6	14	liquid		+ 68.7	− 95.0	0.659	
			heptane	7	16	liquid		+ 98.0	− 91.0	0.684	
			octane	8	18	liquid		+ 126.0	− 57.0	0.703	
			isooctane	8	18	liquid	single bond	+ 99.0	− 22.0	0.706	distillation, fractionation, molecular sieve
			nonane	9	20	liquid	straight chain	+ 150.7	− 54.0	0.717	
			decane	10	22	liquid	(isooctane branched chain)	+ 174.0	− 30.0	0.738	
			cetane	16	34	solid		+ 288.0	+ 18.0	0.774	
			eicosane	20	42	solid		+ 343.0	+ 36.0	0.778	
		alkene (mono-olefin)	ethylene	2	4	dry gas	double bond straight chain	−104.0	−169.4	0.565	cracking of naphtha, ethane and propane
			propylene	3	6	wet/dry gas		− 47.7	−185.0	0.514	cracking of naphtha
			butylene	4	8	wet gas		− 6.3	−185.0	0.595	dehydrogenation of butane
			isobutylene	4	8	wet gas	double bond branched chain	+ 3.7	−138.9	0.621	absorbing in sulphuric acid from butane
		(diolefin)	butadiene	4	6	wet gas	two double bonds	− 18.0	−136.3	0.676	catalytic dehydrogenation of butane and butylene
		higher alkene (mono-olefin)	pentene	5	10	liquid	double bond straight chain	+ 29.2	−138.0	0.641	fractionation and polymerization of lower olefins (eg propylene)
			hexene	6	12	liquid		+ 68.0	−146.0	0.684	
			heptene	7	14	liquid		+ 94.0	−119.0	0.697	
		alkyne (acetylene)	acetylene	2	2	gas	triple bond	− 84.0*	− 81.5	0.618	cracking of methane or by solvent extraction
alicyclic	aliphatic	cyclo-alkane (cyclo-paraffin or naphthene)	cyclopropane	3	6	wet/dry gas		− 33.0	−126.6	0.720	cracking light distillates and chlorination of propane
			cyclopentane	5	10	liquid	single bond ring	+ 49.3	− 93.9	0.740	distillation and solvent extraction
			cyclohexane	6	12	liquid		+ 81	+ 6	0.778	hydrogenation of benzene
			methylcyclohexane	7	14	liquid		+ 101	−106.7	0.769	dehydrogenation of heptane

**sublimes*